P.C.Garratty

Recent Advances in
Animal Nutrition

1988

In the same series:

Recent Advances in Animal Nutrition—1978
Edited by W. Haresign and D. Lewis

Recent Advances in Animal Nutrition—1979
Edited by W. Haresign and D. Lewis

Recent Advances in Animal Nutrition—1980
Edited by W. Haresign

Recent Advances in Animal Nutrition—1981
Edited by W. Haresign

Recent Advances in Animal Nutrition—1982*
Edited by W. Haresign

Recent Advances in Animal Nutrition—1983
Edited by W. Haresign

Recent Advances in Animal Nutrition—1984
Edited by W. Haresign and D.J.A. Cole

Recent Advances in Animal Nutrition—1985
Edited by W. Haresign and D.J.A. Cole

Recent Advances in Animal Nutrition—1986
Edited by W. Haresign and D.J.A. Cole

Recent Advances in Animal Nutrition—1987
Edited by W. Haresign and D.J.A. Cole

Related titles:

Recent Developments in Ruminant Nutrition—2
Edited by W. Haresign and D.J.A. Cole

Recent Developments in Pig Nutrition
Edited by D.J.A. Cole and W. Haresign

Energy Metabolism
Edited by Lawrence E. Mount

Mineral Nutrition of Animals
V.I. Georgievskii, B.N. Annenkov and V.T. Samokhin

Protein Contribution of Feedstuffs for Ruminants
Edited by E.L. Miller and I.H. Pike in association with A.J.M. van Es

Advances in Agricultural Microbiology
Edited by N.S. Subba Rao

Antimicrobials and Agriculture
Edited by M. Woodbine

*out of print

Cover design
The modern domesticated pig has been derived from two species, *Sus scrofa* (the European wild pig) and *Sus vittatus* (the wild pig of eastern and south-eastern Asia). The pig has played an important part in man's farming system as a provider of meat. Before domestication it was hunted. The front cover shows the Altamira boar which is the earliest known picture of a pig and is believed to have been painted 40 000 years ago. It is one of a number of animal paintings by cave dwellers at Altamira in Spain.

Recent Advances in Animal Nutrition

1988

W. Haresign, PhD
D.J.A. Cole, PhD
University of Nottingham School of Agriculture

BUTTERWORTHS
London Boston Singapore Sydney Toronto Wellington

All rights reserved. No part of this publication may be reproduced or
transmitted in any form or by any means, including photocopying and recording,
without the written permission of the copyright holder, application for which
should be addressed to the Publishers, or in accordance with the provisions of
the Copyright Act 1956 (as amended), or under the terms of any licence
permitting limited copying issued by the Copyright Licensing Agency,
7 Ridgemount Street, London WC1E 7AE, England. Such written permission must also
be obtained before any part of this publication is stored in a retrieval system
of any nature.

Any person who does any unauthorized act in relation to this publication may be
liable to criminal prosecution and civil claims for damages.

This book is sold subject to the Standard Conditions of Sale of Net Books and
may not be re-sold in the UK below the net price given by the Publishers in
their current price list.

First published, 1988

© **The several contributors named in the list of contents, 1988**

British Library Cataloguing in Publication Data

Recent advances in animal nutrition.—(Studies in
  the agriculture and food sciences).—1988
  1. Animal nutrition
  I. Series
  636.08′52   SF95

ISBN 0-407-01165 X

Typeset by Latimer Trend & Company Ltd, Plymouth
Printed and bound in Great Britain
by Anchor Brendon Ltd., Tiptree, Essex

PREFACE

This volume, based on the 22nd University of Nottingham Feed Manufacturers Conference, reflects the wide variety of nutritional issues currently of importance to the industry.

The opening section addresses several important nutritional topics. First, the problem of variability of raw materials is examined. While the chapter recognizes the existence of variability within feed ingredients it questions the extent to which it is worth doing much about it. The consideration of European legumes in the diets of non-ruminants reflects the interest in self-sufficiency for high protein materials within the European Economic Community. Considerable attention is given to antinutritional factors in these materials. The mechanism by which Vitamin E acts as a scavenger for, and prevents damage by, free radicals is examined in the last chapter of the opening section, with particular emphasis on how manipulation of dietary Vitamin E content might help reduce the incidence of disease in farm livestock.

There are two specialist pig chapters. The first considers acidification of pig diets and the second examines the various techniques of growth promotion that might be used in pig production. Market opportunities are changing for many agricultural products. This is highlighted in a paper on turkeys which examines nutrition in relation to current market demands. The second paper on poultry looks at responses to energy and amino acids while at the same time considers the determination of feeding strategies to maximize the profitability of laying flocks.

The section on ruminant nutrition presents a progression of ideas on areas of current interest. The chapter on prediction of metabolizable energy of compound feeds is based on work subsequent to the initial Rowett Institute study and the UKASTA/ADAS/COSAC Report. In the chapter on nutrient allowances for ruminants, emphasis is given to animal response to nutrients. The final chapter in this section examines alternative, practical approaches to the characterization of feedstuffs. The suggested approach called MENTOR is largely based on measurements of feed chemistry that are now available to the compounder.

The last part of the book is devoted to three chapters on minority species which reflect the increasing importance of the leisure industry. The chapter on game birds clearly illustrates that, compared with other species, little scientific attention has been paid to their nutrition. The chapters on leisure horses and dogs illustrate that, while their nutrition is based on sound scientific principles, their role is considerably different from that of other farm animals.

The organizers and the University of Nottingham are grateful to BP Nutrition (UK) Ltd for the support they gave in the organization of this conference.

W. Haresign
D.J.A. Cole

CONTENTS

I	**General Nutrition**	**1**
1	PROBLEMS OF DEALING WITH RAW INGREDIENT VARIABILITY M.S. Duncan, *Holly Farms Foods, Wilkesboro, North Carolina, USA*	3
2	EUROPEAN LEGUMES IN DIETS FOR NON-RUMINANTS J. Wiseman and D.J.A. Cole, *University of Nottingham School of Agriculture, Sutton Bonington, UK*	13
3	VITAMIN E AND FREE RADICAL FORMATION: POSSIBLE IMPLICATIONS FOR ANIMAL NUTRITION D.A. Rice and S. Kennedy, *Veterinary Research Laboratories, Stormont, Belfast, Northern Ireland*	39
II	**Pig Nutrition**	**59**
4	ACIDIFICATION OF DIETS FOR PIGS R.A. Easter, *Department of Animal Sciences, University of Illinois, Urbana, Illinois, USA*	61
5	NOVEL APPROACHES TO GROWTH PROMOTION IN THE PIG P.A. Thacker, *Department of Animal Science, University of Saskatchewan, Saskatoon, Canada*	73
III	**Poultry Nutrition**	**85**
6	THE NUTRITIONAL REQUIREMENTS OF TURKEYS TO MEET CURRENT MARKET DEMANDS C. Nixey, *British United Turkeys Ltd, Tarvin, Chester, UK*	87
7	MINERAL AND TRACE ELEMENT REQUIREMENTS OF POULTRY R. Hill, *Royal Veterinary College, University of London, Potters Bar, Hertfordshire, UK*	99

8 RESPONSE OF LAYING HENS TO ENERGY AND AMINO ACIDS	111

R.M. Gous and F.J. Kleyn, *Department of Animal Science and Poultry Science, University of Natal, Pietermaritzburg, South Africa*

IV Ruminant Nutrition 125

9 PREDICTING THE METABOLIZABLE ENERGY (ME) CONTENT OF COMPOUND FEEDS FOR RUMINANTS 127

P.C. Thomas, S. Robertson and D.G. Chamberlain, *Hannah Research Institute, Ayr, Scotland* and R.M. Livingstone, P.H. Garthwaite, P.J.S. Dewey, R. Smart and C. Whyte, *Rowett Research Institute, Aberdeen, Scotland*

10 NUTRIENT ALLOWANCES FOR RUMINANTS 147

J.D. Oldham, *Edinburgh School of Agriculture, West Mains Road, Edinburgh, UK*

11 ALTERNATIVE APPROACHES TO THE CHARACTERIZATION OF FEEDSTUFFS FOR RUMINANTS 167

A.J.F. Webster, R.J. Dewhurst and C.J. Waters, *Department of Animal Husbandry, University of Bristol, Langford, Bristol, UK*

V Nutrition of Alternative Species 193

12 NUTRIENT REQUIREMENTS OF GAMEBIRDS 195

J.V. Beer, *Game Conservancy Ltd, Fordingbridge, Hampshire, UK*

13 NUTRITION OF THE LEISURE HORSE 205

D.L. Frape, *British Horse Feeds Ltd, Rugby, Warks, UK*

14 NUTRITION OF THE DOG 221

J. Corbin, *Department of Animal Sciences, University of Illinois, Urbana, Illinois, USA*

LIST OF PARTICIPANTS 235

INDEX 247

I
General Nutrition

1
PROBLEMS OF DEALING WITH RAW INGREDIENT VARIABILITY

M. S. DUNCAN
Holly Farms Foods, Wilkesboro, North Carolina, USA

The title of this chapter makes two assumptions, (1) that ingredients are variable, and (2) that their variability is of sufficient concern that it must be dealt with. That ingredients are variable goes without saying. That their variability is of concern may be debatable. From a practical point of view, ingredient variability is of concern only when there is a monetary cost associated with that variability and someone incurs those costs.

As Director of Nutrition for the third largest integrated broiler producer in the USA, the unique opportunity was granted of viewing the 'big picture' with respect to the overall economic effects of ingredient variation. Working within a company with an annual production of 1.6 million tonnes of feed, 338 million broilers and 567 million kg of ready-to-cook poultry, it rapidly became apparent that a penny saved here and there could rapidly add up to a million dollars earned. Performance losses which seem trivial to some, can amount to millions of pounds for large companies such as Holly Farms, and this is illustrated by the data in *Table 1.1*. A trial concerned with a small 5% reduction in critical amino acid levels was conducted under commercial conditions (Duncan, 1986). A paired experimental design was employed using commercial houses side by side with capacities of 20 000 or more broilers. Feed conversion was reduced by 1 point, from 2.01 to 2.02 kg feed/kg liveweight gain. This seems like such a small value that it's hardly worth worrying about, and only amounts to about 0.28 pence/kg of dressed poultry. But, multiplied by the annual production figure of 567 million tonnes of oven-ready poultry it amounts to over £660 000 sterling and that is something to worry about.

Before being able to determine if a feed manufacturer will incur any costs from ingredient variability, it is first necessary to determine what goals the feed manufacturer has. The purpose of feed manufacturing is to provide a diet to be fed to animals for some purpose. If the goal is to just keep that animal alive, it is possible just to mix together ingredients of questionable composition in unknown amounts and probably achieve that goal. However, the animal producer generally has a conflicting goal in mind. He wants to buy our feed, feed it to his animals and produce a product which he can market at a profit. He doesn't want to make a profit once in a while, he wants to do it all the time. He therefore demands of us a feed which will consistently provide the level of animal performance which will yield him that profit. And therein lies the major cost of ingredient variability, it is the costs incurred in producing a quality feed from ingredients of variable nutrient composition.

Table 1.1 EFFECT OF A 5% REDUCTION IN DIETARY LYSINE AND TOTAL SULPHUR AMINO ACIDS (TSAA) ON BROILER PERFORMANCE

	Dietary nutrient level (%)	
	100	95
Average weight (kg)	2.12	2.14
Feed conversion (kg feed/kg liveweight gain)	2.01	2.02 ($P < 0.05$)

After Duncan (1986)
22 paired commercial houses—231 700 birds/treatment

For such purposes a quality feed is defined as one which consistently provides a level of nutrition that optimizes animal performance. For feed manufacturers to produce consistency from inconsistency is no small task. It requires that the degree of inconsistency of the raw materials is determined and steps taken to minimize their effect on the consistency of the finished feed. In doing so costs will be incurred from:

(1) analysis of ingredients to determine their nutrient variability,
(2) formulation to achieve minimum nutrient standards for finished feeds.

If feed manufacturers fail to produce a quality feed, they may be faced with the costs of payment of claims filed against feeds that do not meet minimum guarantees and the loss of business due to customer dissatisfaction.

Statistical theory tells us that through sampling and analysis of the nutrient content of an ingredient, it is possible to characterize the variability in terms of a mean or average nutrient content and the degree of deviation from the mean value. Theory also tells us that 50% of the ingredient lots received will have a nutrient content less than the average and 50% will have a nutrient content greater than the average. Of prime importance is the degree of variation, the magnitude of the standard deviation, which measures how closely all of the nutrient values are to the average.

Chung and Pfost (1964) showed that the variation of a nutrient in a feed mixture could be estimated from the nutrient variations of the ingredients by equation (1.1).

$$SD = \sqrt{(X_1 S_1)^2 + (X_2 S_2)^2 + \ldots (X_n S_n)^2} \tag{1.1}$$

where

SD = standard deviation of nutrient in feed mixture
S_n = standard deviation of nutrient in nth ingredient
X_n = fraction of total nutrient contribution by nth ingredient.

The authors noted that half of the feed samples would be expected to contain less than the nutrient level at which the diet was formulated. To overcome this effect, they suggested using the estimated standard deviation of the feed mixture to determine the degree of excess that must be formulated into the diet to insure that only a small percentage of the nutrient content of the feed fell below some minimum level.

Most textbooks on statistical methods will contain a table of the cumulative distribution for the standard normal distribution. From this table, it is possible to determine that if only 20% of the feed manufactured is to fall below a minimum nutrient level, it is essential to formulate that diet to contain an average nutrient content that is 0.84 standard deviations greater than the minimum. If only 10% are to fall below the minimum, then the average must be 1.28 standard deviations greater than the minimum. *Table 1.2* is an analysis of the nutrient variation and cost of variation of a typical broiler diet. From *Table 1.2* the cost of adding safety margins to formulations can be calculated. For instance, increasing the lysine content of the diet by 1 standard deviation would increase the cost of the diet by £0.673/tonne. The difference between having 20% of the feed fall below minimum lysine level and 10% falling below the minimum is £0.30/tonne.

One often overlooked means of minimizing the cost of ingredient variation is the wise expenditure of analysis costs. This should be concentrated on those ingredients and nutrients most likely to be associated with a high cost of variation. From *Table 1.2* it can be seen that variation in the lysine and metabolizable energy content would be expected to impart the greatest cost to any attempt to minimize the effect of variation by overformulation. Unfortunately, these are two nutrients which few feed manufacturers routinely assay for. It therefore becomes necessary to estimate their variation from the assays that are done. One can reasonably estimate the energy content of an ingredient if, at a minimum, the protein, fat and moisture determinations are available. Amino acid variations can be easily estimated from the protein variation if one is willing to accept the premise that amino acids vary in proportion with total protein.

Table 1.3 shows the contribution of the ingredients to dietary nutrient variation of energy and lysine in the diet. If one is worried about maintaining a consistent energy level in the feed, then maize would be the likely candidate for extensive analysis to determine or estimate its energy content and variation. Since the lysine contribution from soyabean meal is by far the greatest, a good estimate of its content and variability would be desirable. By viewing ingredients in this manner, it is possible to determine where analysis costs will be most wisely spent.

Overformulation to insure that the final nutrient level in the feed exceeds some minimum level is a common, but expensive, means of minimizing the effects of ingredient variation. A more desirable alternative would be to reduce ingredient variation itself. By doing so, the degree of excess necessary in any formulations will be reduced and this in turn will reduce the cost of those diets. The most satisfactory method to reduce ingredient variation is by separation. One of the first places to look for the possible benefits of separation is at multiple suppliers of the same ingredient.

Table 1.4 lists some of the characteristics of fish meals from two suppliers. It can be seen that the two products should definitely be kept separate since combining them

Table 1.2 EXPECTED VARIATION IN FINISHED FEED CAUSED BY VARIATION OF NUTRIENTS IN THE INDIVIDUAL INGREDIENTS

Restriction	Units	Restriction cost[a] (£/tonne)	Expected SD[b]	Cost of variation[c] (£/tonne)
ME	(MJ/kg)	0.0891	3.524	0.314
Protein	(g/kg)	0.037	3.676	0.136
Lysine	(g/kg)	2.104	0.320	0.673
TSAA	(g/kg)	1.479	0.100	0.148

[a]Cost/unit/tonne
[b]Standard deviation of nutrient in finished feed estimated by equation (1.1)
[c](SD) × (restriction cost)

6 Problems of dealing with raw ingredient variability

Table 1.3 INGREDIENT CONTRIBUTION TO DIETARY NUTRIENT VARIATION IN AN EXAMPLE FEED

Ingredient	% of diet	Energy (%)		Lysine (%)	
		Total content	Variation	Total content	Variation
Maize	65.340	69.718	67.658	15.684	0.284
Soyabean meal	17.610	13.093	3.824	52.350	93.976
Biscuit meal	5.000	5.904	16.811	1.708	0.010
Fish meal	2.500	2.404	0.208	11.633	5.116
Feather meal	2.500	2.069	0.201	2.354	0.003
Poultry meal	2.000	1.713	2.024	4.690	0.611
Animal fat	2.124	5.098	9.275	0.000	0.000
l-Lysine	0.160	0.000	0.000	11.581	0.000

dramatically increases the variation of the ingredient. In this case, the best way to separate is to find an alternative supplier whose product is more comparable to that of supplier A. There are cases though where it may not be quite so easy to distinguish between suppliers of the same ingredient. Take for example the five suppliers of soyabean meal in *Table 1.5*. It would be difficult to justify the elimination of any of these suppliers, especially if the meal is being purchased on a 490 g/kg protein guarantee. But, because of the effect that combining the meals has on the protein variation, one might consider physical separation.

Physical separation can often be used effectively to reduce the variation of an ingredient. *Table 1.6* shows the effect of physical separation of the soyabean meals listed in *Table 1.5*, based on above and below average protein content. If the separated meal is reblended (that is, used at a 50:50 ratio in the diet), the effective variation can be cut in half. The rapid separation of ingredients based on protein content is technologically feasible. Incoming loads can be rapidly assayed before unloading and then routed to storage depending on protein content. In the soyabean meal example, physical separation yielded an effective reduction in feed costs of £0.46/tonne by reducing the cost of overformulation. If one considers a mill which produces 5000 tonnes of the diet per week, the annual potential savings would be approximately £119 600. If soyabean meal comprised 18% of the diet produced by such a mill then it would require that 36 25-tonne truck loads be separated each six-day working week, or six loads/day.

Table 1.4 COMPARISON OF THE CHARACTERISTICS FOR FISH MEAL FROM TWO SUPPLIERS

Characteristic	Supplier A	Supplier B	Combined
Crude protein (g/kg)			
Mean	663	622	645
SD	8.3	10.8	22.3
Crude fat (g/kg)			
Mean	76	102	89
SD	6.0	13.8	17.2
Est. energy (MJ/kg)			
Mean	12.86	12.93	12.89
SD	0.117	0.431	0.310

Table 1.5 COMPARISON OF PROTEIN CONTENT (g/kg) OF SOYABEAN MEAL FROM FIVE SUPPLIERS

Supplier	Mean	SD
A	504	8.53
B	499	4.47
C	503	4.00
D	501	5.82
E	520	4.80
Combined	505	9.40

Even if the ability to separate ingredients on nutrient content does not exist, the benefits of physical separation can still be gained. If two or more bins are dedicated to an ingredient and equal proportions are removed from each bin during the blending process, the effect on variation is shown by equation (1.2).

$$S' = S/\sqrt{n} \tag{1.2}$$

where

n = number of bins drawn from
S = standard deviation prior to blending
S' = standard deviation of the blend

However, it would require four bins to achieve the same reduction in variation that was gained by separating on nutrient content. Whatever method of separation might be chosen, the expected savings from reducing the cost of minimizing the effect of the

Table 1.6 EFFECT OF SEPARATING SOYABEAN MEAL, BY ABOVE AND BELOW AVERAGE PROTEIN CONTENT AND THEN RECOMBINING IN EQUAL PROPORTIONS, ON NUTRIENT VARIATION

Variation	No separation SD	With separation SD[a]
Soyabean meal		
Protein (g/kg)	9.40	4.01
Lysine (g/kg)	0.60	0.26
TSAA (g/kg)	0.28	0.12
Diet		
Protein (g/kg)	4.42	2.40
Lysine (g/kg)	0.32	0.16
TSAA (g/kg)	0.10	0.06
Diet cost (£/tonne) (at +1 SD)	99.37	98.91

[a]Standard deviation of above and below average loads (SD') is calculated from the truncated normal distribution as:

SD' = 0.603 (SD without separation)

on recombining in equal proportions the standard deviation (SD″) becomes

SD″ = SD'/$\sqrt{2}$

8 *Problems of dealing with raw ingredient variability*

ingredient variation must be weighed against the expected costs incurred from the process of separation.

Thus far the discussion has only addressed the effect of ingredient variation on the feed manufacturer. Of equal importance is the effect that ingredient variation has on the animal producer. Any costs incurred by the feed manufacturer in minimizing the effect of ingredient variation will surely be passed on to the animal producer. But does ingredient variation pose any threat to the producer in terms of reduced animal performance? With today's properly formulated and manufactured diets, one would not expect to encounter gross nutrient deficiencies. But, as already discussed, the likelihood that any given lot of feed will be to some degree deficient in one or more critical nutrients is very real. Could these sporadic, and sometimes small, deficiencies influence animal performance? Intuitively, most people think it would.

Waller and Pfost (1972) thought so too and developed a theoretical model in an attempt to define nutrient variation in terms of expected animal performance. The model, depicted in *Figure 1.1*, requires a response curve of animal performance versus dietary levels of nutrients and the probability distribution of nutrients in the diet. Responses are expressed as % of maximum response and dietary nutrient levels as % of requirement for maximum response. The probability of occurrence of a nutrient level X_i is P_i and the expected response is Y_i. The relative response to a nutrient level X_i is $(P_i)(Y_i)$. When the relative responses are summed over the entire probability distribution function, an estimate of the animal or feed performance to the nutrient variation is obtained.

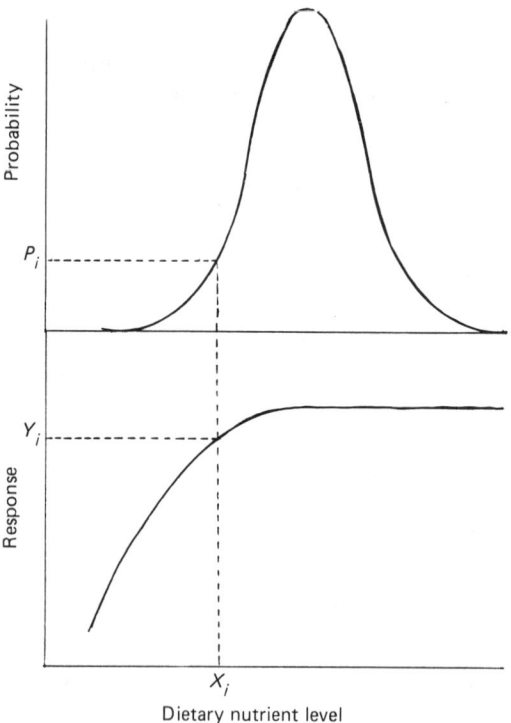

Figure 1.1 Expected frequency distribution of levels of a nutrient in the diet and the average response curve

The theory behind the model was validated experimentally (Duncan, 1973) by showing that day-to-day nutrient variation of the diet would affect animal performance. The experimental design called for feeding broiler chicks diets which simulated the protein variation that might be expected during the manufacture of a typical broiler feed. The protein levels of the experimental diets were formulated to represent the normal distribution of protein in feeds with either a 10% or 20% coefficient of variation (CV). Protein levels of the test diets ranged from 70 to 110% of the expected requirement for optimum growth. Diets were fed randomly for two-day intervals and the number of times a dietary protein level occurred was determined from the frequency distribution function of the treatment.

Table 1.7 presents the comparison of the 10% and 20% CV treatments with a control group which received a constant dietary protein level of 110% of the mean. Growth was reduced in proportion to the degree of nutrient variation. However, growth was not simply a function of protein intake as evidenced by the protein efficiency ratios (PER). The control birds, purposely overfed protein, showed the poorest conversion of protein to bodyweight gain. This was interpreted as a result of protein wastage. The other treatments, being overfed protein only 50% of the time, showed better conversions of protein to bodyweight gain. That the 20% CV treatment converted protein to gain better than did the 10% CV treatment follows the 'law of diminishing returns' which says that as the level of input approaches that required for maximum output, the units of output per unit of input decreases logarithmically until a point is reached where output no longer increases with increasing input.

The results of this trial became the basis for development of an improved model for estimating the effect of nutrient variation on animal performance. Animal responses to dietary nutrient levels were fitted to a logarithmic function in terms of conversion of feed to animal product. A comparison between actual and simulated responses is given in *Table 1.7*.

When evaluating the variation of two or more nutrients in the diet, a multivariate distribution of the nutrient variation is needed (Duncan, Pfost and Waller, 1973). Since animal performance is a function of the most limiting dietary nutrient, the multivariate distribution is a reflection of the probability that at least one of the nutrients will occur at some performance-limiting level. *Table 1.8* represents the multivariate cumulative distribution of the four nutrients in *Table 1.2*. The multivariate cumulative distribution reflects the probability that at least one of the nutrients will occur at a lesser level. In the current case, there is a 98% chance that at least one of the four nutrients will occur at a level (any level) less than 100% of its formulated level. This is a completely different picture from that obtained when only one nutrient

Table 1.7 COMPARISON BETWEEN EXPERIMENTAL AND SIMULATED CHICK RESPONSES TO DIETARY PROTEIN VARIATION OVER THE PERIOD FROM 0 TO 28 DAYS OF AGE

Treatment	Gain (g)	Gain/feed (g/g)	Protein efficiency ratio	Gain/feed, % of maximum	
				Experimental	Simulated
Control	773 [a]	0.575 [a]	2.53 [a]	100	100
10% CV	716 [b]	0.549 [b]	2.64 [b]	95	96
20% CV	703 [b]	0.538 [c]	2.72 [c]	94	92

Values in a column with same letter are not different ($P < 0.05$)

Table 1.8 MULTIVARIATE CUMULATIVE DISTRIBUTION (F) THAT AT LEAST ONE OF FOUR NUTRIENTS WILL OCCUR AT A LEVEL LESS THAN THE REQUIREMENT IN THE DIET (X)

X	F	X	F
87.5	0.00001249	95.0	0.06255716
88.0	0.00002930	95.5	0.09237796
88.5	0.00006198	96.0	0.13433388
89.0	0.00012378	96.5	0.19269837
89.5	0.00023747	97.0	0.27234456
90.0	0.00044084	97.5	0.37655417
90.5	0.00080084	98.0	0.50291349
91.0	0.00141604	98.5	0.64066639
91.5	0.00244299	99.0	0.77900873
92.0	0.00411984	99.5	0.90844219
92.5	0.00680136	100.0	0.98300373
93.0	0.01100517	100.5	0.99897814
93.5	0.01746861	101.0	0.99998298
94.0	0.02721965	101.5	0.99999993
94.5	0.04164035	102.0	1.00000000

was considered where there was only a 50% chance of levels occurring at less than 100% of its formulated level. All is not as bleak as it might seem since there is only a 6% chance that one of the four nutrients will occur at a level less than 95% of its formulated level and only a 1% chance of being less than 93%. A lesson can be learned from the multivariate distribution by those who can't resist adding just one more nutrient guarantee to their feed. As the number of things guaranteed increases, so does the probability that at least one of them will not meet the guarantee.

It is evident from *Figure 1.1* that, if any portion of the nutrient distribution falls below 100% of the requirement for maximum response, performance will be less than 100%. Does this mean that it is necessary to add sufficient excess to formulations to insure that 100% of the feed will contain nutrient levels greater than requirements? The answer is yes and no. When speaking strictly about maximizing animal performance, then the answer is yes. However, when talking about optimizing animal performance, then the answer is no.

Table 1.9 contains data generated by the evaluation of the expected performance of diets formulated using the ingredient variations given in *Table 1.2*. The diet cost represents the increased cost of business incurred by the feed manufacturer in attempting to compensate for ingredient variation by overformulation to guarantee minimum nutrient standards. The diet performance estimates what the customer will receive in terms of conversion of diet to animal product with respect to the maximum

Table 1.9 EFFECT OF ENERGY, PROTEIN, LYSINE AND TSAA VARIATION ON ESTIMATED DIET PERFORMANCE

	Adjustments of formulation specifications			
	+ 0 SD	+ 1 SD	+ 2 SD	+ 3 SD
Diet cost (£/tonne)	98.14	99.37	100.67	103.29
Estimated performance (% maximum attainable)	97.25	98.97	99.73	99.97
Performance cost (£/tonne)	100.92	100.40	100.94	103.32

attainable response. The performance cost is what the formulation is really costing the customer since it represents the extra feed it will require to reach the same level of production that would have been realized had the animals achieved the maximum. The response model estimates that formulating at 2 standard deviations, or greater, over requirements would not yield enough increased performance to offset the increased diet cost. As one can see, the least cost diet does not always yield the least cost animal product. Within an integrated broiler system it has been possible to apply this knowledge toward the formulation of the *best cost* diet. The best cost diet is one that minimizes feed costs while optimizing broiler performance to produce the least cost/kg of broiler produced. The best cost diet, unlike the least cost diet, reduces the cost of doing business for both the feed manufacturer and the animal producer. But, the commercial feed manufacturer cannot take advantage of the best cost formulation unless he is willing to expend the effort to educate his customers; the animal producer must be convinced that the best cost diet will indeed increase his profitability.

Finally, it is important to consider the cost of ingredient variation that befalls the ingredient supplier. In a free marketing system, the value of an ingredient in a diet is a function of the nutrient contribution of the ingredient to the diet and the cost and nutrient contributions of the other ingredients in the diet. Thus, an ingredient tends to seek its own price level within the system. An ingredient with a high degree of nutrient variability will tend to seek a price level lower than the same ingredient with a lesser degree of nutrient variation. This is for two reasons. First, the supplier will be forced to guarantee nutrient contents lower than the average for the ingredient to avoid being deluged with claims. Secondly, the astute feed manufacturer will recognize the price he is paying for the ingredient variation and further devalue the worth of the ingredient in his diets. The ingredient supplier would then be wise to employ the principles of separation to reduce the variation of his product.

The biggest problem of dealing with raw ingredient variability will be determining if it's worth dealing with at all. How important are the restrictions and guarantees placed on feeds? What risk are we willing to accept that guarantees will not be met? Do the benefits offset the costs of dealing with ingredient variation? These are just some of the questions the feed manufacturer must answer. Regardless of the approach we decide to take, the undeniable fact is that there is a cost associated with ingredient variability and someone is going to pay the price. Ingredient variation is something that's not going to go away; but steps can be taken to minimize its effect.

References

CHUNG, D.S. and PFOST, H.B. (1964). Overcoming the effects of ingredient variation. *Feed Age*, **14**(9), 24–27

DUNCAN, M.S. (1973). Nutrient variation: Effect on quality control and animal performance. Doctoral Dissertation. Kansas State University, USA

DUNCAN, M., PFOST, H.B. and WALLER, R.A. (1973). Effect of multivariate inputs on production response. *Proceedings 1973 Winter Meeting of American Society of Agricultural Engineers*

DUNCAN, M.S. (1986). Unpublished research

WALLER, R.A. and PFOST, H.B. (1972). *Statistical Applications in the Feed Industry*, p. 37. American Feed Manufacturers Association

2

EUROPEAN LEGUMES IN DIETS FOR NON-RUMINANTS

J. WISEMAN and D. J. A. COLE
University of Nottingham School of Agriculture, Sutton Bonington, UK

Introduction

Livestock production in Europe has long relied heavily on plant proteins contributing substantially to the diet. Outstanding in this respect is soyabean meal with the European Economic Community (EEC) importing 14.13 million tonnes in 1985/6, of which the UK imported over 1 million tonnes. The importance of soyabean meal relative to similar materials as providers of protein in the UK is shown in *Table 2.1*, which also indicates the demand for plant protein.

Within the EEC, the Common Agricultural Policy was designed in such a way as to encourage the use of home-grown and protein-rich vegetables/materials in animal feeds. However, the success of such a policy depends on other factors as well as the financial incentives. For example, species of plants must be found which are satisfactory as ingredients in animal feeds and also capable of producing good yields in the temperate climate common to much of Europe. In this context, legumes are an obvious group of plants for consideration. Those that are likely to fill such a role are given in *Table 2.2*. However, climatic limitations (latitude, cool temperature, short growing season), harvesting difficulties, disease and low and variable yields on the one hand combined with suitability as an animal feed and economic competitiveness on the other do not make the list a viable proposition for the whole of Europe. For example, in the UK only peas and beans are grown on any scale, and even then contribute only about 10% of the plant proteins used (*Table 2.1*).

Nutritive value of legumes

The decision on whether or not to include a specific raw material in a compound diet is based ultimately upon its price and the relative cost of all the raw materials available, a situation that is influenced considerably by the systems of price support currently in operation. This chapter will be confined to an assessment of the nutritive value of three legume species (field beans, *Vicia faba*; peas, *Pisum sativum* and lupins, *Lupinus* spp.) by examining those components of the crop that contribute positively to nutritive value, those whose presence is detrimental and, finally, how the effects of these antinutritional factors may be mitigated through various processes. An important aspect of this section is the considerable variability in all of these factors within the crops studied.

Table 2.1 PROTEIN IMPORTS INTO UK AS OILSEED CAKES AND MEALS (1985/6) AND CONTRIBUTION OF UK GROWN BEANS AND PEAS

	Quantity ('000 tonnes)	Protein content (%)	Protein yield ('000 tonnes)
(1) *Imported oilseed cakes and meals*			
Soya	1152.5	44.0	507.10
Rape	139.6	31.0	43.28
Sunflower	273.0	31.0	84.63
Cottonseed	222.2	43.0	95.55
Others	397.0	—	127.04
Total imports	2184.3		857.60
(2) *Home-grown beans and peas*			
Beans	158.0	25.0	39.50
Peas	238.0	22.0	52.36
Total beans and peas	396.0		91.86
Total	2580.3		949.46

Based on data from UKASTA (1986)

CHEMICAL COMPOSITION

Protein and amino acid content

Legumes are regarded principally as protein crops, and there is a wealth of data relating to their crude protein composition, some of the more recently published information being presented in *Table 2.3*. It is evident that there is variability in the data which, if not accounted for during assessments of nutritive value, may lead to inaccuracies in formulation. It is difficult to identify trends in the data. This is well illustrated with peas, and although Cousin (1983) reported that wrinkled peas were approximately 2% higher in crude protein than smooth varieties, the range in values for the two groups was 26–33% and 23–31% respectively (fresh weight or dry matter basis not specified), indicating a considerable degree of overlap. A negative correlation between yield and protein content was also observed. However, Edwards, Rogers-Lewis and Fairbairn (1987) suggested that marrowfat peas generally have higher crude protein contents than other types (*Table 2.4*). When other variables such as season and location are included then it is obvious that it is difficult to draw any firm conclusions relating to crude protein content and to specify which cultivars and growing conditions are needed to produce consistently high levels. It is not thought

Table 2.2 POTENTIAL GRAIN LEGUME CROPS IN EUROPE

Pea (*Pisum sativum* L.)
Faba bean (*Vicia faba* L.)
Lupin (*Lupinus* spp.)
Common navy bean (*Phaseolus vulgaris* L.)
Soyabean (*Glycine max* (L.) Merrill)
Lentil (*Lens culinaris* Medic.)
Chick pea (*Cicer arietinum* L.)

Table 2.3 CRUDE PROTEIN (N × 6.25) PERCENTAGE OF VARIETIES OF BEANS, PEAS AND LUPINS (ALL DATA CORRECTED TO 90% DRY MATTER)

Beans		Peas		Lupins		Sample no.
(a) *Whole seed*						
76 (winter)[g]	24.1	Early Dun[f]	20.8–25.6 ($n = 91$)	*L. albus*[b]	35.4	1
Strubes (spring)[g]	25.9	Various[j]	12.0–24.4 ($n = 198$)	*L. albus*[b]	32.3	2
Maris bead (spring)[g]	27.7			*L. albus*[b]	34.5	3
Winter[h] ($n = 28$)	21.9–26.9			*L. angustifolius*[b]	30.0	
Spring[h] ($n = 104$)	23.0–31.9			*L. albus*[b]	31.2	
Various[i] ($n = 29$)	23.9–32.2			*L. albus*[e]	30.9–40.4	
Various[k] ($n = 8$)	27.8–30.0			*L. angustifolius*[e]	25.2–34.1	
(b) *Dehulled seed*						
Maris bead (1977)[a]	31.9	Minerva (1977)[a]	26.2	*L. albus*[d]	43.1	1
Blaze (1977)[a]	31.8	Minerva (1977)[a]	25.3	*L. albus*[d]	41.5	2
Maris bead (1978)[a]	31.7	Maro (1977)[a]	26.6	*L. albus*[d]	40.4	3
Maris bead (1978)[a]	31.8	Minerva (1978)[a]	25.5	*L. albus*[d]	43.1	4
Maris bead (1978)[a]	31.7	Minerva (1978)[a]	25.0	*L. albus*[d]	44.9	5
Blaze (1978)[a]	32.4	Minerva (1978)[a]	27.8	*L. angustifolius*[d]	48.9	
Blaze (1978)[a]	31.8	Maro (1978)[a]	27.8	*L. albus*[e]	37.1–45.6	
Blaze (1978)[a]	32.0	Maro (1978)[a]	30.7	*L. angustifolius*[e]	34.3–32.8	
		Maro (1978)[a]	29.3			
		Filby (1978)[a]	24.7			
		Filby (1978)[a]	23.9			
		Filby (1978)[a]	23.7			

(Date in parentheses for the different varieties is year of harvest)

[a] Evans and Boulter (1980) [c] Hill (1977) [i] Barratt (1982)
[b] Batterham et al. (1986a) [f] Davies (1984) [j] Reichert and Mackenzie (1982)
[c] Batterham et al. (1986b) [g] Bond (1970) [k] Marquardt et al. (1975)
[d] Brillouet and Riochet (1983) [h] Eden (1968)
(recalculated as N × 6.25)

Table 2.4 INFLUENCE OF TYPE, SEASON AND LOCATION ON CRUDE PROTEIN CONTENT OF PEAS (DATA CORRECTED TO 90% DRY MATTER). FIGURES IN PARENTHESES ARE NUMBERS OF SAMPLES

Type	Terrington EHF									Commercial farm	
	1981		1982		1983		1984			1984	
					Crude protein content (%)						
Marrowfat	20.4	(3)	22.2	(4)	22.5	(7)	23.1	(5)		21.5	(5)
Large blue	19.1	(1)	18.5	(2)	20.5	(3)	21.3	(3)		—	
Small blue	17.6	(2)	18.8	(1)	22.5	(3)	22.4	(5)		—	
White	17.9	(5)	19.4	(5)	20.6	(10)	21.8	(9)		19.5	(3)
Coloured	19.7	(2)	20.3	(1)	21.7	(2)	21.2	(2)		—	

Calculated from Edwards *et al.* (1987)

that size of seed is of any significant importance, although the production of smaller pea seeds associated with low yields is accompanied by a greater nitrogen content (Evans and Boulter, 1980). It is evident from *Table 2.3* that dehulling, which is practised frequently with legumes, particularly coloured seeded varieties prior to feeding to non-ruminants, increases the crude protein content by removing a part of the seed with a relatively low value. Similar variability has been shown with beans; both Eden (1968) and Bond (1970) indicated that winter beans had a lower crude protein content than spring sown cultivars, but again it was pointed out that there was considerable overlap between the two.

Traditionally, the factor 6.25 is used to convert nitrogen content to crude protein content. However, there is evidence to suggest that this is far too high. For example, 5.25 was calculated for whole peas for total nitrogen (5.52 when amino acid and amide nitrogen only were the basis for calculation) by Holt and Sosulski (1979), although this was the mean of a number of determinations. Similarly, a figure of 5.80 has been suggested for lupin seed cotyledons (Brillouet and Riochet, 1983), and a range between 5.44 and 6.06, depending on the dominant protein type present, for whole lupin seeds (Hill, 1977). These observations have important implications when comparisons are made between legume species and when estimating protein content. Nonetheless, whichever methods of calculation are employed, crude protein is fundamental when assessing the quality of legumes as animal feeds. However, protein requirements for non-ruminants must be considered in terms of essential amino acids and, more importantly, the relative balance of these amino acids. The content of some essential amino acids is presented in *Table 2.5*. In the examples selected, it is evident that pea protein is of superior quality with respect to sulphur amino acids and lysine, but that lupin has more threonine and tryptophan. Using the data from *Table 2.5*, *Figure 2.1* shows the relative balance of these essential amino acids for the three legume species under consideration compared with the 'ideal' balance for the pig as developed by Cole (1978). This confirms the much reported deficiency of sulphur amino acids in legumes. This deficiency may become more pronounced following dehulling of beans, as Marquardt *et al.* (1975) reported that the concentration of sulphur amino acids in the protein was lower in the cotyledon than in the hulls. The practical relevance of this however must be viewed in the light of the invariable use of legumes with other protein sources and pure amino acids in diets for non-ruminants. For example Cole *et al.* (1971) reported no response to supplementary methionine in diets containing 20% field beans (0.53% methionine + cystine) for growing pigs and

Table 2.5 ESSENTIAL AMINO ACID CONTENT OF WHOLE LEGUMES (g/16 g N).
(FIGURES IN PARENTHESES INDICATE BALANCE RELATIVE TO LYSINE)

	Legume		
Amino acid	White flowered[a] Vienna bean	Progreta[a] pea	Lupin[b]
Lysine	5.4 (100)	7.2 (100)	5.4 (100)
Methionine + cystine	1.6 (30)	2.7 (38)	1.7 (32)
Threonine	3.3 (61)	3.8 (53)	4.2 (78)
Tryptophan	0.7 (13)	1.0 (14)	1.3 (24)
N% (90% DM)	4.83	3.86	

[a]Jagger, Wiseman and Cole (unpublished data)
[b]Hill (1977)

30% (0.49% methionine + cystine) for finishing pigs. The diets were based solely on barley with the addition of white fish meal to the growing pig diet.

It is important to appreciate that amino acid content varies according to cultivar and environment and the above data are to be taken only as an indication. What has been established, however, is that amino acids are not a fixed proportion of total nitrogen content. Thus, an increase in the latter is not associated with a concomitant increase in the former. A considerable number of studies have indicated that, although an increase in legume nitrogen is associated with an increase in the total amount of essential amino acids, it is accompanied by a decrease in the proportion of many essential amino acids relative to nitrogen. In a study of two field bean varieties, Bond (1970) reported a significant negative correlation between nitrogen and essential amino acid content (in g/16 g nitrogen) for lysine, leucine and valine. The negative relationship appeared particularly striking for the sulphur amino acids. Evans and Boulter (1980) obtained the following regression for a mixture of 20 dehulled peas and beans:

$$y \text{ (g methionine + cysteic acid)}/16 \text{ g N} = 3.357 - 0.3255 \text{ N\%} \quad r = -0.86 \quad (2.1)$$

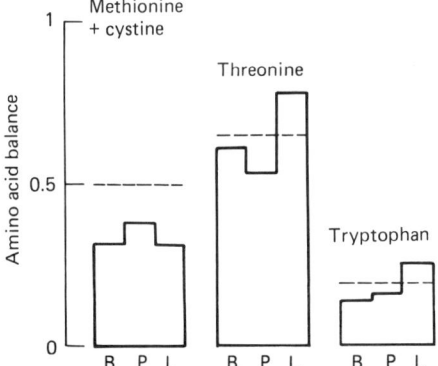

Figure 2.1 Essential amino acid balance of white-flowered field beans (B), peas (P) and lupins (L) relative to a value of 1.0 for lysine. Ideal balance for pigs represented by ---. (Jagger, Wiseman and Cole, unpublished data; Hill, 1977)

However, it is possible that the response of amino acid concentration to increasing nitrogen content is both species and cultivar specific. *Table 2.6* presents data for peas, from which it may be seen that the correlation coefficients were considerably higher when an individual cultivar was considered as opposed to a group of cultivars. It is apparent that there is a general decline in the relative content of most of the amino acids, with the rate of decline being specific to the amino acid in question. The sulphur amino acids, particularly cystine, are among those most affected and this exacerbates an already serious deficiency. The positive correlation between nitrogen content and arginine confirms the observations of Barratt (1982) with field beans, when it was concluded that a large proportion of arginine, an amino acid important in the transport and storage of nitrogen, was unbound.

Similar relationships have been established for field beans (Griffiths, 1984a) where negative correlations between amino acid concentration in the protein and crude protein content were obtained for methionine ($r = -0.433**$), cystine ($r = -0.461**$), phenylalanine ($r = -0.650***$), histidine ($r = -0.670***$) and lysine ($r = -0.592***$) with a positive correlation for arginine ($r = +0.742***$).

The probable reason for these trends is that legumes contain a variety of protein types, all with specific amino acid profiles, and an increase in total nitrogen content of the seed would be accompanied by a change in the relative amounts of these protein types. Breeding for improved protein quality in legumes has been concerned with increasing the proportions of those proteins with a more favourable amino acid profile (e.g. Hill, 1977, for lupins). Up to 80% of the total seed protein is present in the cotyledons of field beans as legumin and vicillin; the former contains considerably more of the sulphur amino acids and tryptophan and consequently there is interest in breeding for greater legumin:vicillin ratios (Hill-Cottingham, 1983).

Carbohydrate

Although livestock do not have a specific requirement for carbohydrates, if digestible

Table 2.6 CORRELATIONS BETWEEN NITROGEN (N) OR CRUDE PROTEIN (CP) CONTENT AND AMINO ACID CONTENT FOR PEAS

Amino acid	$(N)^a$	$(N)^b$	$(CP)^c$	$(CP)^d$
Arginine	0.44	0.93**	0.97**	0.33***
Histidine	-0.46	-0.34	ND	-0.03
Isoleucine	-0.13	-0.59**	ND	-0.32*
Leucine	-0.19	-0.26	ND	-0.23*
Lysine	-0.19	-0.85**	-0.97**	-0.34***
Methionine	0.17	-0.64**	-0.94*	-0.01
Cystine	-0.74**	-0.92**	-0.98**	ND
Phenylalanine	-0.43	-0.22	ND	-0.57***
Tyrosine	0.31	-0.40	ND	ND
Threonine	0.01	-0.80**	-0.97**	-0.36**
Valine	0.05	-0.14	ND	-0.08
Tryptophan	-0.37	-0.32	ND	-0.25**

*$P<0.05$; **$P<0.01$; ***$P<0.001$
ND = No data
[a] Holt and Sosulski (1979)—whole peas, all cultivars ($n = 33$)
[b] Holt and Sosulski (1979)—whole peas, one cultivar at several locations ($n = 16$)
[c] Reichart and Mackenzie (1982)—dehulled pea flour, one cultivar ($n = 198$)
[d] Davies (1984)—whole peas, one cultivar ($n = 91$)

they represent the principal energy-yielding components of legumes. In addition the many polymeric forms are associated with the indigestible 'plant fibre' fraction generally regarded as being of low or negligible nutritive value for non-ruminants. Carbohydrates are an extremely diverse group of compounds and it is outside the scope of this chapter to consider them or the various fractionation procedures in detail. There have been a number of recent studies characterizing the major carbohydrate types in legumes and, from a nutritional point of view, it is customary to divide them into two major groups according to whether or not they are of any value to the animal to which the legumes are fed. Such a distinction is however animal specific as it is the enzyme system present in the animal that determines this value. Additionally, it should be noted that there are various methods employed in the determination of carbohydrates which may make comparisons difficult.

Starch as a storage polysaccharide is generally assumed to be the carbohydrate of greatest potential nutritive value to non-ruminants, and estimates of the levels commonly found in legumes are presented in *Table 2.7*. It is evident that there is considerable variability both between and within species. The negligble starch content of lupins has been well established. Smooth seeded peas appear to have significantly more starch than wrinkled seeded varieties. However, this is accompanied possibly by a relatively lower level of sugar and a lower level of amylose in the starch (Davies and Downey, 1983). There seems to be little difference between white-flowered and coloured-flowered varieties of field beans, although there is a marginal superiority of winter over spring cultivars.

Cell wall related material (non-storage polysaccharides—fibre) is generally assumed to be of negligible nutritive value for poultry (Carre and Leclercq, 1985) although it may undergo considerable breakdown in the hindgut of pigs. Extensive studies of this fraction by Carre and Brillouet (1986) have revealed that its content on a dry matter basis for whole seeds is of the order of 30.9% for lupins, 17.4% for field beans and 13.1% for smooth peas. Somewhat lower figures were obtained for cotyledons (lupins 17.1–22.8%, smooth pea 6.9%, broad bean 7.2%; Brillouet and Carre, 1983; Brillouet and Riochet, 1983). What was of particular interest in these studies was that legume cell walls, in contrast to those from cereals, are associated with significant amounts of pectic substances. These are solubilized during the neutral detergent fibre (NDF) procedure for analysing cell walls, which suggests that NDF considerably underestimates the 'unavailable' carbohydrate content of legumes.

A third category of carbohydrate within legumes that has attracted attention is the α-galactosides. Levels, on a dry matter basis for whole seeds, are of the order of 3.9 to 5.3% for field beans, 5.0 and 7.2% for smooth and wrinkled peas respectively and 6.7

Table 2.7 STARCH CONTENT (% WHOLE SEED ON A DM BASIS) OF LEGUMES. (FIGURES IN PARENTHESES ARE THE NUMBER OF SAMPLES ANALYSED)

Field beans	*Peas*	*Lupins*
32.9–34.8 winter ($n = 2$)[a]	47.9 smooth ($n = 2$)[c]	< 1.0[c]
19.3–30.0 spring ($n = 6$)[a]	32.9 wrinkled ($n = 2$)[c]	
30.0–42.3 ($n = 16$)[b]		
41.2 white flowered ($n = 2$)[c]		
41.3 coloured flowered ($n = 2$)[c]		

[a]Pritchard, Dryburgh and Wilson (1973)
[b]Cerning, Saposnik and Guilbot (1975)
[c]Cerning-Beroard and Filiatre (1976)

Table 2.8 OIL (% AIR DRY SEED) AND MAJOR FATTY ACIDS (% OF OIL) OF WHOLE LUPIN SEEDS

	L. albus			L. angustifolius	L. luteus
	A	B		A	A
		1	2		
Oil[a]	—	8.3–10.6	8.6–11.3	—	—
Fatty acid[b]					
C16:0	7.6	6.9– 9.4	5.4– 1.5	7.6–11.7	5.3
C18:0	1.7	1.1– 2.1	0.4– 1.4	4.6– 7.0	1.5
C18:1	52.6–60.6	45.0–54.4	43.6–53.5	28.5–47.5	23.8–39.1
C18:2	16.5–23.4	17.2–22.5	19.4–26.9	33.7–48.3	45.0–48.5
C18:3	2.5– 8.1	6.7–15.2	9.2–13.6	1.8– 6.7	0.9– 7.6

[a]Determined by near infra-red spectroscopy
[b]Chain length followed by number of double bonds
A Hill (1977)
B Green and Oram (1983); 1 and 2 refer to different sites

to 8.5% for lupins (Cerning, Saposnik and Guilbot, 1975; Cerning-Beroard and Filiatre, 1976).

Lipid

Of the three legume species under consideration, only lupins have a significant content of lipid (*Table 2.8*). This fraction is important because of its high energy-yielding potential, and also its content of the essential fatty acid, linoleic acid. There would appear to be little variability in total oil content. Variations in fatty acid profiles would be unlikely to influence significantly the dietary energy values since the ratios of unsaturated to saturated fatty acids are considerably higher than those which are associated with impaired fat availability (e.g. Wiseman and Lessire, 1987). Differences in essential fatty acid content may however be important if diets are formulated to a specific essential fatty acid content. Furthermore, with the considerable interest in the adverse effects of soft oily carcass fat associated with high dietary levels of polyunsaturated fatty acids (e.g. Wood, 1984), together with the specific relationship between linolenic acid (C18:3) and the production of off-odours due to oxidative deterioration, it may be advisable to analyse for fatty acid content if lupins are to be used in diets. The oil content of field beans and peas has been studied by Welch and Griffiths (1984). Values of 1.50–1.93% and 1.37–2.80% were obtained respectively for field beans and peas. Despite relatively low levels, the oil content and, additionally, the fatty acid profile which indicated combined values of linoleic and linolenic acids of 45.2–62.9% and 50.3–69.7% respectively for field beans and peas, may have an important influence on stability of the seeds during physical processing prior to incorporation into compound diets where the development of rancidity could have an adverse effect on product quality. There would however appear to be negligible problems associated with storage of the whole seed.

NUTRIENT AVAILABILITY IN LEGUMES

Although gross chemical composition is a useful guide to the nutritive value of dietary

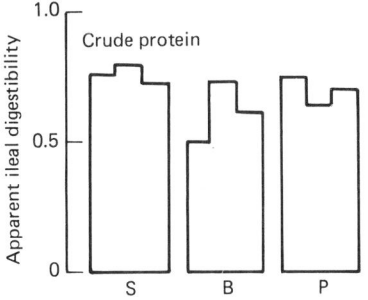
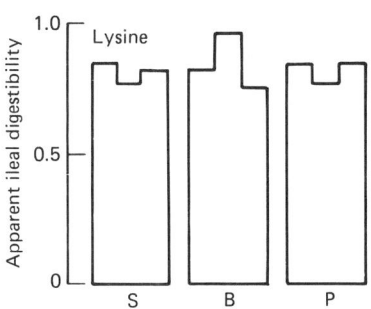
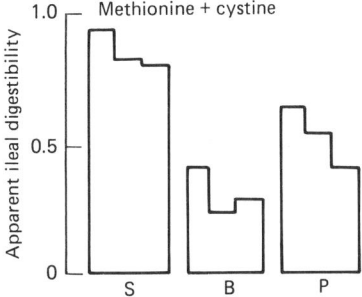

Figure 2.2 Coefficient of apparent ileal digestibility of crude protein, lysine and methionine + cystine for three varieties of soyabean meal (S), field beans (B) and peas (P). (Jagger, Wiseman and Cole, unpublished data)

raw materials, nutrient availability as measured by digestive and absorptive efficiency together with utilization *in vivo* is a far more precise assessment. Available nutrients, whatever system of measurement is used, are not frequently the basis for diet formulation. However the considerable variability which exists both between and within classes of raw materials in terms of nutrient availability indicates that such measurements should be considered. For example, *Figure 2.2* shows apparent ileal digestibilities of protein and some essential amino acids determined with the growing pig. The possible reasons for such differences will be discussed later.

Antinutritive factors

It is well known that legumes contain a number of naturally occurring factors which interfere with nutrient availability and whose presence is responsible for reports of suboptimal levels of animal performance arising from their use. It is important to appreciate that antinutritive factors are often difficult to measure, not least because of their diversity, and have complex and, frequently, imperfectly understood modes of action. Their content in legumes is likely to be variable, which may be real or attributable to a considerable lack of standardization in the methods of analysis between laboratories as well as their destruction during assays. Additionally, the use of a specific name for a certain antinutritive factor may in fact merely describe a group of related compounds with similar modes of action. Furthermore, in assessments of their biological effects, it is probable that sensitivity to antinutritive factors is influenced by species (with pigs being less sensitive than poultry—Aherne, Lewis and Hardin, 1977; chickens more sensitive than rats—Marquardt *et al.*, 1974) and age (with hens being more sensitive than chicks—Davidson, 1973). Finally it is apparent that, because legumes contain varying amounts of many different groups of antinutritional factors, it is often difficult to separate their biological effects. Some genotypes of the three legume species under consideration do contain specific antinutritive factors which result in them not being suitable for inclusion into diets for non-ruminants. An example would be lupin alkaloids to which pigs are particularly susceptible at levels of the order of 0.3 g/kg diet (Godfrey *et al.*, 1985).

Tannins Tannins are a complex group of phenolic compounds that are widely distributed in the plant kingdom. Those of nutritional relevance to non-ruminants that are found in legumes are predominantly condensed polyphenols of molecular weight 500–5000 and of flavonoid origin. Generally they are confined to the seed coat.

There is considerable variability in the tannin content of field beans. In an analysis of 55 samples, Cabrera and Martin (1986) reported a range in condensed tannin content from 0 to 3.54% (air dry basis). The study confirmed the well-established view that tannin content is related to flower colour rather than to seed colour, with white-flowered lines having negligible amounts. Griffiths and Jones (1977) produced evidence to suggest that the testa of winter varieties tended to have a lower tannin content than spring varieties, although the results were expressed as total phenols relative to a tannic acid standard rather than as condensed tannins. This is regarded as of questionable value when attempting to identify tannins of nutritional relevance (Deshpande, Cheryan and Salunkhe, 1986).

Griffiths (1981) confirmed the low condensed tannin content of white-flowered field beans, and also reported the same relationship between flower colour and condensed tannin content for peas (*Table 2.9*). As with field beans, it would seem that seed colour of peas was of little value in predicting condensed tannin content.

The adverse biological effects of the presence of condensed tannins in diets for non-ruminants are associated with their ability to form complexes with proteins, including enzymes, although they are reported to have an astringent taste which may adversely influence palatability.

Studies of enzyme systems have revealed that the presence of condensed tannins reduces their activity. Thus, Griffiths (1979) reported an *in vitro* inhibition of trypsin,

Table 2.9 CONDENSED TANNINS (AS 'CATECHIN EQUIVALENTS') IN FIELD BEANS AND PEAS

	Flower colour	Testa	Cotyledons	Whole seeds
Field beans				
Maxime	Coloured	4.20	0.08	0.59
Minden	Coloured	4.20	0.08	0.70
Triple White	White	0.04	0.07	0.06
Peas				
Marathon	Coloured	2.86	0.08	0.23
Minerva	Coloured	3.75	0.10	0.35
Filby	White	0.03	0.08	0.06

(Griffiths, 1981)

α-amylase and lipase activities of 36.4, 94.5 and 68.6% respectively when water extracts of testa from coloured-flowered field beans were added. This compared with inhibitions of 7.6, 26.2 and 10.6% respectively when tannin-free testa from white-flowered varieties were used. Similar reductions in α-amylase and trypsin activities were reported in *in vitro* studies with water-soluble extracts from the testa of coloured-flowered peas (Griffiths, 1981). These responses were confirmed *in vivo* with rats (Griffiths and Moseley, 1980), where a reduction in both trypsin and α-amylase activity was reported when diets including testa from a coloured-flowered field bean were fed. Interestingly, lipase activity was increased under these conditions, and it was argued that this was attributable to an increase in pancreatic lipase secretion coupled with a lower affinity of condensed tannins for lipase. The combined effects of enzyme inhibition were to reduce performance of rats when fed testa from a coloured-flowered field bean cultivar compared with those from a white-flowered genotype (*Table 2.10*). The lack of response for biological value indicates that tannins present in the water extracts of testa do not have an adverse metabolic role *in vivo*. Ford and Hewitt (1979a and b) reported that methionine availability as measured with a microbiological assay was consistently higher and that there was a marginal superiority in terms of true digestibility with chicks for white-flowered over coloured-flowered field bean cultivars.

It has already been stated that tannins are diverse compounds, and there have been studies attempting to assess the relative antinutritive activity of some tannin classes. The situation is further complicated by the conclusions of Mitaru, Reichart and Blair (1984) who, working with sorghum tannins, considered that the affinity of condensed tannins for proteins may be dependent upon protein structure.

The degree of polymerization of condensed tannins was investigated by Martin-Tanguy, Guillaume and Kossa (1977) who fractionated the total condensed tannins, which ranged from 1.85 to 5.46% of the testa of coloured-flowered cultivars of field beans (with a white-flowered variety giving a value of zero) into four categories (A, B, C, and D) of decreasing molecular weight (assumed to represent decreasing degrees of polymerization). Attempts were then made to correlate total condensed tannins with a variety of biological measurements with poultry (including growth rate, egg weight, apparent nitrogen digestibility) with a limited degree of success, except that the white-

Table 2.10 EFFECT OF FEEDING TESTA FROM DIFFERENT FIELD BEAN CULTIVARS ON PERFORMANCE IN RATS

	Cultivar flower colour		
	Red	*White*	*SE*
Liveweight gain (g)	47.6	55.9	1.1
Food conversion efficiency (g gain/g feed DM)	0.311	0.348	0.007
Protein efficiency ratio	2.55	2.93	0.06
Net protein utilization	57.7	64.7	1.0
Biological value of protein	77.1	76.0	1.0
Apparent coefficient of digestibility of:			
organic matter	0.843	0.904	0.009
crude protein	0.724	0.827	0.007
total soluble carbohydrates	0.955	0.990	0.008
lipid	0.979	0.989	0.001

(Moseley and Griffiths, 1979)

flowered variety was always superior. However, when responses were assessed in terms of the sum of fractions B + C + D or C + D, correlations were markedly improved, the overall result being illustrated in *Figure 2.3*. This suggests that high polymeric forms of condensed tannins are relatively inert biologically.

It is evident that tannins present in testa will have an adverse, although variable, effect on the nutritive value of legumes in which they are contained. This explains the considerable interest in breeding white-flowered genotypes. However, these do not appear to have as favourable agronomic characteristics as coloured-flowered cultivars, which has led to the suggestion that breeding for low tannin content may be more appropriate (D. A. Bond, personal communication). However it should be borne in mind that other antinutritional factors, discussed subsequently, may also be important; for example the three field bean cultivars in *Figure 2.2* with poor apparent availability of amino acids did include a white-flowered, zero tannin variety.

Processing of high tannin cultivars would be an alternative, and dehulling of field beans is frequently practised, producing a raw material of higher nutritive value. However, an important feature of this procedure, apart from the costs incurred, is the ability to dispose of the hulls. This appears not to present problems, particularly as Barry and Manley (1986) reported that the condensed tannin/protein complex present within the hull of the legume *Lotus pedunculatus* confers a degree of protein protection to ruminants. The complex appears to dissociate post-ruminally where, presumably, protein is available for digestion. However, the free condensed tannins would probably be available to form complexes with digestive enzymes and also with proteins of the gut wall. The net effect of the presence of condensed tannins may therefore be negligible. The use of agents such as polyethylene glycol and polyvinyl pyrolidine to inactivate free condensed tannins by forming complexes stronger than those with protein and extracting them with water has been studied (e.g. Griffiths, 1981; Barry and Manley, 1986; Ford and Hewitt, 1979b). The commercial applicability of such procedures is unclear.

Protease inhibitors It has been shown that condensed tannins may act as enzyme inhibitors, but they are probably not specific for any one enzyme. A further category of antinutritive factors present in legumes are those with specific protease inhibitory characteristics, usually antitrypsin and antichymotrypsin activity. Although there is variability between legume cultivars and the animal species to which they are fed, the general effect of these protease inhibitors is a reduction in protein digestibility, and an

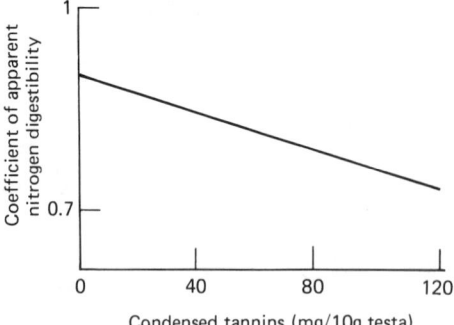

Figure 2.3 Coefficient of apparent nitrogen digestibility (ND) in horse beans as influenced by level of low molecular weight condensed tannins (CT). (Martin-Tanguy, Guillaume and Kossa, 1977) (*see* text for further details)

increase in pancreatic secretion of proteases. These proteases are relatively rich in sulphur amino acids, as are the inhibitors themselves (Leiner, 1976). Thus, an inherent relative deficiency in these amino acids within legumes is seriously compounded with the presence of protease inhibitors.

The content of trypsin and chymotrypsin inhibitor activity is shown in *Table 2.11*. The general conclusions to be drawn from these data is that there is little variability in protease inhibitor content amongst cultivars of field beans. This, however, is at variance with the conclusions of Marquardt et al. (1975) who considered that a range in trypsin inhibitor activity of 2.7 to 4.3 (smaller in fact than the range of data in *Table 2.11*) was significant, and reported that it had been possible to select lines with high and low trypsin inhibitor activity. Furthermore, Sjodin, Martensen and Magyarosi (1981) reported that an increase in antitrypsin factor activity from 2.16 to 3.46 units/mg was associated with a significant reduction in biological value of field beans fed to rats.

Absolute values are greater and variability among pea cultivars is more evident than for beans, with a tendency for winter and smooth-seeded cultivars to have higher levels of protease inhibitor activity. However there are exceptions to this as illustrated by the greater variability in wrinkled peas (Griffiths, 1984b) attributable to considerably larger values for the two marrowfat varieties. Protease inhibitor activity appears to be concentrated in the cotyledons, which may have implications if seeds are dehulled prior to feeding, although Marquardt et al. (1975) reported that trypsin inhibitor levels were lower in cotyledons than hulls, and Wilson, McNab and Bentley (1972) considered that not all the trypsin inhibitor activity was located in the cotyledon. It is possible that some antiprotease activity is associated not with specific antitrypsin factors but with non-specific agents such as tannins. This would explain some of the anomalous observations relating to hulls. There was also a significant positive correlation between trypsin and chymotrypsin inhibitor activity ($r = 0.986$, Griffiths, 1984b). Trypsin inhibitor activity appears not to fall with time following harvest (D. A. Bond, personal communication).

Despite the well-established adverse effect of protease inhibitors naturally present in field beans and peas on trypsin and chymotrypsin activity, together with pancreatic hypertrophy, it is difficult to assess whether these effects are of any real significance in practice when considering overall animal performance. Protease inhibitors are present in raw soyabean meal at considerably higher levels than in field beans, peas and lupins (*see Table 2.11*), and heat processing of these thermolabile factors is considered essential if this raw material is to be included in diets for non-ruminants (e.g. Wiseman, 1980). Indeed, the levels of protease inhibitors in field beans and peas are often below those which would be considered perfectly acceptable for adequately processed soyabean meal.

An important issue is that a safe dietary level of protease inhibitors, as opposed to levels in any raw material, has yet to be established. Consequently it is not possible to recommend levels for raw materials. Furthermore, the contribution of protease inhibitor activity in a specific raw material to overall dietary levels will depend upon the rate of inclusion of that raw material.

Notwithstanding these points, heat treatment of peas and beans does result in a reduction in protease inhibitor activity. The extent of this is variable and depends upon the precise heating conditions employed, the use of moist as opposed to dry heat allegedly being far more effective (Griffiths, 1984b). Such treatment results in an improvement in animal performance when processed peas and beans are subsequently incorporated into compound diets (McNab and Wilson, 1974; Marquardt,

Table 2.11 TRYPSIN INHIBITOR ACTIVITY (MEASURED AS TRYPSIN UNITS INHIBITED/mg) AND CHYMOTRYPSIN INHIBITOR ACTIVITY (MEASURED AS UNITS INHIBITED/mg) OF LEGUMES. (NUMBERS IN PARENTHESES INDICATE THE NUMBER OF SAMPLES ANALYSED)

	Trypsin inhibitor activity			Chymotrypsin inhibitor activity
	Whole seed	Hull	Cotyledon	Whole seed
Field bean[a]	3.3–6.2 ($n = 26$)	0 ($n = 1$)	6.9 ($n = 1$)	
Winter	4.1 ($n = 8$)			
Spring	4.5 ($n = 18$)			
Field bean[b]	1.41–1.56 ($n = 5$)			0.38– 0.77 ($n = 5$)
Peas[a]	2.9–10.8 ($n = 16$)	0.6 ($n = 5$)	7.8 ($n = 5$)	
Peas[a]				
Winter smooth	10.3 ($n = 2$)			
Winter wrinkled	7.9 ($n = 1$)			
Spring smooth	4.9 ($n = 2$)			
Spring wrinkled	2.7–3.7			
Peas[b]				
Smooth	0.15–1.07 ($n = 5$)			0.74– 3.86 ($n = 5$)
Wrinkled	0.66–4.62 ($n = 4$)			2.44–10.24 ($n = 4$)
Lupins[a]	<1 ($n = 3$)			
Raw soyabean meal, defatted[a]	59.4 ($n = 1$)			

[a]Valdebouze et al. (1980) data on DM basis
[b]Griffiths (1984b) data on fresh weight basis

Campbell and Ward, 1976; Moseley and Griffiths, 1979; Johns, 1987). However, the reduction in protease inhibitor activity following heat treatment did not fully account for the improvement in animal performance recorded. Coupled with these observations are those of Abbey, Neale and Norton (1979a,b) who illustrated that purified extracts of protease inhibitors did not account fully for the depression in performance of rats fed raw bean diets. Protease inhibitor–trypsin or –chymotrypsin complexes were not secreted by rats and it was concluded that protease inhibitors are hydrolysed by enzymes, although not by trypsin or chymotrypsin. Amino acids thus liberated would therefore be available to the animal, unless of course such hydrolytic activity was confined to the large intestine.

Phytohaemagglutinins Phytohaemagglutinins, otherwise referred to as lectins, are compounds which agglutinate red blood cells. In purified forms they may be extremely toxic materials and, when injected into mice at levels as low as 0.001 µg/g body weight, the phytohaemagglutinin from castor bean, which is one of the most toxic known, results in death. Levels of phytohaemagglutinins in the diet as low as 0.5% from *Phaseolus* beans depress growth rates (Liener, 1974), or may even result in considerable mortality depending upon the source of material (Liener, 1976). Phytohaemagglutinins are widely distributed in the plant kingdom; they are extremely diverse in character and assessments of their biological role are complicated by their subdivision into a toxic protein fraction with no agglutinating activity and an agglutinin which may not be toxic when fed. Jaffe (1980) suggested that when fed their adverse effects are linked to them forming complexes with the gut wall. Under such conditions, nutrient absorption is impaired. Presumably the ability of phytohaemagglutinins to agglutinate red blood cells is of little importance in influencing nutritive value as they are high molecular weight compounds and would probably not be absorbed.

In spite of these observations however, agglutinating potential is still widely used as a means of measuring phytohaemagglutinin content of legumes, which is then linked with nutritive value. The situation is further complicated by differences in sensitivity of red blood cells to phytohaemagglutinin action. Marquardt *et al.* (1975) illustrated that rabbit erythrocytes were considerably more sensitive to phytohaemagglutinins from field beans than those from chickens, sheep or cattle (which showed negligible agglutination). Erythrocytes from mice, pigs, turkeys and rats showed intermediate responses between the two extremes. Results such as these are used frequently to justify the use of rabbit erythrocytes in determinations of phytohaemagglutinin content because they are the most sensitive. However, the considerable variability in the responses determined by choice of erythrocyte may make this a questionable procedure. Furthermore, it is evident that the mechanism of agglutination varies according to the origin of the phytohaemagglutinin in question (Marquardt, Campbell and Ward, 1976). It may not be valid therefore to employ a common analytical procedure to compare different legume species. Finally, Huisman and Van der Poel (1987) reported that some so-called phytohaemagglutinins only have one binding site and are not therefore able to agglutinate erythrocytes, although they are able to complex with the gut wall. This suggests that the term phytohaemagglutinin is inappropriate.

The phytohaemagglutinin content of field beans, peas and legumes measured using rabbit erythrocytes are presented in *Table 2.12*. The values indicate that peas have higher levels than field beans, with no consistent trend within species, while lupins appeared to have negligible levels. However, in all cases the content was much lower

Table 2.12 PHYTOHAEMAGGLUTININ CONTENT (EXPRESSED AS UNITS/mg DETERMINED WITH RABBIT ERYTHROCYTES) IN LEGUMES

Field beans	($n = 26$)	25– 100
Peas	($n = 3$)	100– 400
Lupins	($n = 3$)	negligible
Raw defatted soyabean meal		1600–3200

(Valdebouze et al., 1980)

than that found in extracted soyabean meal. In terms of distribution throughout the seed, Marquardt et al. (1975) reported that phytohaemagglutinins were confined to the cotyledon in field beans.

The discussion above indicates that it would be extremely difficult to assess the biological significance of phytohaemagglutinins from the three legume species under consideration, not least because Marquardt, Campbell and Ward (1976) reported that they are readily inactivated below pH 4, and to recommend supposedly safe levels in compound diets. As with protease inhibitors, they are destroyed by heat, with moist heat conditions proving more effective. However, a reduction or destruction of phytohaemagglutinin content and the elimination of protease inhibitor activity does not fully account for the improvement in nutritive value of field beans following heating (Marquardt, Campbell and Ward, 1976), suggesting that other antinutritive factors are present.

Favism factors   Haemolytic anaemia, characterized by the disruption of the integrity of erythrocytes, following ingestion of field beans is a condition in susceptible humans referred to as favism. The two factors that have been implicated in producing this condition are vicine and convicine, which are pyrimidine glucosides, although in fact it is the aglycones that are thought to be responsible (Hill-Cottingham, 1983). Similar disorders would appear not to have been reported in livestock. However, vicine and convicine have been implicated as causative agents in reducing egg weight in laying hens (Olaboro, Marquardt and Campbell, 1981; Olaboro et al., 1981). The data presented in *Table 2.13* appear to indicate that there is a negative correlation

Table 2.13 INFLUENCE OF VICINE ON EGG WEIGHT IN LAYING POULTRY. A–G REFER TO EXTRACTS FROM FIELD BEANS ASSUMED TO CONTAIN VICINE AT VARIOUS RATES

Diet		Vicine (%)	Feed intake (g/hen/day)	Mean egg weight (g)	Egg production (%)	Liver lipid (%)	Yolk weight (g)
(a)	Control	0	98	52.8	—	30.0	—
	A	0.02	100	51.0	—	28.0	—
	B	0.37	100	51.8	—	22.0	—
	C	0.59	93	49.8	—	20.0	—
	D	1.17	94	47.7	—	22.0	—
(b)	Control	0	114	58.4	—	—	—
	E	0.40	107	54.6	—	—	—
	F	0.40	109	54.6	—	—	—
(c)	Control	0	113.6	64.2	82	—	18.4
	G	1.00	98.0	60.0	72	—	17.0

Olaboro, Marquardt and Campbell, 1981

between dietary vicine and egg weight. It was argued that the mechanism by which vicine reduces egg weight was through its adverse influence on hepatic lipogenesis, in a fashion similar to the alleged role of vicine in the development of favism, which resulted in smaller amounts of lipid present in the yolk. However, it would also be possible to conclude that the effect of vicine was mediated predominantly through its influence on feed intake, particularly as egg numbers were also reduced. An adverse metabolic role of vicine on lipid metabolism and egg weight of the sort implicated in producing favism would require the action of β-glucosidase. As already stated, it is the aglucone rather than the pyrimidine glucoside that is active. The presence of this enzyme *in vivo* is unlikely, plant sources probably have little effect and the aglucones themselves are unstable. All of these factors prompted Olsen and Andersen (1978) to question whether vicine and convicine are major causative agents in favism. Finally, studies with poultry have indicated that heat stable vicine is probably not of significant importance in chicks, as heating field beans produced a marked improvement in performance (Marquardt, Campbell and Ward, 1976).

Nevertheless, vicine and convicine are still implicated in reducing the biological value of protein for rats, and Bjerg *et al.* (1984) concluded that breeding for low or negligible levels would be of some considerable value. Pitz, Sosulski and Rowland (1981) reported a range of vicine and convicine levels respectively of 0.51–0.75 and 0.19–0.60% (DM basis) for 36 cultivars and concluded that, despite the confounding effects of environment and year, breeding for low pyrimidine glucoside levels was possible.

Processing of legumes

It has been shown that a number of antinutritive factors which have accepted adverse biological effects are present in field beans, peas and lupins, but that their practical significance is difficult to assess. There are numerous reports to the effect that heating legumes will inactivate the thermolabile proteases and phytohaemagglutinins but that the improvement in nutritive value is frequently greater than could be explained by the absence of these antinutritive factors. However, it must be said that the nutritive value of legumes is frequently lower than could be predicted from a knowledge of their chemical analysis, even allowing for the denaturation of heat labile antinutritive factors. It would be possible to conclude that other hitherto unidentified antinutritive heat labile factors are present in raw legumes, as has been proposed for example by Marquardt, Campbell and Ward (1976). Moseley and Griffiths (1979) suggested that an antigenic factor in field beans similar to those found in soyabean meal could be of importance in interfering with protein metabolism *in vivo*.

However, other studies have concentrated on the structure of proteins and carbohydrates present in legumes, as it is possible that they are inherently less digestible than similar nutrients found in cereals. This section will consider the effects of processing on nutritive value, and will concentrate on improving nutrient availability and dietary energy value *per se* rather than on reducing antinutritive factor activity. Assessments of the effect of heat processing are invariably confined to the biological consequences. However, both the practicality and economic feasibility of treatments should be considered. Thus, for example, autoclaving has been used frequently during experiments, but this is not a practical procedure.

Peas Extensive studies on the digestion of components of peas have recently been reported by Carre et al. (1987) and Longstaff and McNab (1987). Both groups subjected peas to a number of physical processes including grinding, steam pelleting and a variety of heat treatments. Selected results are presented in *Table 2.14*. It is evident that it is possible to enhance the nutritive value of peas through processing, and the improvements in dietary energy value closely matched those for coefficients of both starch and protein digestibility. The digestibility of starch is influenced by a number of factors including size of the starch granule, accessibility to enzymes and time of exposure (and, incidentally, by the analytical procedures adopted). Processes which reduce particle size will probably improve digestibility, particularly with poultry where rate of passage of food is relatively rapid. Additional heating processes may be of benefit, possibly because of disruption of intermolecular bonds in starch and its gelatinization, although Longstaff and McNab (1987) considered that the formation of retrograde starch with heating would, because this form is not readily digested, offset any advantages due to gelatinization if indeed such a development itself conferred any advantage on the utilization of starch by poultry.

What was evident from both studies was that none of the treatments imposed raised the coefficients of digestibility of pea starch to the value of 0.99 reported for wheat (Longstaff and McNab, 1987). The presence of an α-amylase inhibitor in peas was not considered likely (Longstaff and McNab, 1987).

Protein digestibility was improved following processing, which, it was argued, was attributable to lower anti-trypsin factor activity or to protein denaturation. Moran, Summers and Jones (1968) observed that moist heat treatment (autoclaving or steam pelleting) substantially improved the nutritive value of peas and concluded that this effect was probably a consequence of both protein denaturation and rupture of the starch granule. Dry heat processing was ineffective. It is, however, interesting to note the conclusions of Liener (1976) that even legume proteins fully denatured following cooking are incompletely attacked by digestive enzymes, and it should be appreciated that numerous studies have indicated that excessive heat treatment may, through the formation of a number of complexes, render protein less available to the animal.

Field beans In studies with broiler chicks, Marquardt et al. (1974) concluded that there was no evident improvement in growth performance following heat treatment of starch isolates from field beans, and Aherne, Lewis and Hardin (1977) observed that autoclaving field beans failed to improve their nutritive value for pigs. The effect of heat treatment on the nutritive value of field beans was investigated by McNab and Wilson (1974) who obtained an improvement in AME_N from 10.59 to 11.67 MJ/kg DM when raw beans were micronized, and this was accompanied by an increase in *in vitro* carbohydrate availability. Studies undertaken by Lacassagne et al. (1988) indicate that, as with peas, mechanical processing of field beans improves their nutritive value when fed to poultry by enhancing both starch and protein digestibility and, accordingly, dietary energy values (*Table 2.15*). The rate of improvement was greater than with peas (Carre et al., 1987) because the study with field beans was based on young chicks as opposed to adult cockerels. It is probable that the benefits of processing are greater in young stock. The AME_N values of raw field beans in the two studies described compare favourably with those obtained by Edwards and Duthie (1970, 1972) of 10.51 MJ/kg DM ($n = 11$), and 10.83 and 10.01 MJ/kg DM ($n = 6$) for both winter and spring cultivars.

The white-flowered tannin-free variety (*Table 2.15*) had, as expected, a greater protein digestibility than the coloured-flowered cultivars, but starch digestibility and AME_N were seemingly unaffected by tannin content. This would suggest that tannins

Table 2.14 INFLUENCE OF PROCESSING ON THE APPARENT AND TRUE METABOLIZABLE ENERGY VALUES CORRECTED TO ZERO NITROGEN RETENTION (AME$_N$, TME$_N$) AND COEFFICIENTS OF DIGESTIBILITY OF PROTEIN AND STARCH IN PEAS DETERMINED WITH ADULT COCKERELS

(a)

	Process						AME$_N$ (MJ/kg DM)	Coefficient of digestibility	
	Ground (2 mm screen)	Steam pellet (4 mm die)	Ground (2 mm screen)	Steam pellet (2 mm die)	Ground (2 mm screen)			True protein	Starch
Maize	+						12.79	0.809	0.921
Basal	+	+					13.02	0.831	0.968
	+	+	+				13.28	0.825	0.973
Wheat	+						12.11	0.735	0.909
Basal	+	+					12.67	0.772	0.950
	+			+	+		12.59	0.767	0.968

(b)

Process	TME$_N$ (MJ/kg)	Coefficient of starch digestion (two extractions)
Whole	9.91	0.756
Ground 0.5 mm	11.38	0.881
Heated, ground 0.5 mm	11.66	0.904
Autoclaved, ground 0.5 mm	11.06	0.912
Dehulled, ground 0.5 mm	12.39	0.928
Ground 1 mm	11.68	0.923
Cooked, ground 1 mm	12.20	0.918
Ground 1 mm + 'cellulase'	11.59–12.09	0.906

(a) Carre *et al.*, 1987
(b) Longstaff and McNab, 1987

Table 2.15 INFLUENCE OF MECHANICAL TREATMENT OF FIELD BEANS ON THE APPARENT METABOLIZABLE ENERGY CONTENT (MJ/kg DM) CORRECTED TO ZERO NITROGEN RETENTION (AME_N) AND COEFFICIENTS OF DIGESTION OF STARCH (S) AND PROTEIN (P) DETERMINED WITH YOUNG BROILERS

Process			Cultivar								
			Winter			Spring			White-flowered		
Ground (2 mm screen)	Steam pelleted (2.5 mm die)	Ground (2 mm screen)	AME_N	S	P	AME_N	S	P	AME_N	S	P
+			10.30	0.834	0.694	10.64	0.856	0.669	10.17	0.751	0.826
+	+		11.49	0.935	0.723	11.50	0.935	0.706	11.80	0.897	0.872
+	+	+	11.61	—	0.721	11.57	—	0.706	12.03	—	0.865

Lacassagne et al. (1988)

exert their antinutritive factor activity principally through binding with dietary protein rather than as anti-enzyme factors.

Lupins Since it would appear that lupins are comparatively free of the heat labile antinutritive factors no significant benefits arising from heat treatment would be expected. However, it is evident that the nutritive value of lupins is not as high as would be predicted from a knowledge of chemical composition. In a study of the site of digestion of components of lupin seed meal, Taverner, Curic and Rayner (1983) concluded that the apparent ileal availability of amino acids from lupin seed meal was high and of the order of 90%. However, comprehensive studies of lysine availability, as determined by slope assay techniques which therefore include assessments of metabolic as well as digestive and absorptive efficiency, indicated that availability was considerably lower in raw samples than those that had been heated (Batterham *et al.*, 1986a,b). It was suggested that low digestibility was not the problem and that therefore lysine metabolism was impaired either through the presence of a heat stable antinutritive factor or because it was present in a form that could not be utilized. Interestingly the poor response appears to be confined to pigs as it was reported that performance levels with rats and chicks were higher. Aguilera *et al.* (1986) reported that autoclaving had no effect on protein digestibility with rats, but that there was an improvement in biological value. They speculated that a heat labile antinutritive factor was present which interfered with protein metabolism, and discussed the possible role of a substance which increased the requirement for vitamin B_{12}. As with field beans, it may therefore be unwise to extrapolate from one animal species to another when evaluating lupins or feedstuffs generally.

The disappearance of energy from the hind gut of pigs fed lupins was also studied by Taverner, Curic and Rayner (1983) who observed that 32% of ingested gross energy was removed during passage from the ileum to the anus. This suggests a considerable amount of bacterial activity in the hind gut, and in fact the fibre fraction in lupins was reported to be fairly digestible. However, the efficiency of utilization of the volatile fatty acids thus produced is a contentious issue (*see* Low, 1985). Batterham *et al.* (1986b) concluded that energy deficiency was not responsible for the depressed growth responses obtained, which suggests a high efficiency of utilization. However, the allegedly low potential contribution of volatile fatty acids to energy balance in pigs (Pond, 1981) suggests that efficiency of utilization is comparatively poor. An additional problem with hind gut fermentation, which is more specific to legumes, is the potential for flatus production arising from the particular oligosaccharides present. The α-galactosides are probably not digested in the foregut of non-ruminants due to the absence of α-galactosidase but are broken down in the hind gut of pigs. They have been implicated as causative agents in flatus production (Fleming, 1981). Flatus would not of course be assessed in a simple balance experiment.

The consequences of excessive hind gut fermentation are that the ratio of net energy:digestible energy is likely to be considerably lower than anticipated. It is possible that this may be a reason for the lower than anticipated nutritive value of lupins and legumes in general.

Conclusion

It is evident that there is still a reluctance to incorporate field beans, peas and lupins routinely in large amounts into compound diets for non-ruminants, for which there may be a number of reasons.

Evidence has been presented relating to the considerable variability in nutrient content within cultivars of the same legume species, a situation which tends not to be reflected in pricing structure. Although the content of antinutritive factors in European legumes is generally low, it may be high enough within certain cultivars or samples to reduce nutritive value. This of course would present a few problems if legumes were routinely processed to denature heat labile factors. However, the efficacy of processing is questionable when antinutritive factor content is variable. A decision will always need to be made as to whether the savings associated with not having to process a sample which fortuitously contains low levels of antinutritive factors are greater than the financial loss associated with not processing and feeding a sample where antinutritive factor content may be high. Finally, variability in supply both between years and on a seasonal basis may present problems.

It is perhaps pertinent to note that soya, which is the raw material against which European legumes must compete, is traded on strict quality control criteria which is reflected in price. It is always processed prior to inclusion into diets for non-ruminants because it is known to contain levels of antinutritive factors which are dangerously high, and is also freely available year-round.

References

ABBEY, B., NEALE, R.J. and NORTON, G. (1979a). *British Journal of Nutrition*, **41**, 31–38
ABBEY, B., NEALE, R.J. and NORTON, G. (1979b). *British Journal of Nutrition*, **41**, 39–45
AGUILERA, J.F., PRIETO, C., FONOLLA, J. and GIL, F. (1986). *Animal Feed Science and Technology*, **15**, 33–40
AHERNE, F.X., LEWIS, A.J. and HARDIN, R.T. (1977). *Canadian Journal of Animal Science*, **57**, 321–328
BARRATT, D.H.P. (1982) *Journal of the Science of Food and Agriculture*, **33**, 603–608
BARRY, T.N. and MANLEY, T.R. (1986). *Journal of the Science of Food and Agriculture*, **37**, 248–254
BATTERHAM, E.S., ANDERSEN, L.M., BURNHAM, B.V. and TAYLOR, G.A. (1986a). *British Journal of Nutrition*, **55**, 169–177
BATTERHAM, E.S., ANDERSEN, L.M., LOWE, R.F. and DARNELL, R.E. (1986b). *British Journal of Nutrition*, **56**, 645–659
BJERG, B., EGGUM, B.O., OLSEN, O. and SORENSEN, H. (1984). In *Vicia faba: Agronomy, Physiology and Breeding*, pp. 287–296. Ed. Hebblethwaite, P.D., Dawkins, T.C.K., Heath, M.C. and Lockwood, G. Commission of the European Communities, Martinus Nijhoff/Dr W. Junk, The Hague.
BOND, D.A. (1970). *Proceedings of the Nutrition Society*, **29**, 74–79.
BRILLOUET, J-M. and CARRE, B. (1983). *Phytochemistry*, **22**, 841–847
BRILLOUET, J-M. and RIOCHET, D. (1983). *Journal of the Science of Food and Agriculture*, **34**, 861–868
CABRERA, A. and MARTIN, A. (1986). *Journal of Agricultural Science (Cambridge)*, **106**, 377–382
CARRE, B. and BRILLOUET, J-M. (1986). *Journal of the Science of Food and Agriculture*, **37**, 341–351
CARRE, B., ESCARTIN, R., MELCION, J.P., CHAMP, M., ROUX, G. and LECLERCQ, B. (1987). *British Poultry Science*, **28**, 219–229

CARRE, B. and LECLERCQ, B. (1985). *British Journal of Nutrition*, **54**, 669–680
CERNING, J., SAPOSNIK, A. and GUILBOT, A. (1975). *Cereal Chemistry*, **52**, 125–138
CERNING-BEROARD, J. and FILIATRE, A. (1976). *Cereal Chemistry*, **53**, 968–978
COLE, D.J.A. (1978). In *Recent Advances in Animal Nutrition*, pp. 59–72. Ed. Haresign, W. Butterworths, London
COLE, D.J.A., BLADES, R.J., TAYLOR, R. and LUSCOMBE, J.R. (1971). *Experimental Husbandry*, **20**, 6–11
COUSIN, R. (1983). In *Perspectives for Peas and Lupins as Protein Crops*, pp. 146–164. Ed. Thompson, R. and Casey, R. Commission of the European Communities, Martinus Nijhoff, The Hague
DAVIDSON, J. (1973). *British Poultry Science*, **14**, 557–567
DAVIES, R.L. (1984). *Australian Journal of Experimental Agriculture and Animal Husbandry*, **24**, 350–353
DAVIES, D.R. and DOWNEY, C. (1983). In *Perspectives for Peas and Lupins as Protein Crops*. Eds. Thompson, R. and Casey, R. Commission of the European Communities, Martinus Nijhoff, The Hague
DESHPANDE, S.S., CHERYAN, M. and SALUNKHE, D.K. (1986). *CRC Critical Reviews in Food Science and Nutrition*, **24**, 401–449
EDEN, A. (1968). *Journal of Agricultural Science (Cambridge)*, **70**, 299–301
EDWARDS, D.G. and DUTHIE, I.F. (1970). *Journal of Agricultural Science (Cambridge)*, **76**, 257–259
EDWARDS, D.G. and DUTHIE, I.F. (1972). *Journal of Agricultural Science (Cambridge)*, **79**, 169–170
EDWARDS, S.A., ROGERS-LEWIS, D.S. and FAIRBAIRN, C.B. (1987). *Journal of Agricultural Science (Cambridge)*, **108**, 383–388
EVANS, I.M. and BOULTER, D. (1980). *Journal of the Science of Food and Agriculture*, **31**, 238–242
FLEMING, S.E. (1981). *Journal of Food Science*, **46**, 794–798
FORD, J.E. and HEWITT, D. (1979a). *British Journal of Nutrition*, **42**, 317–323
FORD, J.E. and HEWITT, D. (1979b). *British Journal of Nutrition*, **42**, 325–340
GODFREY, N.W., MERCY, A.R., EMMS, Y. and PAYNE, H.G. (1985). *Australian Journal of Experimental Agriculture*, **25**, 791–795
GREENE, A.G. and ORAM, R.N. (1983). *Animal Feed Science and Technology*, **9**, 271–282
GRIFFITHS, D.W. (1979). *Journal of the Science of Food and Agriculture*, **30**, 458–462
GRIFFITHS, D.W. (1981). *Journal of the Science of Food and Agriculture*, **32**, 797–804
GRIFFITHS, D.W. (1984a). In *Vicia faba: Agronomy, Physiology and Breeding*, pp. 271–278. Ed. Hebblethwaite, P.D., Dawkins, T.C.K., Heath, M.O. and Lockwood, G. Commission of the European Communities, Martinus Nijhoff/Dr W. Junk, The Hague
GRIFFITHS, D.W. (1984b). *Journal of the Science of Food and Agriculture*, **35**, 481–486
GRIFFITHS, D.W. and JONES, D.I.H. (1977). In *Protein Quality from Leguminous Crops*, pp. 105–115. Commission of the European Communities
GRIFFITHS, D.W. and MOSELEY, G. (1980). *Journal of the Science of Food and Agriculture*, **31**, 255–259
HILL, G.D. (1977). *Nutrition Abstracts and Reviews*, **47**(8), 511–529
HILL-COTTINGHAM, D.G. (1983). In *The Faba Bean*, pp. 159–180. Ed. Hebblethwaite, P.D. Butterworths, London
HOLT, N.W. and SOSULSKI, F.W. (1979). *Canadian Journal of Plant Science*, **59**, 653–660
HUISMAN, J. and VAN DER POEL, A.F.B. (1987). *Proceedings of the XXXVIII European Association for Animal Production*. Lisbon

JAFFE (1980). In *Toxic Constituents of Plant Foodstuffs*. Ed. Liener, I.E. Academic Press, London
JOHNS, D.C. (1987). *New Zealand Journal of Agricultural Research*, **30**, 169–175
LACASSAGNE, L., FRANCESCH, M., CARRE, B. and MELCION, J.P. (1988). *Animal Feed Science and Technology* (in press)
LIENER, I. (1974). *Journal of Agricultural and Food Chemistry*, **22**, 17–22
LIENER, I. (1976). *Journal of Food Science*, **41**, 1076–1081
LONGSTAFF, M. and MCNAB, J. (1987). *British Poultry Science*, **28**, 261–285
LOW, A.G. (1985). In *Recent Advances in Animal Nutrition*, pp. 87–112. Ed. Cole, D.J.A., and Haresign, W. Butterworths, London
MARTIN-TANGUY, J., GUILLAUME, J. and KOSSA, A. (1977). *Journal of the Science of Food and Agriculture*, **28**, 757–765
MARQUARDT, R.R., CAMPBELL, L.D. and WARD, T. (1976). *Journal of Nutrition*, **106**, 275–284
MARQUARDT, R.R., CAMPBELL, L.D., STOTHERS, S.C. and MCKIRDY, J.A. (1974). *Canadian Journal of Animal Science*, **54**, 177–182
MARQUARDT, R.R., MCKIRDY, J.A., WARD, T. and CAMPBELL, L.D. (1975). *Canadian Journal of Animal Science*, **55**, 421–429
MCNAB, J.M. and WILSON, B.J. (1974). *Journal of the Science of Food and Agriculture*, **25**, 395–400
MITARU, B.N., REICHERT, R.D. and BLAIR, R. (1984). *Journal of Nutrition*, **114**, 1787–1796
MORAN, E.T., SUMMERS, J.D. and JONES, G.E. (1968). *Canadian Journal of Animal Science*, **48**, 47–55
MOSELEY, G. and GRIFFITHS, D.W. (1979). *Journal of the Science of Food and Agriculture*, **30**, 772–778
OLABORO, G., MARQUARDT, R.R. and CAMPBELL, L.D. (1981). *Journal of the Science of Food and Agriculture*, **32**, 1074–1080
OLABORO, G., MARQUARDT, R.R, CAMPBELL, L.D. and FROLICH, A.A. (1981). *Journal of the Science of Food and Agriculture*, **32**, 1163–1171
OLSEN, H.S. and ANDERSEN, J.H. (1978). *Journal of the Science of Food and Agriculture*, **29**, 323–331
PITZ, W.J., SOSULSKI, F.W. and ROWLAND, G.G. (1981). *Journal of the Science of Food and Agriculture*, **32**, 1–8
POND, W.G. (1981). *Journal of Animal Science*, **65**, 497–499
PRITCHARD, P.J., DRYBURGH, E.A. and WILSON, B.J. (1973). *Journal of the Science of Food and Agriculture*, **24**, 663–668
REICHERT, R.D. and MACKENZIE, S.L. (1982). *Journal of Agricultural and Food Chemistry*, **30**, 312–317
SJODIN, J., MARTENSEN, R. and MAGYAROSI (1981). *Zeitschrift fur Pflanzenzuchtung*, **86**, 231–247
TAVERNER, M.R., CURIC, D.M. and RAYNER, C.J. (1983). *Journal of the Science of Food and Agriculture*, **34**, 122–128
UKASTA (1986). In *Feed Facts 1986*. United Kingdom Agricultural Supply Trade Association, London
VALDEBOUZE, P., BERGERON, E., GABORIT, T. and DELORT-LAVAL, J. (1980). *Canadian Journal of Plant Science*, **60**, 695–701
WELCH, R.W. and GRIFFITHS, D.W. (1984). *Journal of the Science of Food and Agriculture*, **35**, 1282–1289
WILSON, B.J., MCNAB, J.M. and BENTLEY, H. (1972). *British Poultry Science*, **13**, 521–523

WISEMAN, J. (1980). *Feed Forum 80*, Society of Food Technologists, Dublin
WISEMAN, J. and LESSIRE, M. (1987). *British Poultry Science*, **28**, 663–676.
WOOD, J. (1984). In *Fats in Animal Nutrition*, pp. 407–435. Ed. Wiseman, J. Butterworths, London

3

VITAMIN E AND FREE RADICAL FORMATION: POSSIBLE IMPLICATIONS FOR ANIMAL NUTRITION

D. A. RICE and S. KENNEDY
Veterinary Research Laboratories, Stormont, Belfast, Northern Ireland

The enzymatic and chemical production of free radicals in biological systems is a normal ongoing process. For example, free radicals are produced by neutrophils during the process of phagocytosis and the killing of invading pathogens. They are produced in subcellular membranes during the enzymatic production of eicosanoids, e.g. prostaglandins, prostacyclins and leukotrienes. They are also produced in the mitochondria of cells, where electron transfer is utilized as a means of energy utilization. Thus, free radical production is a normal and necessary part of normal cell function.

There are, however, circumstances where the rate of production of free radicals or the type of radicals produced exceeds the body's capacity to maintain them, or restrain them within the limits of the cell where their effect is required. Under these circumstances, the normal body defence mechanisms cannot prevent the highly active free radicals from reacting with other molecules within the cell which are then destroyed or damaged in the process. Cell death, loss of function or genetic mutation can then ensue.

To prevent or at least minimize such damage, nature has provided a variety of defence systems. Many of these are enzymatic and most are under nutritional control. One of these nutritional defence systems is vitamin E. It acts as a lipophilic free radical scavenger minimizing free radical induced damage in the lipid environment of subcellular membranes. It is not the only mechanism available to the body for detoxifying free radicals, but its unique property lies in its ability to scavenge free radicals in a lipid environment, thus quenching the potential formation of lipid peroxides at source.

The purpose of this chapter is therefore to review current knowledge on the mechanism whereby vitamin E prevents free radical damage, to outline problems which result from faulty prevention and to suggest how these problems have been and are being overcome.

Lipid peroxidation

Auto-oxidation or peroxidation of lipids is a phenomenon well known to lipid chemists involved in the food industry. This oxidative breakdown of polyunsaturated fats leads to rancidity and all of its accompanying changes in the colour, taste and odour of food. As a consequence, there have been many years of research concerned with understanding the mechanism of this instability, all aimed at finding antioxidants capable of preventing it (Wiseman, 1986). Unsaturated fats, when they occur in biological systems, particularly in the subcellular membranes of cells, are no less susceptible to auto-oxidation. The major difference, however, between the biological and non-biological system is that peroxidation of lipids in the biological environment *cannot and should not be completely prevented*. The reason for this is that certain types of lipid peroxidation are beneficial. For example, the enzyme-catalysed peroxidation of lipid which precedes the formation of eicosanoids is absolutely essential for life, and any attempt to prevent this will lead to physiological upsets in normal cell function. Consequently, the objective of animal nutritionists is to feed the right balance of nutrients which will permit essential lipid peroxidation but prevent the unwanted co-oxidation of subcellular lipid which would result in cell death.

ESSENTIAL LIPID PEROXIDATION

The essential fatty acids, linoleic, linolenic and arachidonic, are the principal polyunsaturated fatty acids (PUFAs) required by mammals. The reason that they are essential relates to their requirement in subcellular membranes, e.g. lysosomal and mitochondrial membranes. In the membrane they have two functions: (1) they ensure that the membrane remains fluid and not rigid; and (2) they are the primary building block for eicosanoid synthesis.

Eicosanoids are enzyme-catalysed derivatives of polyunsaturated fatty acids. The principal enzymes involved in this process are cyclo-oxygenase and lipoxygenase. The principal substrate on which they act is arachidonic acid, although other 20-carbon fatty acids which contain at least three double bonds can also act as substrates. These fatty acids are released from the subcellular membrane by the action of phospholipase. The released fatty acid is converted by cyclo-oxygenase to an endoperoxide. This is subsequently enzymatically converted to a variety of eicosanoids, e.g. the prostaglandins PGA_1, PGA_2, PGE_1, PGE_2, $PGF_{2\alpha}$. Alternatively the endoperoxide may be converted to prostacyclin (PGI_2) and thromboxane (TXA_2). The released arachidonic acid may be converted by lipoxygenase to a hydroperoxy fatty acid which is a precursor for the leucotrienes.

All of these potent tissue hormones are essential for normal cell function. If the cell does not produce a balanced profile of these, then pathological abnormalities will result. For example, PGI_2 prevents the aggregation or clotting of platelets in blood whereas TXA_2 is pro-aggregatory. Should the production of these products cease, or an adequate balanced profile not be produced then clotting would ensue, a microthrombosis would form and the animal would die as a result of damage to blood capillaries. This subject has been reviewed recently (Holman, 1986).

Certain drugs can modify this profile, e.g. aspirin is a cyclo-oxygenase inhibitor. Aspirin therefore decreases the production of eicosanoids with the associated beneficial effects, felt by many humans, in reducing headaches. It also decreases the PGI_2/TXA_2 ratio thus minimizing the risk of thrombosis. Currently many millions of pounds are being spent on a search for lipoxygenase inhibitors, suitable for pharmacological use in humans. It has been predicted that such inhibitors will have marked beneficial effects on the progress of diseases resulting from chronic inflammatory diseases, e.g. emphysema, asthma, arthritis, etc. (*see* review by Higgins, 1985).

The eicosanoid profile can also be modified nutritionally. The principal method of achieving this is by altering the dietary input of fatty acids. Both the n-6 (linoleic family) and the n-3 (linolenic family) fatty acids act as substrates for cyclo-oxygenase and lipoxygenase. Eicosopentaenoic acid (C20:5, n-3), often referred to as eskimo oil, produces eicosanoids with decreased and often opposite physiological effects from those of its arachidonic based counterparts. Nutritional modification of the eicosanoid profile by increased dietary concentration of n-3 fatty acids is currently receiving a lot of research effort (Higgs, Moncada and Vane, 1986).

Thus, it can be seen that lipid peroxidation is a normal essential ongoing body process which cannot be terminated. It must however be controlled in its own micro-environment within the cell, because if not, then disease may ensue. It would appear from chemical studies of these reactions that under normal circumstances the stereochemistry of the PUFA/enzyme interaction will not permit the PUFA free radical to inadvertently attack contiguous PUFA not involved in the reaction (Ingold, 1983). However, it has been hypothesized that this system may be 'leaky', i.e. that free radicals may on occasion leak from the controlled environment (McKay and King, 1980). The most likely reason why vitamin E is an essential nutrient is because it acts as a free radical scavenger, mopping up 'leaked' PUFA radicals that escape during the arachidonic acid cascade.

Since vitamin E is a natural ingredient of feedstuffs and since eicosanoid synthesis is a normal ongoing process, one would imagine that there would normally be sufficient vitamin E in the diet to prevent a chain reaction occurring as a result of 'leaking' PUFA radicals and disease would not normally ensue. This has probably been true during the evolutionary process, as animals became adapted to their normal diets. However, with modern farming methods vitamin E requirements of animals now exceed the levels present in the diet under many circumstances. For example, animals which evolved on grass and legume diets, high in vitamin E, are now obliged to eat large quantities of cereals and by-products which are low in this vitamin. In addition, the vitamin E requirement of livestock is not constant but can be increased, depending on the 'rate' of lipid peroxidation. If the lipid peroxidation challenge is initiated by a high 'oxygen radical' attack, then there may be a higher requirement for vitamin E to prevent 'auto-oxidation' of membrane PUFA than that which is required to maintain an optimum PUFA/cyclo-oxygenase/vitamin E ratio in the eicosanoid producing environment.

Auto-oxidation initiated by oxygen radicals

At this stage it is necessary to define what is meant by a 'free radical'. It is a molecule, or part of a molecule containing a single unpaired electron. This unpaired electron confers on the molecule a considerable degree of chemical reactivity. Not all free radicals have the same degree of reactivity, some being virtually unreactive, others being highly reactive and thus potentially dangerous. These free radicals can have either reducing or oxidizing properties which can vary, for the same radical, depending on the experimental or biological conditions. Oxidizing radicals are the most important from the standpoint of vitamin E because they can abstract a hydrogen from PUFA thus initiating lipid peroxidation.

Free radicals are produced in biological systems: (a) enzymatically, (b) by the action of metal ions, (c) by ionizing radiation, or (d) by the presence of xenobiotics.

OXYGEN FREE RADICALS

Until the point in history when blue-green algae evolved, oxygen radicals presented no problems. At that time, however, nature developed a method of using oxygen in aerobic metabolism, which increased energy utilization from food. As humans, we are in no doubt of the essentiality of oxygen. For the most part, however, we are unaware of the potential problems associated with the molecule. These problems apparently relate to the triple ground state, electron spins and the conservation of angular momentum of the molecule which together produce a variety of oxygen free radicals.

A number of oxygen species known to cause disease are shown in *Table 3.1*. They share the same property in that they are radicals or have the potential to cause radical reactions in biological systems (Pryor, 1986).

These oxygen radicals can initiate lipid peroxidation, which, once initiated will self-propagate. As with lipid peroxidation, not all oxygen radical production is detrimental. It occurs in aerobic environments as a host defence mechanism. In this process phagocytes such as neutrophils produce superoxide and hydrogen peroxide, by an enzymatic process in which electrons are transferred from NADPH to O_2. This production of active oxygen species is referred to as the 'respiratory burst'. The superoxide and hydrogen peroxide or derivatives of these, are used by the phagocyte to kill pathogens which they have engulfed (Forman and Thomas, 1986). Macrophages which do not produce or produce low concentrations of superoxide are incapable of killing invading pathogens and leave the animal more susceptible to disease. However, this point will be discussed later in more detail. At this stage we are more concerned with the other role which these oxygen radicals have in the host animal, a detrimental role of initiating lipid peroxidation.

The process by which these radicals are produced in the animal are as follows:

$$O_2 + e^- \rightarrow O_2\cdot \text{ (superoxide radical)} \tag{3.1}$$

Superoxide dismutase (a copper dependent enzyme) converts superoxide to hydrogen peroxide:

$$2O_2\cdot^- + 2H^+ \rightarrow H_2O_2 + O_2 \tag{3.2}$$
$$\text{(hydrogen peroxide)}$$

Table 3.1 EXAMPLES OF SOME OXYGEN FREE RADICALS AND OXYGEN SPECIES IMPORTANT IN FREE RADICAL PRODUCTION

Symbol[a]	Description of radical
$^1O_2$	Singlet oxygen
NO_2	Nitrogen dioxide
H_2O_2	Hydrogen peroxide
HO·	Hydroxyl radical
$O_2\cdot^-$	Superoxide
HOO·	Hydroperoxyl radical
RO·	Aloxyl radical
ROO·	Peroxyl radical

[a]The free radical is symbolized by a dot (·)

Although superoxide is not particularly active it can be rapidly converted *in vivo* to much more active radicals such as its protonated form ($HO_2\cdot$) and the hydroxyl radical ($OH\cdot$). This latter radical is formed from a Haber Weiss–Fenton reaction involving an iron catalyst as indicated by the two stages:

$$H_2O_2 + Fe^{++} \rightarrow OH\cdot + Fe^{+++} + OH^- \qquad (3.3)$$

$$O_2\cdot^- + Fe^{+++} \rightarrow Fe^{++} + O_2 \qquad (3.4)$$

The sum effect of these reactions is represented by:

$$O_2\cdot^- + H_2O_2 \xrightarrow{\frac{Fe^{+++}}{Fe^{++}}} OH^- + OH\cdot + O_2 \qquad (3.5)$$

There is currently much research into the toxicity and particularly the carcinogenicity of oxygen radicals. Examples of diseases where oxygen radicals are thought to play important aetiological roles are: inflammation, rheumatoid arthritis, atherosclerosis, lung disorders, alcoholism, toxic liver damage and in cardiac injury as a result of reperfusion (*see* reviews by Slater *et al.*, 1987, Halliwell, 1987; Dianzani, 1987; Golden and Ramdath, 1987; Jackson, 1987). It is beyond the scope of this chapter to discuss these diseases in detail. However, it is important to realize that these oxygen radicals, which can be formed by a variety of means, can initiate lipid peroxidation by abstracting a hydrogen atom from a specific methylinic carbon of PUFA. The principal culprit is hydroxyl radical.

The process by which this progresses is:

$$LH - H \rightarrow L\cdot \qquad \qquad \textit{initiation} \quad (3.6)$$

$$L\cdot + O_2 \rightarrow LOO\cdot$$
(lipid peroxy free radical)

Once initiated, the lipid peroxidation chain reaction is self-perpetuating until it is chemically terminated:

$$LOO\cdot + LH \rightarrow LOOH + L\cdot \qquad \textit{propagation} \quad (3.7)$$
(lipid)

Termination of this reaction can be achieved by reaction of the lipid radical with other cell components including protein and DNA. It can also, of course, be terminated by interaction with vitamin E which is a free radical scavenger.

While enzymatically initiated lipid peroxidation is a natural beneficial continuous process in the body and oxygen free radicals are also continuously produced in the body for host defence purposes, the detrimental effects of these two processes are kept under control by a variety of trace nutrient-dependent mechanisms. However, under certain circumstances both enzymatic and oxygen free radicals may leak from their well controlled micro-environment and attack the peroxidation-susceptible PUFAs in subcellular membranes. This produces a lipid peroxidation chain reaction and results in disease.

Problems caused by uncontrolled lipid peroxidation

The cell damage and pathological changes occasioned by a lipid peroxidation chain reaction will depend on the magnitude of the reaction, the function of the damaged membrane and associated molecules and the function of the cell.

As previously stated, lipid peroxidation is mainly a problem in subcellular membranes. The resulting pathology can therefore be related to the effects of this damage on individual components of the membrane or its environs or the collective effects of the summed damage. Each organelle within the cell, e.g. golgi complex, lysosomes, mitochondria, endoplasmic reticulum, has its own particular membrane keeping it as a separate compartment within itself. This membrane not only has a physical restraining function, but also maintains within the membrane many essential molecules, e.g. proteins, required for regulating cell metabolism. It also regulates ion concentrations within the compartment. The membrane also has in its environment a variety of additional molecules which will interact with molecules within the membrane.

Consequently, lipid peroxidation chain reactions can lead to (a) reduction of the PUFA concentration in the membrane, (b) cross-linking of peroxidized PUFA, (c) fragility of the membrane, (d) increased membrane permeability, e.g. to calcium, (e) damage to membrane proteins, and (f) inability of substrates to recognize their target enzyme in the altered 'laterally segregated lipid domains' of the membrane.

In addition, the lipid free radicals may react with nucleic acids within the cell. This can cause somatic mutation and lead to carcinogenesis. The lipid peroxides may also act as eicosanoid precursors, leading to the production of an unwanted profile of prostaglandins, protacyclins and thromboxanes within the cell. They could also lead to further upsets in cell metabolism.

The above relates only to the toxic effects of the lipid peroxides *per se*. However, the hydroperoxides, once formed, are unstable. These are further metabolized to a variety of cell-toxic products, e.g. aldehydes.

MECHANISM BY WHICH VITAMIN E PREVENTS LIPID PEROXIDATION

Dam (1957), subsequent to a series of elegant experiments in which he and his co-workers demonstrated the interactions between vitamin E and high levels of PUFA in the diet, proposed that vitamin E acted as an *in vivo* anti-oxidant. This was supported by his ability to demonstrate peroxides in tissues of deficient animals. This hypothesis was subsequently built on by Tappel (1962) and, in spite of alternative proposals, it is still the most tenable hypothesis. The most predominant alternative theory proposed by Diplock and co-workers (*see* review by Diplock, 1983) suggested that vitamin E became an integral component of the subcellular membrane in a stabilizing form. This proposal is considered by many to be untenable because of the fluid nature of membranes where such 'lock and key' interactions are unlikely, and because there is no theoretical evidence that there would be sufficient chemical energy available to maintain such a rigid bond.

More recently, Infante (1986) has reviewed the aspects of vitamin E function which are not readily explained by the anti-oxidant function. He has proposed that vitamin E deficiency results in a loss of function of n-3 and n-6 fatty acid desaturases. Whereas many of his arguments to sustain this proposal are hypothetical, such reviews indicate that not all branches of science are convinced by the anti-oxidant theory and most scientists would not be surprised if additional, non-anti-oxidant, roles for the vitamin were found.

The mechanism whereby vitamin E exerts its anti-oxidant effect is related to the phenol hydroxyl group (*Figure 3.1*). This can donate a hydrogen to the lipid free radical; the hydrogen reacts with the free electron, quenching that free radical and thus preventing a chain reaction. The tocopherol thus formed, is apparently continuously regenerated at the expense of vitamin C (Packer, Slater and Wilson, 1979). Thus:

$$\begin{matrix} LOO\cdot \\ LOOH \end{matrix} \begin{matrix} \alpha\text{-tocopherol OH} \\ \alpha\text{-tocopherol O}\cdot \end{matrix} \begin{matrix} \text{vitamin C}\cdot \\ \text{vitamin C} \end{matrix} \quad (3.8)$$

By this means vitamin E terminates the ongoing or potential lipid peroxidation chain reaction.

Vitamin E, in nature, occurs as α-, β-, γ- and δ-tocopherol and the same isomers of tocotrienols. In certain feedstuffs, α-tocopherol may occur at a relatively lower concentration than some of these other isomers. This presumably relates to a higher requirement of the other tocopherol isomers by these plants. In livestock, however, the major requirement appears to be for α-tocopherol. Although all isomers appear to be absorbed (Peake and Beirl, 1971), they also appear to be rapidly excreted. Studies in cattle and in pigs have shown that, although α-tocopherol did not predominate in feed, it was the only vitamin E isomer to occur in substantial quantities in blood and tissues (McMurray and Rice, 1982). While the critical structure component of vitamin E as a free radical scavenger is the hydroxyl group, which is common to all tocopherols and tocotrienols, the methyl groups on the chromanol ring and the structure of the side chain have obvious essential qualities for localization of this important radical quenching hydroxyl group in the subcellular membrane where it is required.

Molennaar, Hulstaert and Hardonk (1980) proposed a mechanism whereby α-tocopherol exerted its anti-oxidant function. They suggested that α-tocopherol becomes an integral component of the membrane in that area of the membrane where mixed function oxidases could potentially react with polyunsaturated fatty acids (*Figure 3.2*). The methyl groups on the phytol chain of the α-tocopherol would fit into

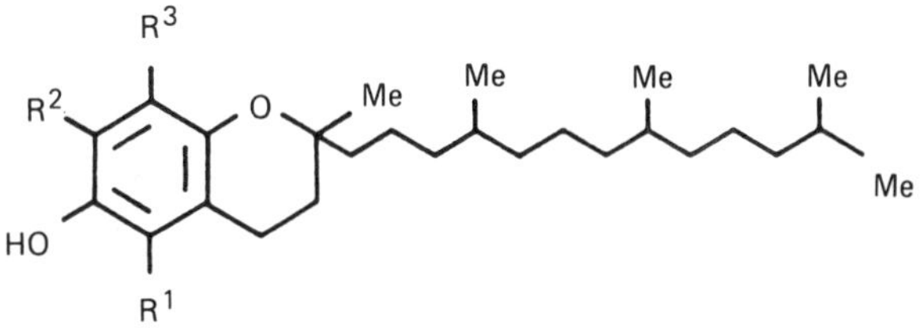

Compound	R^1	R^2	R^3
α-Tocopherol	Me	Me	Me
β-Tocopherol	Me	H	Me
γ-Tocopherol	H	Me	Me
δ-Tocopherol	H	H	Me

Figure 3.1 Formula of tocopherol showing the important radical-scavenging hydroxyl (OH) group. The biological activity of the different tocopherol compounds is dependent on the number and positioning of methyl groups on the chromanol ring at R^1, R^2 and R^3

46 Vitamin E and free radical formation

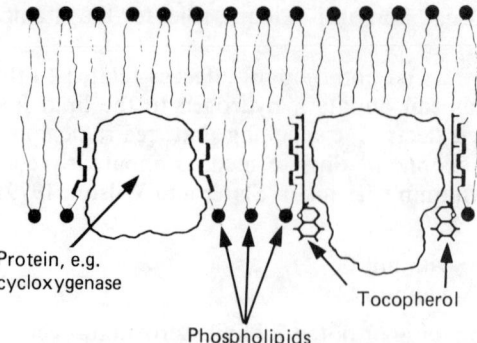

Figure 3.2 A diagrammatic representation of the localization of α-tocopherol in the phospholipid environment of the subcellular membrane. It prevents the co-oxidation of phospholipid PUFA adjacent to but not actually involved in eicosanoid synthesis. ⌐⌐ indicates the *cis* double bonds of the arachidonic acid in the phospholipid membrane. (After Molennaar, Hulstaert and Hardonk, 1980)

the cavities made by the *cis*-double bonds of the arachidonic acid. The theory of Diplock (1983) referred to earlier has been ridiculed because the ratio of α-tocopherol to arachidonic acid in biological membranes is 1:500 to 1:1000 and thus not all fatty acid molecules within the membrane could be protected by a tocopherol molecule. However, it is possible that since the long-chain PUFA most susceptible to peroxidative attack is that in close opposition to a 'leaky' radical producing enzyme (protein), then these may be the only ones requiring a contiguous α-tocopherol molecule.

Pathological consequences of lipid peroxidation

The most interesting aspect of lipid free radical disease in livestock, from the comparative pathology viewpoint, is the diversity of tissues affected by the disease and the great variety of pathological changes observed. This is in contrast to humans where vitamin E, having become the *in vogue* vitamin of the 1930s–1950s, fell into disrepute. The reasons for this are that although it showed such promise as an anti-sterility vitamin and an anti-myopathic vitamin in rats, rabbits and guinea pigs, the original experimental animals used by early workers in this field (*see* review by Mason, 1980), it failed quite miserably when tested in humans for similar functions.

More recently, certain diseases, particularly of neonatal children, have been associated with vitamin E deficiency and can be prevented by vitamin E supplementation (*see* review by Muller, 1987). From the beginning, however, a number of vitamin E responsive conditions of farm livestock were identified, and this list has continued to expand until today.

The most common manifestations of deficiency are myopathy of both skeletal and cardiac muscle; the vascular system is also affected resulting in testicular degeneration in the cock, dietetic microangiopathy in pigs, exudative diathesis of chicks and embryonic degeneration in hen and turkey eggs; the liver is affected by telangiectasis in the bovine and lipoid degeneration in fish; the pancreas is fibrosed in the chick; the cerebellum is affected causing encephalomalacia in the chick and adipose tissue becomes necrotic causing steatitis in the chick and pig. This list is by no means comprehensive; it merely presents some of the principal diseases associated with deficiency and probably reflects the tissues which are most susceptible to peroxidative attack in individual species of farm livestock. It should also be pointed out that if the peroxidative attack is particularly high in a particular species, unusual manifestations of vitamin E deficiency can be observed. For example, dietetic microangiopathy and exudative diathesis have been observed in deficient cattle, although very rarely.

It is not proposed to discuss all of the vitamin E responsive diseases which occur in farm animals; rather three particular examples will be selected in an attempt to consider whether they still occur, and if so why, using the experiences of the Veterinary Research Laboratories at Stormont in Northern Ireland. The three diseases in question are: (a) nutritional degenerative myopathy in steers, (b) pancreas disease/myopathy of farmed Atlantic salmon, and (c) dietetic microangiopathy of pigs. The effect of vitamin E on the immune response of livestock and on the porcine stress syndrome will also be discussed.

NUTRITIONAL DEGENERATIVE MYOPATHY

Whereas myopathy in cattle was traditionally a disease associated with pre-ruminant calves (Blaxter, 1955), in the early and mid-1970s it became prevalent in cattle 6–24 months old. This was shown by Allen *et al.* (1975) to be associated with the feeding of vitamin E and selenium-deficient diets. Since many parts of the UK and Ireland are selenium deficient, livestock eating homegrown, unsupplemented rations are often selenium deficient, or more precisely, their tissues are deficient in the selenium-dependent enzyme glutathione peroxidase. This enzyme detoxifies lipid peroxides in tissues. Such livestock therefore depend on the α-tocopherol they absorb from feeds, for their defence against lipid peroxidation. In the 1970s it became customary to treat cereal grains with propionic acid at harvesting, to prevent fungal growth during storage. This organic acid destroys the α-tocopherol (vitamin E) in cereals within weeks (Rice, Blanchflower and McMurray, 1985). Livestock therefore became deficient in both of these nutrients and as a consequence nutritional myopathy ensued (Rice and McMurray, 1986).

In addition, if subclinically deficient cattle were turned onto spring pasture, they often became severely clinically diseased. This was due to the surge of linolenic acid from within the grass into plasma and tissues. This work has clearly demonstrated a cause and effect relationship between the linolenic acid surge and the onset of myopathy. It has also overturned many currently held tenets on PUFA hydrogenation in the rumen (McMurray, Rice and Kennedy, 1983).

A summary of the stages in the development of degenerative myopathy is as follows:

(1) Many cattle are deficient in glutathione peroxidase due to low levels of selenium in their foodstuffs (McMurray and Rice, 1982).
(2) Preservatives such as propionic acid and sodium hydroxide will destroy the α-tocopherol in stored cereal grains (Rice, Blanchflower and McMurray, 1985).
(3) Cattle eating such diets will become deficient in both vitamin E and selenium, because of the synergistic effects between these two nutrients, this will result in subclinical myopathy (Rice and McMurray, 1986).
(4) Grass contains high levels of PUFA, particularly C18:3, (n-3) (linolenic acid). Much of this PUFA escapes rumen hydrogenation at turnout onto spring pasture, causing increased concentrations of linolenic acid in the phospholipids, triglycerides and cholesterol esters of plasma (McMurray, Rice and Kennedy, 1983).

(5) This plasma linolenic acid is incorporated into tissues, particularly into tissue phospholipids (Rice et al., 1986).
(6) The increase in arachidonic acid levels in tissues, resulting from vitamin E and selenium deficiency *per se*, together with the additional increase in tissue linolenic acid at turnout, are sufficient to explain the chronic type of myopathy which occurs indoors and the acute myopathy which occurs at turnout onto spring pasture (Rice et al., 1986).

These observations on n-3 fatty acid surges in tissues are obviously of much importance in relation to explaining the aetiology of myopathy at turnout of calves onto pastures. More recently, Horta, Chassagne and Brochart (1986) have made similar observations in relation to the linolenic acid surges in tissues of cows at turnout onto spring pasture. They also observed that this surge changes the ratio of n-3 to n-6 fatty acids in tissues of cows and have shown that this alters the profile of prostaglandins produced by the reproductive tract. They propose that this, in turn, may have a bearing on fertility.

DIETETIC MICROANGIOPATHY

This disease, sometimes referred to as mulberry heart disease, was shown to be caused by vitamin E and selenium deficiency by Grant (1961). The disease which causes sudden death in weaned/fattening pigs results from the formation of microthrombi in small capillaries and arterioles of cardiac muscle. This leads to haemorrhage in the heart and sudden death. The incidence of the disease has decreased considerably in recent years because of the higher levels of vitamin E and selenium being used in pig diets although cases of the disease do still occur.

In a study carried out in 1986–87 (Rice and Kennedy, 1988c), it was found that the current field disease is histologically similar but not identical to the experimental disease described by Grant (1961), Nafstad and Tollersrud (1970) and Van Vleet, Ferrans and Ruth (1977). Of more interest was the finding of lower concentrations of vitamin E in the tissues of affected pigs, particularly in heart tissue (*Table 3.2*). The glutathione peroxidase and selenium concentrations of tissues were not different.

Table 3.2 VITAMIN E (AS α-TOCOPHEROL) AND GLUTATHIONE PEROXIDASE (GSH Px) LEVELS OF TISSUES OF PIGS WITH DIETETIC MICROANGIOPATHY AND OF CONTROL PIGS

		Vitamin E (μg/g)		GSH-Px (μg/mg protein)	
		Liver	Heart	Kidney	Heart
Pigs with dietetic microangiopathy	Mean	2.02	2.63	363	216
	SD	0.84	1.34	93	162
	n	25	23	27	20
Controls	Mean	3.12	4.93	449	168
	SD	1.78	2.29	156	77
	n	26	24	26	24
Significance (unequal variance *t*-test)		$P < 0.01$	$P < 0.001$	$P < 0.05$	NS

After Rice and Kennedy (1988c)

The lower levels of vitamin E in affected pigs was not due to lower dietary supply, since the affected pigs received 44 mg/kg α-tocopherol compared with 40 mg/kg for controls. The oil content of the rations and the PUFA content were also similar in affected and control pigs.

Whatever the reason for the lower tissue levels of vitamin E, it is reasonable to assume that higher dietary levels of the vitamin would prevent even more cases of this disease. Whether farmers will be prepared to pay the extra cost of the higher levels of vitamin E supplementation required to eliminate it or whether they are prepared to live with the current incidence of mulberry heart disease is not known. However, if pigs are still developing mulberry heart disease because of a high peroxidative challenge (e.g. a large production of superoxide from activated macrophages), is their vitamin E status optimal? Moreover, is the vitamin E status of the tissues of their littermates optimal particularly in relation to host defence mechanisms? It may be that these questions will only be answered when high levels (200 mg/kg) of vitamin E as α-tocopherol, to boost the host's defence mechanisms, becomes more widespread under field conditions. Such levels may also eliminate dietetic microangiopathy.

PORCINE STRESS SYNDROME

Pigs have long been known to have higher levels of plasma creatine kinase than other species of livestock (Moss and McMurray, 1979), indicating a continuing release of this enzyme from muscle. Previous attempts to reduce these circulating creatine kinase levels in stress susceptible pigs by additional dietary vitamin E have been unsuccessful. However, recent work by Arthur and his co-workers has shown that the levels of creatine kinase can be reduced by supplementation of the diet with 235 mg/kg α-tocopherol. This decreased the circulating levels of lipid peroxidation by-products and prevented the disease (Duthie, Arthur and Hoppe, 1988). These results indicate that previous workers aiming to prevent the disease were not using high enough levels of vitamin E. These data, together with those presented earlier for dietetic microangiopathy, suggest that the vitamin E requirement of pigs in the field are not currently being fully met. It is also possible that other vitamin E deficiency diseases, as yet undiagnosed, also exist in pigs.

PANCREAS DISEASE OF SALMON

This disease of farmed Atlantic salmon is observed during their first season in sea cages, and causes a high mortality rate. Recovering fish are severely emaciated and thrive poorly. The name 'pancreas disease' was shown to be a misnomer by Ferguson *et al.* (1986) who demonstrated that affected fish had severe degenerative myopathy, affecting principally heart and red skeletal muscle although the oesophagus and other skeletal muscles are also affected. Subsequent studies demonstrated that the principal lesions were associated with vitamin E deficiency (Ferguson, Rice and Lynas, 1986). It is still not clear, however, whether this deficiency is primary or induced, i.e. whether it causes pancreatic malfunction or results from it. It was, however, quite clear from those studies that the vitamin E status of plasma and liver was severely depleted. The selenium and glutathione peroxidase status of the tissues, although marginally lower in diseased fish appeared to be responding to the reduced food intake rather than having any aetiological significance (*Table 3.3*).

Table 3.3 PLASMA AND TISSUE DATA OF ATLANTIC SALMON AFFECTED BY PANCREAS DISEASE

Group			Vitamin E (as α-tocopherol)		Selenium		Glutathione peroxidase Erythrocytes (U/g Hb)	Creatinine kinase Plasma (U/l)
			Liver (μg/g)	Plasma (μmol/l)	Liver (μg/g)	Plasma (μmol/l)		
(1)	Clinically affected fish	Mean	20.8	4.9	1.4	0.32	54	81 770
		SD	8.2	3.9	0.5	0.21	11	122 000
		n	10	6	10	10	10	10
(2)	Cage mates of clinically affected fish	Mean	64.7	14.4	2.3	0.28	58	84 300
		SD	34	11.7	0.4	0.09	16	59 464
		n	10	7	10	10	10	
(3)	Healthy fish from an unaffected site	Mean	280	68.3	3.9	1.10	64	4 475
		SD	148	17.5	0.7	0.18	9	3 072
		n	10	8	10	9	9	9
Significance								
Unequal variance *t*-test		1v2	$P < 0.001$	$P < 0.05$	$P < 0.001$	NS	NS	NS
		1v3	$P < 0.001$	$P < 0.001$	$P < 0.001$	$P < 0.001$	$P < 0.05$	$P < 0.05$
		2v3	$P < 0.001$	$P < 0.001$	$P < 0.001$	$P < 0.001$	NS	$P < 0.001$

From Ferguson, Rice and Lynas (1986)

The feed from the affected site contained 98 mg/kg α-tocopherol, 1.26 mg/kg selenium and 14.5% fat. The polyunsaturated fatty acid content of the fat was high, with > 50% represented as C18:2 (n-6) or longer chain length acids, but the peroxide value was low at less than 1.0 mEq of oxygen/kg of lipid. The problem did not therefore appear to relate to rancid fat in the ration. There were no differences in the polyunsaturated fatty acid content of tissues, with particular reference to liver. It would therefore appear that either the level of vitamin E in the ration was inadequate or that there was a conditioned deficiency related to lack of uptake. The precise aetiology has still not been satisfactorily resolved. However, until it is resolved, there appears to be sufficient evidence to recommend that the α-tocopherol levels of the diets of such salmon should be increased substantially to a value in the region of 200–400 mg/kg. This vitamin E concentration is already used under field conditions by rainbow trout producers (Ferguson, Rice and Lynas, 1986).

VITAMIN E AND THE IMMUNOCOMPETENCE OF LIVESTOCK

Earlier in this chapter, discussion centred on the mechanisms of free radical production with particular reference to the initiation of lipid peroxidation. It is now widely believed that the major mechanisms involve oxygen radical production in phagocytes and eicosanoid production in subcellular membranes. Because both of these processes are integral components of the animal's host defence mechanisms, it is not surprising that recent research has demonstrated a direct effect of vitamin E deficiency on these host defence mechanisms *per se*. This effect of vitamin E deficiency has two manifestations, firstly, on the phagocytic cell which is rendered less able to engulf and kill invading pathogens and secondly, in decreasing the host animal's ability to mount specific B- and T-cell immune responses to challenge by an alien antigen. Thus, both antibody production and cell mediated immunity are compromised.

What is perhaps more surprising are reports that much higher than 'normal' levels of either parenteral or oral vitamin E improve the immunocompetence still further. Examples of some of these effects are summarized in *Table 3.4*. It would appear from examination of these results that the addition of 100–300 mg/kg α-tocopherol to the feed will certainly boost the animal's resistance to disease. It will also boost its humoral antibody response to vaccination which in turn improves the resistance of vaccinated animals to disease.

It has been shown that vitamin E affects eicosanoid synthesis in the animal and it has been proposed that this is the principal mechanism by which it modulates the immune system of livestock. Lafuze *et al.* (1983) have demonstrated that as a result of antigenic challenge, neutrophils are activated, free radicals are thus produced and the neutrophils adhere to endothelial surfaces. With high doses of vitamin E, the neutrophil did not adhere so readily to endothelium but remained in the circulation. There was also a lowered production of H_2O_2 by the neutrophil. This would result in a decreased risk of both H_2O_2 or hydroxyl/radical attack on endothelial cells and direct peroxidative attack on the endothelium. There would also be an enhanced production of PGI_2 by the endothelium, resulting in a decreased adherence of neutrophils to endothelium.

Other workers have shown similar modifications of the prostanoid profile in a variety of tissues, as a result of vitamin E supplementation. In relation to the prevention of clotting and thrombosis, Toivenen (1987) has demonstrated beneficial

Table 3.4 EFFECT OF DIETARY VITAMIN E (AS α-TOCOPHEROL) ON THE IMMUNE RESPONSE

	Level of vitamin E in diet (mg/kg)	Challenge	Effect of the higher dose(s)
Chickens	150, 300	E. coli	↓ mortality ↑ antibody titres[a]
Turkeys	100	E. coli	↓ mortality, no change in antibody titre[b]
	300	E. coli	↓ mortality ↑ antibody titre[b]
Hens	150	Brucella abortus vaccine	↑ maternal transfer of antibody to chicks[c]
Hens	450	Brucella abortus	↑ antibody titres in chicks vaccinated at 2 weeks of age[c]
Pigs	110	E. coli vaccine	↑ antibody titre[d]
Lambs	300	Clostridial vaccine	↑ antibody titres[e]
Sheep	300	Chlamydia induced pneumonia	↑ weight gain and decreased[f] isolation of organism

[a] Heinzerling, Nockels and Tengerdy (1974)
[b] Julseth (1974)
[c] Jackson, Law and Nockels (1978)
[d] Ellis and Vorshies (1976)
[e] Tengerdy et al. (1983)
[f] Stephens, McChesney and Nockels (1979)

roles for vitamin E as well as selenium and vitamin C. These nutrients decrease the production of thromboxane (TXA_2) which promotes clotting and increase the production of anti-aggregatory PGI_2 (*Table 3.5*). Similarly, Lawrence et al. (1985) have demonstrated that vitamin E decreases PGE_2, PGF_2 and TXB_2 concentrations in the bursa of chicks infected with *E. coli*. They also showed that alteration to the eicosanoid profile induced by vitamin E was organ specific. This has led to the hypothesis that vitamin E enhances antibody production by depressing PGE_2 which is a known T-lymphocyte suppressing agent (Lawrence et al., 1985).

Those working in the livestock feed industry will realize that in spite of these demonstrated beneficial effects of vitamin E at levels of 100–300 mg/kg as α-tocopherol in the feed on the immune status of animals, compounders have not increased dietary vitamin E to anywhere near these levels. This indicates that the feed industry is either not convinced by the arguments that vitamin E produces a consistent beneficial effect on the host response or feels that the cost/benefit of the higher levels of inclusion is suspect. Certainly not all studies where vitamin E has been included in the ration at low levels have shown evidence of a decreased immunocompetence of deficient livestock (Anderson, Hartley and Bennett, 1986) nor have all experiments with a high inclusion level of vitamin E demonstrated an enhancement of immunocompetence (Marsh, Dietert and Combs, 1981).

When those involved in deciding 'inclusion levels' of micronutrients, or indeed any nutrient, in a compound feed read these contradictory statements, it cannot enhance their opinion of the usefulness of the research process. They need information which will demonstrate a consistent beneficial effect of the nutrient in a particular species before they can justify recommending increased expenditure. At the very least they need to be able to predict when a beneficial response is likely to be obtained.

The inconsistencies in experimental findings may reflect the presence of other *unmeasured* variables which were not recorded by the researchers. It is possible that

Table 3.5 EFFECTS OF VITAMIN E, SELENIUM AND VITAMIN C ON THE SYNTHESIS OF EICOSANOIDS[a] WHICH MODULATE BLOOD CLOTTING IN VASCULAR ENDOTHELIUM

	Endothelial prostacyclin	Platelet thromboxane
Vitamin E	No effect	Marked decrease
Selenium	No effect	Decrease
Vitamin C	Marked increase	Slight decrease

The study was carried out *in vitro* using human vascular endothelial cells in tissue culture (after Toivenen, 1987)

[a]Prostacyclin inhibits whereas thromboxane promotes platelet aggregation

the beneficial effects of vitamin E supplementation will be obtained only when certain selected variables are in operation. As already stated, the production of free radicals is a complex process with many modulating effects. In the experiments described above, the lack of a beneficial effect may be due to an inadequate free radical challenge in controls or an excessive challenge in the treated groups. There are many factors which may increase free radical production and elimination, and therefore influence the likelihood of a vitamin E benefit being obtained:

(1) *Intercurrent disease* The presence of disease, clinical or subclinical will increase the production of oxygen radicals, as the host's phagocytes attempt to kill invading pathogens. Similarly, an antigenic response will stimulate eicosanoid synthesis, the produced prostaglandins regulating the immune response. This prostanoid synthesis may therefore additionally increase the likelihood of 'leakage' of lipid free radicals from their enzyme microenvironment in the subcellular membrane. Thus, under different experimental conditions, inapparent subclinical disease in control animals may increase the peroxidative challenge and enhance the beneficial effect of vitamin E supplementation observed in treated animals.

(2) *Toxins* Many toxins actually initiate oxygen free radical production, e.g. carbon tetrachloride, ozone, nitrogen dioxide, ethanol (Slater *et al.*, 1987; Pryor, 1986). Consequently, toxins in the feed or environment may also increase the peroxidative challenge and have a similar effect to that mentioned above for intercurrent disease.

(3) *Other nutritional factors* Dietary concentrations of vitamin A, vitamin C, copper, iron and selenium are examples of nutrients which at adequate or high levels in the diet will have a modulating effect on free radical production. Vitamin A is known to quench free radical production under certain circumstances (Burton and Ingold, 1984) and can also boost the immune response at high levels of intake (Vyas and Chandra, 1984). Vitamin C at low concentrations initiates lipid peroxidation and at high concentrations quenches it (McCay, 1985), probably by recycling the vitamin E radical. This would allow a single vitamin E molecule to scavenge many free radicals, thereby potentiating its usefulness (Packer, Slater and Wilson, 1979). Copper, as an integral component of superoxide dismutase, is essential for converting the superoxide radical to hydrogen peroxide, so preventing the formation of the highly active hydroxyl radical. Iron and selenium, through their roles in catalase and glutathione peroxidase, respectively, remove hydrogen peroxide from the cell, pre-empting

its conversion to the hydroxyl radical. All of these factors, if in optimum balance within the animal's tissues, will tend to minimize free radical damage by eliminating free radicals as they are produced. Conversely, diets that are suboptimal will increase risk and increase the likelihood of a vitamin E responsive condition.

Thus, as previously stated (Rice and Kennedy, 1988a) the likelihood of disease occurring or of the immune response suffering as a result of vitamin E deficiency will depend on other factors within the cell:

(1) The inherent or genetically predetermined degree of unsaturation in the membrane.
(2) The concentration of PUFA within the membrane resulting from dietary input. High levels of dietary PUFA will increase the PUFA concentration of membranes.
(3) The level of desaturation of PUFA in the membrane, as longer-chain PUFA with more double bonds is a higher peroxidative risk.
(4) The intensity of the initiation of the peroxidative challenge, whether oxygen radical mediated or related to enzyme driven prostanoid synthesis, i.e. this may be mediated by intercurrent disease.
(5) The inherent ability of the subcellular membrane, under attack, to maintain high concentrations of peroxidation antagonists, e.g. vitamin E, glutathione, peroxidase, superoxide dismutase, catalase, etc. within the cell.
(6) The concentration of these antagonists in the cell as a result of dietary input and of their bio-availability to the animal.

Those experiments which demonstrate the beneficial effect of vitamin E on the immune response may reflect the ability of high levels of vitamin E to prevent self-inflicted injury to lymphoreticular and other host defence system cells, as they mount a free radical mediated attack on microbial or toxin induced insult. Under field conditions, high dietary inputs of vitamin E will be required only where there is a specific challenge to the host's defence systems. Therefore, an individual producer may only expect a response when certain stresses occur on his livestock. This may in turn mean that there may be an inconsistent effect on any one farm, benefit occurring only when challenge occurs. This hypothesis can only be tested in the field using sophisticated epidemiological techniques involving cohort comparisons. To date no such trials have been done.

To carry out such studies, large numbers of flocks/herds will have to be studied and accurate records kept for comparative purposes. An example of such production, management, disease computerized recording system has recently been published (McIlroy, Goodall and McMurray, 1988). In that recording system, the profitability of broilers has been shown to be inversely related to factors such as subclinical Gumboro disease and the incidence of skin lesions (breast burns) which, in turn, has been related to climatic variables. Testing the usefulness of high levels of vitamin E in large field trials, using large numbers of birds and producers, it should be possible to define the circumstances under which an improved response to supplementation will be obtained in terms of management, nutrition and disease variables.

The additional benefit of utilizing these higher levels of vitamin E under such conditions will demonstrate whether the appearance of the classic vitamin E responsive diseases such as dietetic microangiopathy of pigs and skeletal myopathy of turkeys or more recent ones such as the porcine stress syndrome can be totally eliminated. The sudden deaths of pigs resulting from dietetic microangiopathy may be eliminated if producers can be convinced that the cost of the higher levels of vitamin E will reduce not only the sporadic cases of sudden death but also boost the immune response.

WELFARE IMPLICATIONS

There is currently a tide of opinion among animal behaviourists, veterinarians and indeed the public at large (motivated by animal rights pressure groups), that livestock must be housed, managed and fed according to their 'physiological and ethological needs'. This opinion resulted in the production of a 'European Convention for the protection of animals kept for farming purposes' by the Council of Europe in 1976. As a signatory of this, all current and future UK legislation must comply with the spirit of this Convention.

Few people within the industry have any doubt that in the future the consumer and legislation will demand that animals shall not be knowingly maintained in a suboptimal nutritional state which could affect their well-being. Producers may well be forced in the future to feed levels of all nutrients which are compatible with optimum animal welfare. They may not be permitted to 'take a chance', as is currently the case with vitamin E, that the number of deaths or lack of productivity resulting from deficiency, will be less than the cost of inclusion of these beneficial higher levels.

References

ALLEN, W.M., PARR, H.W., BRADLEY, R., SWANNOCK, K., BARTON, C. and TYLER, R. (1975). *Veterinary Record*, **94**, 373–375

ANDERSON, P.H., HARTLEY, P. and BENNETT, S. (1986). In *Proceedings of the Sixth International Conference on Production Disease in Farm Animals*, pp. 203–206. Ed. McMurray, C.H., Rice, D.A., Kennedy, S. and McLoughlin, M. Veterinary Research Laboratories, N. Ireland

BAXTER, K.L. (1962). *Vitamins and Hormones*, **20**, 633–643

BLAXTER, K.L. and MCGILL, R.F. (1955). *Veterinary Reviews and Annotations*, **1**, 91–114

BURTON and INGOLD (1984). *Science*, **224**, 569–573

DAM, H. (1957). *Pharmacological Reviews*, **24**, 1–16.

DIANZANI, M.U. (1987). *Proceedings of the Nutrition Society*, **46**, 43–52

DIPLOCK, A.T. (1983), In *Biology of Vitamin E*, pp 45–55. Ciba Foundation Symposium—101. Pitman, London

DUTHIE, G., ARTHUR, J. and HOPPE, P. (1988). In *Oxygen Radicals in Biology and Medicine*. Ed. Simic, M. Plenum, New York (in press)

ELLIS, R.P. and VORHIES, M.V. (1976). *Journal of the American Veterinary Medical Association*, **168**, 231–232

FERGUSON, H.W., ROBERTS, R.J., RICHARDS, R.H., COLLINS, R.O. and RICE, D.A. (1986). *Journal of Fish Diseases*, **20**, 95–98

FERGUSON, H.W., RICE, D.A. and LYNAS, J.K. (1986). *Veterinary Record*, **119**, 297–299

FORMAN, H.J. and THOMAS, M.J. (1986). *Annual Review of Physiology*, **48**, 669–680

GOLDEN, M.H.N. and RAMDATH, D. (1987). *Proceedings of the Nutrition Society*, **46**, 53–68

GRANT, C.A. (1961). *Acta Veterinaria Scandanavica*, **2** (Suppl) 3
HALLIWELL, B. (1987). *Proceedings of the Nutrition Society*, **46**, 13–26
HEINZERLING, R.H., NOCKELS, C.L. and TENGERDY, R.P. (1974). *Proceedings of the Society for Experimental Biology and Medicine*, **146**, 279–283
HIGGINS, A.J. (1985). *Veterinary Quarterly*, **7**, 44–59
HIGGS, E.A., MONCADA, S. and VANE, J.R. (1986). *Progress in Lipid Research*, **25**, 5–11
HOLMAN, R.T. (Ed.) (1986). *Essential Fatty Acids, Prostaglandin and Leucotrienes: the Second International Congress.* (*Progress in Lipid Research*, vol. 25). Pergamon, Oxford
HORTA, A.E.M., CHASSAGNE, M. and BROCHART, M. (1986). *Annales de Recherche Veterinaire*, **17**, 353–359
INFANTE, J.P. (1986). *Molecular and Cellular Biochemistry*, **69**, 93–108
INGOLD, K.U. (1983). In *Biology of Vitamin E*, pp. 242. Ciba Foundation Symposium–101. Pitman, London
JACKSON, D.W., LAW, G.R.J. and NOCKELS, C.F. (1978). *Poultry Science*, **57**, 70–73
JACKSON, M.J. (1987). *Proceedings of the Nutrition Society*, **46**, 77–80
JULSETH, D.R. (1974). Evaluation of vitamin E and disease stress in turkey performance. MS thesis, Colorado State University, Fort Collins, USA
LAFUZE, J.E., WEISMAN, S.J., INGRAHAM, L.M., BUTTERICK, C.J., ALPERT, L.A., BAEHNER, R.L. (1983). In *Biology of Vitamin E*, pp. 130–146. Ciba Foundation Symposium–101. Pitman, London
LAWRENCE, L.M., MATHIAS, M.M., NOCKELS, C.F. and TENGERDY, P. (1985). *Nutrition Research*, **5**, 497–509
MCCAY, R.B. (1985). *Annual Review of Nutrition*, **5**, 323–340
MCILROY, G., GOODALL, E.A. and MCMURRAY, C.H. (1988). *Agricultural Systems*, **27**, 11–22
MCKAY, P.B. and KING, M.M. (1980). In *Vitamin E: A Comprehensive Treatise*. Ed. Machlin, L.J. Marcel Dekker Inc., New York
MCMURRAY, C.H. and RICE, D.A. (1982). *Irish Veterinary Journal*, **36**, 57–67
MCMURRAY, C.H., RICE, D.A. and KENNEDY, S. (1983). In *Biology of Vitamin E*, pp. 201–233. Ciba Foundation Symposium–101. Pitman, London
MARSH, J.A., DIETERT, R.R. and COMBS, G.F. (1981). *Proceedings of the Society for Experimental Biology and Medicine*, **166**, 228–236
MASON, K.E. (1980). In *Vitamin E: A Comprehensive Treatise*, pp. 1–6. Ed. Machlin, L.J. Marcel Dekker Inc., New York
MOLENNAAR, I., HULSTAERT, C. and HARDONK, M. (1980). In *Vitamin E: A Comprehensive Treatise*, pp. 372–390. Ed. Machlin, L.J. Marcel Dekker Inc., New York
MOSS, B. and MCMURRAY, C.H. (1979). *Research in Veterinary Science*, **26**, 1–6
MULLER, D.P.R. (1987). *Proceedings of the Nutrition Society*, **46**, 69–75
NAFSTAD, I. and TOLLERSRUD, S. (1970). *Acta Veterinaria Scandinavica*, **11**, 1–29
NOCKELS, C.F. (1987). In *Recent Advances in Animal Nutrition—1986*, pp. 177–192. Ed. Haresign, W. and Cole, D.J.A. Butterworths, London
PACKER, J.E., SLATER, T.F. and WILSON, R.L. (1979). *Nature*, **278**, 737–738
PEAKE, I.R. and BEIRL, J.G. (1971). *Journal of Nutrition*, **101**, 1615–1619
PRYOR, W.A. (1986). *Annual Review of Physiology*, **48**, 657–666
RICE, D.A., BLANCHFLOWER, W.J. and MCMURRAY, C.H. (1985). *Journal of Agricultural Science (Cambridge)*, **105**, 15–19
RICE, D.A. and KENNEDY, S. (1988a). *Proceedings of the Nutrition Society*, **47**, 43–50
RICE, D.A. and KENNEDY, S. (1988b). *British Veterinary Journal*, 144

RICE, D.A. and KENNEDY, S. (1988c). In *Trace Elements in Man and Animals—6*. Ed. Hurley, L.S. and Keen, G.L. Lonnerdal, B. and Rucker, R. Plenum, New York

RICE, D.A., KENNEDY, S., MCMURRAY, C.H. and BLANCHFLOWER, W.J. (1986). In *Proceedings of the Sixth International Conference on Production Disease in Farm Animals*, pp. 229–233. Ed. McMurray, C.H., Rice, D.A., Kennedy, S. and McLoughlin, M. Veterinary Research Laboratories, N. Ireland

RICE, D.A. and MCMURRAY, C.H. (1986). *Veterinary Record*, **118**, 173–176

SLATER, T.F., CHEESEMAN, K.H., DAVIES, M.J., PROUDFOOT, K. and XIN, W. (1987). *Proceedings of the Nutrition Society*, **46**, 1–12

STEPHENS, L.C., MCCHESNEY, A.E. and NOCKELS, C.F. (1979). *British Veterinary Journal*, **135**, 291–293

TAPPEL, A.L. (1962). *Vitamins and Hormones*, **20**, 493–510

TENGERDY, R.P., MEYER, D.L., LAUERMAN, L.H., LUEKER, D.C. and NOCKELS, C.F. (1983). *British Veterinary Journal*, **139**, 147–152

TOIVENEN, J.L. (1987). *Prostaglandins, Leucotrienes and Medicine*, **26**, 265–280

VAN FLEET, J.F., FERRANS, V.J. and RUTH, G.R. (1977). *Laboratory Investigations*, **37**, 201–211

VYAS, D. and CHANDRA, R.K. (1984). In *Nutrition, Disease Resistance and Immune Function*, pp. 325–343. Ed. Watson, R.R. Marcel Dekker Inc., New York

WILSON, R.L. (1987). *Proceedings of the Nutrition Society*, **46**, 27–34

WISEMAN, J. (1986). In *Recent Advances in Animal Nutrition—1986*, pp. 47–76. Ed. Haresign, W. and Cole, D.J.A. Butterworths, London

II

Pig Nutrition

4

ACIDIFICATION OF DIETS FOR PIGS

R. A. EASTER
Department of Animal Sciences, University of Illinois, Urbana, Illinois, USA

Introduction

The evolution of a biologically and economically satisfactory strategy for feeding piglets weaned at an early age is not yet complete. Progress is being made on new research providing fundamental information on the functional capabilities, and limitations, of the gastrointestinal tract during the transition from sow's milk to dry, cereal-based diets. A key discovery was the recognition that the young piglet may not be able to maintain appropriate gastric pH during the period. The discussion that follows provides an overview of the research leading to that conclusion and the experimental evidence to establish the basis for the use of acidification in weanling pig diets.

Physiological difficulties for the early-weaned piglet

Growth failure is a well established phenomenon in early-weaned pigs (Okai, Aherne and Hardin, 1976). Undoubtedly, this is a manifestation of an array of interacting environmental, social and physiological factors, not the least of which is digestive immaturity. The data reported by Etheridge, Seerley and Huber (1984) and presented in *Table 4.1* serve to illustrate this problem. Pigs were either provided cereal-based diets, beginning at day 7, and weaned at 21 days of age or allowed to suckle the sow for 35 days post partum without access to dry feed. Faecal samples which were collected and analysed provided an indication that fermentation in the lower bowel was significantly greater in pigs receiving the cereal-based diet. Faecal osmolarity and volatile fatty acid concentrations were both increased.

These results are not particularly surprising. It has been adequately demonstrated that the weanling pig is ill-prepared, enzymatically (Becker *et al.*, 1954; Corring, Aumaitre and Durand, 1978), to digest the complex carbohydrates found in most cereal-based weaner diets. Attempts to cause precocious maturation of digestive functions by treatment with hydrocortisone and/or ACTH have met with only limited success (Chapple, Cuaron and Easter, 1983). Incomplete digestion results in passage of fermentable substrate into the lower bowel. The osmotic upset observed by Etheridge, Seerley and Huber (1984) supports the notion that there is a relationship between diarrhoea and digestive development. Early-weaned pigs also lack the

Table 4.1 EFFECT OF DIET-TYPE ON MEASURED OSMOLARITY AND OSMOTICALLY ACTIVE CONSTITUENTS IN FAECAL CONTENTS OF PIGS AT 35 DAYS OF AGE (VALUES, EXCEPT FOR pH, ARE EXPRESSED AS mosmol/litre)

	Dietary treatment		
	Maize–soyabean meal	Oats–casein	Sow's milk
pH	5.9	6.5	7.1
Lactic acid$^-$	14.1	5.5	0.3
Total VFA$^-$	8.7	3.9	2.7
Na$^+$	13.8	10.8	2.0
K$^+$	15.0	10.0	20.0
Cl$^-$	4.7	3.0	1.1
Ca$^+$	1.4	0.5	0.3
R$^-$	12.2	10.3	11.3
Measured	149.8	88.5	49.0

After Etheridge, Seerley and Huber (1984)

capacity to produce gastric acid and it is probable that this has a negative effect on digestion. This chapter will provide a review of the information that leads to the hypothesis that digestion and, consequently, growth efficiency can be enhanced by diet acidification.

The need for dietary acid presumes that there is a deficiency in the pig's ability to maintain proper gastric pH. Mature pigs adjust stomach pH by secretion of hydrochloric acid from the parietal cells. Although there is variation due to diet, time after the meal at which the pH is measured and sampling site, stomach of mature pigs can reach very acid values, i.e. pH 2.0–3.5 (Slivitskii, 1975 as cited by Kidder and Manners, 1978).

The situation in young pigs is quite different. Although a subject of debate for some time, it is now evident that the newborn pig does produce some hydrochloric acid (Forte, Forte and Machen, 1975; Cranwell, Noakes and Hill, 1976). Initially, the production is low but increases with advancing age (Cranwell, 1985). This is apparently the consequence of limited secretory capacity and not lack of stimulation. Cranwell (1985) reported that maximal acid output in response to intravenous betazole hydrochloride infusion averaged 3.4 mmol H^+/h for pigs at 9–12 days and increased to 7.6 mmol H^+/h when pigs reached an age of 27–38 days.

The suckling pig employs several strategies to solve the problem of limited acid secretion. First, the primary carbohydrate in sow's milk is lactose which can be converted to lactic acid by the *Lactobacillus* bacteria normally resident in the stomach. This, in fact, appears to be the primary method of gastric acidification in suckling pigs. Secondly, nursing pigs reduce the need for momentary secretion of copious amounts of acid by consuming frequent, relatively small, meals (Pond and Maner, 1984). Finally, diets differ greatly in buffering capacity (Manners, 1970). Sow's milk is undoubtedly easier to acidify than is a high-protein weanling diet that is richly supplemented with calcium carbonate. Maner *et al.* (1962) found that stomach pH values dropped below 2.0 within 2 h after feeding a casein–dextrose diet while more than 4 h were required for a similar pH to be reached when pigs were fed a soyabean protein–dextrose diet.

What are the consequences of failure to maintain a low gastric pH? There are two working hypotheses. First, stomach pH has a role in preventing the movement of viable bacteria from the environment into the upper small intestine (Stevens, 1977).

Second, hydrochloric acid is involved in the activation of pepsinogens. Additionally pepsin has two pH optima, one at pH 2.0 and another at pH 3.5 (Rerat, 1981). Thus, it is likely that pigs having an elevated gastric pH also experience a net reduction in efficiency of protein digestion.

Response of piglets to organic acids

It isn't at all surprising that the first attempts to use acidification in pig farming were directed at the alleviation of post-weaning diarrhoea (*cf.* Kershaw, Luscombe and Cole, 1966). These workers found that 1% lactic acid addition to the drinking water would improve growth rate and feed efficiency. They also reported reductions in the *Escherichia coli* count in the duodenum and jejunum of pigs fed acids in the drinking water. In another experiment, Cole, Beal and Luscombe (1968) found that growth rate and feed efficiency were significantly improved by the addition of 0.8% lactic acid to the drinking water (*Table 4.2*). Moreover, there was a reduction in haemolytic *E. coli* counts in both the duodenum and jejunum.

The hypothesis that acidification may reduce the incidence of scours in young pigs was tested in an experiment by White *et al.* (1969). Pigs were separated from (or left with) the sow 48 h post partum. Those removed from the sow were given a standard rearing diet or that diet with sufficient lactic acid to reduce the diet pH to 4.8. Pigs fed the diets with lactic acid had lower stomach pH values than either those fed the standard diet or those that were allowed to suckle the sow. Acidification provided a prophylaxis against scouring (*Table 4.3*). These results imply an effect of lactic acid therapy on stomach pH values. This was confirmed by Thomlinson and Lawrence (1981). They also reported that the multiplication of *E. coli* 0141:K85(B) was reduced by acidification with a corresponding reduction in piglet mortality.

Table 4.2 EFFECT ON GROWTH PERFORMANCE OF ADDING SEVERAL MATERIALS AT 0.8% TO THE DRINKING WATER OF POST-WEANLING PIGLETS

	Control	Lactic acid	Propionic acid	Calcium propionate	Calcium acrylate
Feed intake (kg/day)	1.009	1.027	0.927	0.931	0.868
Liveweight gain (kg/day)[a]	0.372	0.409	0.354	0.345	0.310
Gain: feed ratio (kg gain/kg feed)	0.367	0.395	0.373	0.370	0.355

After Cole, Beal and Luscombe (1968)
[a] Improvement due to lactic acid ($P < 0.05$)

Table 4.3 EFFECT OF ACIDIFICATION OF THE DIET ON INCIDENCE OF SCOURING

	Nursed by sow	Post-weaning diet treatment		
		'Normal' diet	Lactic acid diet	High casein diet
No. pigs	6	7	6	6
No. scouring	4	7	3	6
Mean days scouring/pig	1.2	6.0	1.0	3.5

After White *et al.* (1969)

The ultimate value of organic acids in the prophylaxis of post-weaning diarrhoea is yet to be fully established. It is of passing interest that 'folk' remedies in tropical regions of the world include the use of lime juice to treat scours in young piglets (Costa Rican swine producer, personal communication).

Research attention in the 1970s turned from reduction of diarrhoea to more general effects on growth rate and efficiency of feed utilization. A report by Kirchgessner and Roth (1982) summarized several experiments involving the use of fumaric acid. These studies showed improved liveweight gain, feed intake and feed conversion efficiency when weaner pigs were fed diets supplemented with 1.5–2.0% fumaric acid. They showed that older pigs also responded to fumaric acid but the magnitude of the response was less. The effects were attributed in part to improved digestibilities of nutrients. Nitrogen balance was improved by 5–7% and the metabolizable energy values of the diets were increased by 1.5–2.1% (Kirchgessner and Roth, 1980).

Attempts to replicate the European work in North America have met with mixed success. In an early experiment, Lewis (1981) fed pigs, weaned at four weeks of age, diets containing graded levels of fumaric acid. These diets were based on maize and soyabean meal and contained small amounts of dried whey, fat and fish solubles. Gain was improved but evidence of an improvement in feed intake or feed conversion efficiency was lacking. Moreover, the incidence of scours was unaffected.

Using relatively more complex diets, i.e. containing barley, wheat, oat groats, soyabean meal, fish meal, dried skim milk, tallow, vitamins and minerals, Falkowski and Aherne (1984) reported that grain tended to be improved by 1 or 2% addition of either fumaric acid or citric acid but the effect was not significant. Feed conversion efficiency was improved ($P < 0.05$) by 5–10%, depending on treatment. The response was similar for both fumaric and citric acid. The trend for feed conversion efficiency to respond more consistently than growth has been confirmed by Giesting and Easter (1985).

The response to acid appeared to occur independently of effects due to other growth promotants. Edmonds, Izquierdo and Baker (1985) conducted a series of factorial experiments to evaluate the interaction of citric acid, copper sulphate and antibiotics relative to effects on growth performance of weanling pigs. Acidification improved performance in the presence and absence of both copper and antibiotic. As in previous studies, the feed conversion efficiency response tended to be of greater magnitude than the growth response.

Similar positive benefits were noted when 3.0% citric acid was added to the diets of pigs weaned at ten days of age (Henry, Pickard and Hughes, 1985). In these experiments the response to citric acid was greater than the response to fumaric acid. Interestingly, when pigs were given a choice there was evidence of a selection preference for non-acidified diets.

Giesting and Easter (1985) compared the response of weaner pigs to 2% additions of propionic, fumaric and citric acid to simple diets formulated with maize, soyabean meal, vitamins and minerals. The pigs were weaned at an average age of 30 days and were fed the assigned diets for a four-week period. Addition of each acid improved efficiency of gain but propionic acid caused a reduction in food intake and a depression in gain. In a second experiment, fumaric acid additions of 1, 2 or 3% were made to the diet. There was a linear improvement (*Table 4.4*) in both growth rate and feed conversion efficiency with the maximum response at 3% acid addition.

Table 4.4 EFFECT OF GRADED LEVELS OF FUMARIC ACID ON GROWTH-PERFORMANCE OF WEANER PIGS (10.0–18.7 kg)

	Fumaric acid level (%)					
	0	1	2	3	4	PSE[a]
Diet pH	5.96	4.77	4.33	3.98	3.80	
Liveweight gain (g/day)	261	261	257	296	297	14.6
Feed intake (g/day)	501	484	445	493	493	23.4
Gain: feed ratio (kg gain/kg feed)	0.52	0.54	0.57	0.60	0.60	0.02

After Giesting and Easter (1985)
[a] Pooled standard error

Relationship of diet type to the acidification response

The fact that Giesting and Easter (1985) were able to obtain a response to higher levels of acid than those used previously led to the hypothesis of a possible interaction between diet type and the acid response. Prior to the work by Giesting and Easter (1985), acid additions had been made to diets containing some milk product. The lactose contained in these diets would have been available for the formation of lactic acid which may have ameliorated the need for dietary acid.

The results from an experiment designed to evaluate the relationship between diet type and the acid response are presented in *Table 4.5* (Giesting, 1986). Diets were formulated with either maize and soyabean meal or maize with 11.95% soyabean meal and 25% dried skim milk. The level of acid addition to each type was 0, 2 or 3%. The maximum response was obtained with 2% acid when the diet containing dried skim milk. In contrast, there was a linear gain and efficiency response with up to 3% acid in the simple, maize–soyabean meal diet.

It has been suggested (Kidder, 1982) that vegetable-protein diets are more difficult for the weaner pig to digest than are diets having milk protein as the supplemental amino acid source. If proteolysis is enhanced by acidification, then it might be expected that pigs fed diets formulated with soya–protein concentrate would exhibit a

Table 4.5 EFFECT OF FUMARIC ACID ADDITION ON PERFORMANCE OF PIGS FED DIETS FORMULATED WITH SOYABEAN MEAL OR DRIED SKIM MILK[a]

Protein source:	Soyabean meal			Dried skim milk			
% Fumaric acid:	0	2	3	0	2	3	PSEM[b]
	Weeks 0–2						
Liveweight gain (g/day)	133	152	171	195	223	195	23.3
Feed intake (g/day)	306	296	304	312	332	310	22.1
Gain: feed ratio (kg gain/kg feed)	0.430	0.510	0.560	0.600	0.670	0.630	0.0486
	Weeks 0–4						
Liveweight gain (g/day)	289	320	311	327	359	350	22.3
Feed intake (g/day)	540	549	533	532	565	536	33.1
Gain: feed ratio (kg gain/kg feed)	0.540	0.580	0.580	0.610	0.640	0.650	0.137

[a] After Giesting (1986). The average initial weight was 8.2 kg and the duration of the experiment was four weeks
[b] Pooled standard error of the mean

greater response to acid than pigs fed diets prepared with casein. The results of an experiment to test that hypothesis are shown in *Table 4.6* (Giesting, 1986). Zero or 3% fumaric acid was added to diets of similar lysine content. Both liveweight gain and feed conversion efficiency were improved by acid addition. There was, however, an interaction with diet type. Pigs fed the diet formulated with soya–protein concentrate responded more to acidification than did pigs fed the diet containing a large amount of casein.

This experiment included a fifth treatment combination wherein sodium bicarbonate was added to the fumaric acid-supplemented, casein-based diet in an attempt to demonstrate that the effect of the acid could be negated by neutralization of the acid. There was an intriguing response to bicarbonate addition. Pigs fed the diet containing both bicarbonate and fumaric acid grew more efficiently ($P < 0.05$) than did those consuming the other diets. This response to bicarbonate was unexpected, but has been confirmed (Roos, Giesting and Easter, 1987) by subsequent experiments. Giesting (1986) has proposed that the bicarbonate response results from correction of a metabolic acid load, i.e. H^+, arising from ingestion of large quantities of fumaric acid. Kirchgessner and Roth (1982) reported that rats fed high-energy, low-protein diets supplemented with fumaric acid had elevated activities of glutamate–oxaloacetate transaminase and glutamate–pyruvate transaminase in the liver, along with reduced serum urea levels. The increased enzyme levels may be an indicator of increased deamination of amino acids to produce ammonia to buffer H^+ in the urine.

In an attempt to establish that diet acidification does, indeed, improve protein digestion, Giesting (1986) used surgically modified pigs to obtain ileal samples on which to base digestion estimates. A simple 'T' cannula was installed in the terminal ileum at two weeks of age using a modification of the procedure first described by Funderburke *et al.* (1982). Following surgery, the pigs were returned to the sow and allowed to suckle normally until four weeks of age. They were weaned, placed in individual cages, fed test diets and ileal samples were obtained. The pigs were fed *ad libitum* and digestion coefficients were calculated using chronic oxide as an indigestible marker.

The results of an experiment (Giesting, 1986) designed to examine the effects of diet type, i.e. milk-based versus soya-based, and fumaric acid addition on ileal digestibility values are presented in *Table 4.7*. Not unexpectedly, there was substantial variation in the data. It was clear, however, that both nitrogen and dry matter digestibility increased linearly with age. Both nitrogen and dry matter digestibility were greater for diets containing skim milk than for diets formulated with soyabean meal. Trends for improved digestibility with fumaric acid addition were present but not significant.

These data also suggest that the acid response may decline with maturation of the gastrointestinal tract. To test this hypothesis, Giesting and Easter (1985) fed diets containing fumaric acid to finishing pigs. There was no response in either growth rate or efficiency of feed conversion.

Alternatives to organic acids

In view of economic considerations, alternatives to the organic acids discussed above have been investigated for their potential as diet acidifiers. Giesting (1986) attempted to demonstrate growth responses to the addition of hydrochloric, phosphoric and sulphuric acids in amounts calculated to provide acidification similar to that obtained with 3% fumaric acid. Concentrated hydrochloric acid addition to weaner-pig diets

Table 4.6 EFFECT OF FUMARIC ACID ADDITION ON PERFORMANCE OF PIGS FED DIETS BASED ON SOYA–PROTEIN CONCENTRATE OR CASEIN[a,b]

Protein source:	Soya–protein concentrate	Soya–protein concentrate	Casein	Casein	Casein	PSEM[c]
Fumaric acid:	–	+	–	+	+	
Sodium bicarbonate:	–	–	–	–	–	
			Weeks 0–2			
Liveweight gain (g/day)	123	174	148	169	198	13.3
Feed intake (g/day)	306	322	330	352	348	16.4
Gain: feed ratio (kg gain/kg feed)	0.410	0.530	0.460	0.480	0.570	0.0279
			Weeks 0–4			
Liveweight gain (g/day)	298	311	292	330	315	11.7
Feed intake (g/day)	580	598	578	621	592	19.5
Gain: feed ratio (kg gain/kg feed)	0.510	0.520	0.500	0.530	0.530	0.0104

[a]After Giesting (1986). The average initial weight was 7.4 kg and the duration of the experiment was four weeks
[b]Fumaric acid addition was 3.0%, sodium bicarbonate addition was 2.74%
[c]Pooled standard error of the mean

Table 4.7 EFFECT OF DIET TYPE, TIME AFTER WEANING, AND FUMARIC ACID ADDITION ON ILEAL DIGESTIBILITY VALUES FOR NITROGEN AND DRY MATTER[a]

	By week				
Item	1	2	3	4	PSEM[c]
No. of observations	18	19	17	14	
Liveweight gain (g/day)	142	207	418	573	28.1
DM digestibility (%)	68.5	69.8	75.9	71.8	1.55
N digestibility (%)	63.5	69.7	77.8	75.2	1.90
	By diet				
Fumaric acid:	−	+	−	+	
Protein:[b]	SBM	SBM	DSM	DSM	
Liveweight gain (g/day)	231	265	421	423	28.1
DM digestibility (%)	68.8	69.4	73.8	74.0	1.55
N digestibility (%)	67.7	70.5	74.0	74.0	1.90

[a] After Giesting (1986). Pigs were fitted with ileal cannulae at 14 days of age, allowed to suckle until weaned at day 25 of life. Samples were obtained during the following four weeks
[b] Abbreviations: SBM = simple, maize–soyabean meal diet, DSM = maize diet with 25% dried skim milk
[c] Pooled standard error of the mean

resulted in a severe depression in growth. This is not surprising in view of the fact that this treatment resulted in a diet with about 1.3% Cl^-. Dietary electrolyte balance affects animal performance. The calculated index ($Na^+ + K^+ - Cl^-$ expressed in mEq/100 g of diet) was − 6.7. Data from experiments conducted by Patience, Austic and Boyd (1987) provide evidence that an index value in the negative range is consistent with dramatic reductions in growth. Sulphuric acid addition also depressed performance, probably for the same reason.

Of the three inorganic acids tests, only phosphoric did not result in a growth depression. However, there was no indication of improvement in performance. This is particularly disappointing in view of the fact that this acid could serve a dual role as a source of both acidity and inorganic phosphorus.

Other organic acids have also been investigated. For example, Schutte and van Weerden (1986) found positive effects from feeding calcium formate in combination with either fumaric or propionic acid to pigs during a 35-day period beginning at an initial weight of 12.5 kg. The results are presented in *Table 4.8*. This comprehensive experiment also included the use of copper, added as copper sulphate. Growth rate and feed conversion efficiency were significantly improved by calcium formate addition but not by fumaric acid addition. Calcium formate in combination with fumaric acid gave a significant response in feed conversion efficiency but not in growth rate. The best performance was obtained when calcium formate was used in combination with the copper. This agrees nicely with the earlier observation by Edmonds, Izquierdo and Baker (1985) regarding the additivity of the organic acid and copper sulphate responses.

A logical extension of the diet acidification research is the hypothesis that performance of young pigs can be enhanced by the addition of lactic acid-producing miocrobes to the diet on the assumption that this will result in increased formation of lactic acid in the stomach. Pollmann, Danielson and Peo (1980a) fed weaner pigs having an average initial weight of 7 kg, diets supplemented with *Lactobacillus acidophilus* or *Streptococcus faecium*. There was a significant improvement in daily

Table 4.8 EFFECT OF CALCIUM FORMATE ON GROWTH-PERFORMANCE OF PIGS

Diet description	Liveweight[a] gain	Gain: feed ratio (kg gain/kg feed)	Feed intake (g/day)
Control[b]	16.6	0.51	927
+ 1.5% fumaric acid	16.4	0.52	901
+ 1.5% calcium formate	18.6	0.53	995
+ 1.0% fumaric acid + 0.5% calcium formate	17.6	0.53	954
+ 1.0% calcium formate + 0.5% propionic acid	19.0	0.56	976
+ 165 ppm copper	18.9	0.56	1070
+ 165 ppm copper + 1.5% calcium formate	20.0		1013

After Schutte and von Weerden (1986)
[a]Pigs were assigned to treatment at an average weight of 12.5 kg and remained on the test for a total of 35 days
[b]The control diet was of practical composition and contained no antibiotics

gain and feed conversion efficiency in the first experiment with the addition of either organism, but in the second trial only a response to *Lactobacillus* was detected. In the second experiment, lactic acid (DL-lactic acid, 220 mg/kg) was also included and gave a significant (3.09 versus 2.61) improvement in feed conversion ratio.

In a subsequent series of experiments Pollmann, Danielson and Peo (1980b) demonstrated that colonization of the gastrointestinal tract by *Lactobacillus acidophilus* is enhanced by dietary lactose. The utility of lactose in enhancement of growth in weaner pigs has also been demonstrated by Giesting (1986). Pigs were fed diets supplemented with 25% dried skim milk, or diets wherein the carbohydrate and protein components in the first diet were simulated by addition of 13% lactose and 10% casein. Replacement of lactose by either cornstarch or a variety of hydrolysed cornstarch products (*Table 4.9*) depressed performance. This observation is consistent with the notion that gastric acidification in the young pig is mediated through the formation of lactic acid from lactose. The starch or hydrolysed starch diets did not provide the substrate for this function.

A recent experiment at the University of Kentucky (Cromwell and Burnell, 1987) tested the hypothesis that an 'acidification response' could be obtained by feeding a combination of organic acids along with a substantial number of bacterial cells (*Lactobacillus acidophilus* and *Streptococcus faecium*), enzymes and flavouring agents. The results are presented in *Table 4.10*. A growth response was evident, both when

Table 4.9 EFFECT OF VARYING PROTEIN AND CARBOHYDRATE SOURCES ON PERFORMANCE OF STARTER PIGS (8.09–19.54 kg BODY WEIGHT)[a,b]

Item	SBM	DSM	LAC CAS	HMS CAS	LAC ISP	HMS ISP	PSEM[c]
Liveweight gain (g/day)	345	454	457	422	406	367	15.1
Feed intake (g/day)	618	641	659	645	630	570	19.9
Gain: feed ratio (kg gain/kg feed)	0.560	0.710	0.690	0.660	0.650	0.650	0.0129

After Giesting (1986)
[a]Abbreviations used: SBM = soyabean meal, DSM = dried skim milk, LAC = lactose, CAS = casein, HMS = hydrolysed maize starch, ISP = isolated soya protein
[b]The negative control diet was formulated with maize and soyabean meal. The second diet contained 25% dried skim milk and the remaining diets were formed by substituting LAC or HMS for the carbohydrate in the skim milk and CAS or ISP for the protein.
[c]Pooled standard error of the means

Table 4.10 PERFORMANCE RESPONSE OF PIGS TO AN ACIDIFYING AGENT[a,b]

Basal diet:	Maize–soya		Maize–soya–whey	
Acidifier:	0	1.0	0	1.0
Number of pigs	48	48	48	48
Liveweight gain (g/day)	291	318	327	341
Feed intake (g/day)	514	527	577	559
Gain: feed ratio (kg gain/kg feed)	0.564	0.602	0.561	0.595

Cromwell and Burnell (1987)
[a] Data are averaged over two experiments. Pigs weighed 6.8 kg initially and were treated for a 28-day period
[b] The acidifier was a commercial product containing citric acid, sorbic acid and benzoate along with small amounts of *Lactobacillus acidophilus*, *Streptococcus faecium*, enzymes and flavouring agents

pigs were fed simple maize–soyabean meal diets or diets formulated with maize, soyabean meal and dried whey. Additionally, though not shown in the table, the reponse to the acidifier was additive with antibiotic and copper responses. Additional research will undoubtedly be forthcoming in this area.

Diet acidification is not the complete answer to the post-weaning growth check. However, a substantial body of published literature does support the general conclusion that pigs will respond to reduced diet pH in the weeks immediately following weaning. The magnitude of the response is likely related to the nature of the diet with the greatest benefit evident when diets are formulated with cereal grains supplemented with plant proteins and are devoid of lactose.

References

BECKER, D.E., ULLREY, D.E., TERRILL, S.W. and NOTZOLD, R.A. (1954). *Science*, **120,** 345
CHAPPLE, R.P., CUARON, J.A. and EASTER, R.A. (1983). *Journal of Animal Science*, **57** (Suppl. 1), 94
CLEMENS, E.T., STEVENS, C.E. and SOUTHWORTH, M. (1975). *Journal of Nutrition*, 105
COLE, D.J.A., BEAL, R.M. and LUSCOMBE, J.R. (1968). *Veterinary Record*, **83,** 459–464
CORRING, T., AUMAITRE, A. and DURAND, G. (1978). *Nutrition and Metabolism*, **22,** 231–243
CRANWELL, P.D. (1985). *British Journal of Nutrition*, **54,** 305–320
CRANWELL, P.D., NOAKES, D.E. and HILL, K.J. (1976). *British Journal of Nutrition*, **36,** 71–86
CROMWELL, G.L. and BURNELL, T.W. (1987). *Animal Nutrition and Health*, **42,** (4) 14–16
EDMONDS, M.S., IZQUIERDO, O.A. and BAKER, D.H. (1985). *Journal of Animal Science*, **60,** 462–469
ETHERIDGE, R.D., SEERLEY, R.W. and HUBER, T.L. (1984). *Journal of Animal Science*, **58,** 1403–1410
FALKOWSKI, J.F. and AHERNE, F.X. (1984). *Journal of Animal Science*, **58,** 935–938
FORTE, J.G., FORTE, T.M. and MACHEN, T.W. (1975). *Journal of Physiology*, **244,** 15–31
FUNDERBURKE, D.W., KVERAGAS, C.L., VANDERGRIFT, W.L. and SEERLEY, R.W. (1982). *Journal of Animal Science*, **64,** 457–466
GIESTING, D.W. (1986). *Utilization of Soy Protein by the Young Pig*. PhD Thesis. University of Illinois, Urbana

GEISTING, D.W. and EASTER, R.A. (1985). *Journal of Animal Science*, **60,** 1288–1293
HENRY, R.W., PICKARD, D.W. and HUGHES, P.E. (1985). *Animal Production*, **40,** 505–509
KERSHAW, G.F., LUSCOMBE, J.R. and COLE, D.J.A. (1966). *Veterinary Record*, **79,** 296
KIDDER, D.E. (1982). *Pig News and Information*, **3,** 25–28
KIDDER, D.E. and MANNERS, M.J. (1978). *Digestion in the Pig*, Scientechnica, Bristol
KIRCHGESSNER, N. and ROTH, F.X. (1980). *Zeitschrift fur Tierphysiologie, Tierenahrung und Futtermittelkunde*, **44,** 239–246
KIRCHGESSNER, M. and ROTH, F.X. (1982). *Pig News and Information*, **3,** 259–264
LEWIS, A.J. (1981). *Nebraska Swine Research Report*, University of Nebraska, Lincoln
MANER, J.H., POND, W.G., LOOSLI, J.K. and LOWREY, R.S. (1962). *Journal of Animal Science*, **21,** 49–52
MANNERS, M.J. (1970). *Journal of the Science of Food and Agriculture*, **21,** 333–340
OKAI, D.B., AHERNE, F.X. and HARDIN, R.T. (1976). *Canadian Journal of Animal Science*, **56,** 573–587
PATIENCE, J.F., AUSTIC, R.E. and BOYD. R.D. (1987). *Journal of Animal Science*, **64,** 457–466
POLLMANN, D.S., DANIELSON, D.M. and PEO, E.R. JR (1980a). *Journal of Animal Science*, **51,** 577–581
POLLMANN, D.S., DANIELSON, D.M. and PEO, E.R. JR (1980b). *Journal of Animal Science*, **51,** 638–644
POND, W.G. and MANER, J.H. (1984). *Swine Production and Nutrition*, Avi Publishing Company, Westport, CT
RERAT, A.A. (1981). *World Review of Nutrition and Dietetics*, **37,** 229–287
ROOS, M.A., GIESTING, D.W. and EASTER, R.A. (1987). *Journal of Animal Science*, **65,** (Suppl. 1), 245
SCHUTTE, J.B. and VON WEERDEN, E.J. (1968). *CAFO as a Feed Additive in Diets for Young Pigs*. ILOB Report 569, Wageningen, The Netherlands
SLIVITSKII, M.G. (1975). *Vest. Sel'Khoz. Nauk.*, **7,** 75–80
STEVENS, C.E. (1977) In *Dukes' Physiology of Domestic Animals*, pp. 216–232. Ed. Swenson, M.J. Comstock, London
THOMLINSON, J.F. and LAWRENCE, T.L.J. (1981). *Veterinary Record*, **109,** 120–122
WHITE, F., WENHAM, G., SHARMAN, G.A.M., JONES, A.S., RATTRAY, E.A.S. and MCDONALD. I. (1969). *British Journal of Nutrition*, **23,** 847–857

5

NOVEL APPROACHES TO GROWTH PROMOTION IN THE PIG

P. A. THACKER
Department of Animal Science, University of Saskatchewan, Saskatoon, Canada

Introduction

Antibiotics have played a major role in the growth and development of the pig industry for more than 30 years. Their efficiency in increasing growth rate, improving feed utilization and reducing mortality from clinical disease is well documented (Hays and Muir, 1979). However, consumers are becoming increasingly concerned about drug residues in meat products (Lindsay, 1984). In addition, it has been suggested that the continuous use of antibiotics may contribute to a reservoir of drug-resistant bacteria which may be capable of transferring their resistance to pathogenic bacteria in both animals and humans (Solomons, 1978). Thus it is possible that the future use of antibiotics in animal feeds may be restricted.

Alternative methods of growth promotion must be made available in order to allow the continued development of a viable pig industry. Recently, several new methods of growth promotion have been developed which may have potential for use with pigs. These include the use of growth hormone injections, somatostatin immunization, repartitioning agents, probiotics as well as various enzyme preparations.

Growth hormone injection

The endocrine system plays an important role in the regulation of growth and in the partitioning of nutrients between muscle and adipose tissue (Schanbacher, 1984; Etherton and Kensinger, 1984). The hormones which are known to exert a significant effect on growth rate in pigs include insulin, growth hormone, thyroxine, glucocorticoids, oestrogen, testosterone and a variety of peptides loosley referred to as growth factors (Welsh, 1985). Manipulation of the endocrine system would therefore seem to have considerable potential as a method of increasing growth rate and improving the efficiency of feed utilization in commercial pig operations.

The principal hormone involved in stimulating growth is called somatotrophin or growth hormone (Spencer, 1985). Growth hormone is a protein, produced in the pituitary gland of the pig, which has been shown to stimulate hepatic synthesis of DNA, RNA and protein (Welsh, 1985). Growth hormone has also been shown to increase the concentration of free fatty acids in the blood, while decreasing amino acid breakdown (Welsh, 1985). These changes alter the way that the pig partitions the nutrients contained in the feed, with the end result being an increase in protein synthesis and a decrease in fat synthesis (Etherton *et al.*, 1987).

Growth hormone also stimulates the hepatic synthesis of a group of compounds called somatomedins which are a family of small peptides that mediate bone and muscle growth (Phillips and Vassilopoulou-Sellin, 1979). They exert their effect by causing an increase in the rate of multiplication of bone and muscle cells which is eventually translated into an increase in body size (Daughaday, 1982).

Since endogenously produced growth hormone has been shown to stimulate growth rate, a considerable amount of effort has been made to isolate growth hormone for use as a growth promoter (Turman and Andrews, 1955; Henricson and Ullberg, 1960; Lind *et al.*, 1968; Machlin, 1972). However, progress has been relatively slow due to problems in producing large quantities of growth hormone, the high cost of its production, as well as a lack of knowledge as to the most effective dosage, the correct time for treatment and its proper duration.

Recently, large amounts of pure, species-specific growth hormone have been produced by genetically altered bacteria, through recombinant-DNA technology (Goeddel *et al.*, 1979). The low cost, apparent abundance and purity of this biosynthetic growth hormone has rekindled interest in utilizing exogenous growth hormone as a growth promoter.

Daily injections of growth hormone (0–70 µg/kg body weight) have been shown to produce a 12.6% increase in average daily gain, a 17.5% improvement in feed conversion efficiency, a 24.7% decrease in fat content and a 14.7% increase in muscle mass (Etherton *et al.*, 1987; *Table 5.1*). Similar results have been reported by other workers (Baile, Della-fera and McLaughlin, 1983; Chung, Etherton and Wiggins, 1985; Etherton *et al.*, 1986; Boyd *et al.*, 1987).

In terms of growth rate, it would appear that the best response is obtained with a dosage of approximately 90 µg/kg body weight (Boyd *et al.*, 1987). At higher levels, feed intake is depressed with a concomitant decrease in growth rate. However, improvements in fat content and muscle mass continue up to the maximum levels of growth hormone tested.

The use of growth hormone as a growth promoter has many advantages. Unlike other growth promoters, such as antibiotics or anabolic steroids, growth hormone is not deposited in the body tissues. However, even if it was deposited in animal tissue, it would be broken down during cooking and digested in the gut just like any other protein. As a consequence, its use should find favour with consumers concerned with drug residues in meat products. In addition, since the half-life of growth hormone in plasma is estimated to be 8.9 min (Althen and Gerrits, 1976), a withdrawal period

Table 5.1 EFFECT OF GROWTH HORMONE INJECTION ON PIG PERFORMANCE

	Dosage (µg/kg bodyweight/day)			
	0	10	30	70
Liveweight gain (g/day)	900	980	950	1030
Feed conversion ratio (kg feed/kg liveweight gain)	2.86	2.72	2.58	2.36
Carcass fat (%)	28.7	28.7	24.4	21.6
Muscle mass (%)	26.0	27.7	28.4	30.5

After Etherton *et al.* (1987)

should not be necessary, allowing producers to take advantage of its growth promoting effects right through to market weight.

Unfortunately, the use of growth hormone to increase growth rates in pigs has practical limitations which must be overcome before it can be utilized under commercial conditions. At present, daily injections must be given and fairly large amounts of the hormones must be injected. The increased labour required to inject pigs on a daily basis would probably prevent most producers from taking advantage of the benefits of growth hormone treatment. Further research is required to develop a delivery system which would allow pigs to be injected at less frequent intervals.

Somatostatin immunization

The benefits of manipulation of the immune system to provide protection against disease are universally recognized (Quirke, 1985). The availability of effective vaccines against a wide variety of infectious organisms has facilitated the development of many modern intensive and highly efficient systems of animal production. However, it is only recently that attention has been focused on the possibility of utilizing the immune response to influence endocrine function and hence alter the growth rate of domestic animals (Spencer, 1986).

The secretion of growth hormone from the pituitary gland is controlled by hormones secreted from the hypothalamus. Growth hormone releasing factor (GH-RF), a 44 amino acid peptide, stimulates pituitary growth hormone secretion (Lance *et al.*, 1984), while somatostatin (SR-IF), a 14 amino acid peptide, inhibits growth hormone secretion (Brazeau *et al.*, 1973). The amount of growth hormone secreted depends on the balance between the degree of stimulation or inhibition by these two peptides (*Figure 5.1*). Therefore, a reduction in the levels of circulating somatostatin could lead to an increase in the secretion of growth hormone and thereby stimulate growth.

Inhibition of growth hormone secretion can be reduced by inducing an immune response in the animal against its own circulating somatostatin (Spencer, 1986). By coupling somatostatin to a foreign carrier protein and injecting it with a suitable adjuvant, it is possible to trick the animal's immune system into believing that the somatostatin is a foreign compound. Therefore, the body attempts to remove the somatostatin from the blood, with a concomitant increase in the levels of circulating growth hormone.

The preliminary results of an experiment conducted to determine the effects of somatostatin immunization on weaner pig performance are shown in *Table 5.2*. In this experiment, pigs were immunized (0–100 µg somatostatin) at five weeks of age and their performance was monitored for four weeks. During the four week experiment, pigs immunized against somatostatin gained approximately 12% faster than did control pigs while the efficiency of feed utilization was unaffected (Thacker and Laarveld, 1988).

The development of somatostatin immunization as a growth promoting technique is attractive from a practical standpoint. Unlike growth hormone injection which requires daily injections in order to obtain a response, somatostatin immunization requires only two injections spaced several weeks apart. Therefore, it may be possible to fit somatostatin immunization into a regular vaccination schedule.

The technique of somatostatin immunization appears to have a considerable amount of potential as a practical means of enhancing the growth rate of pigs. The most limiting aspect of the technique is the ability of inducing a consistent immune

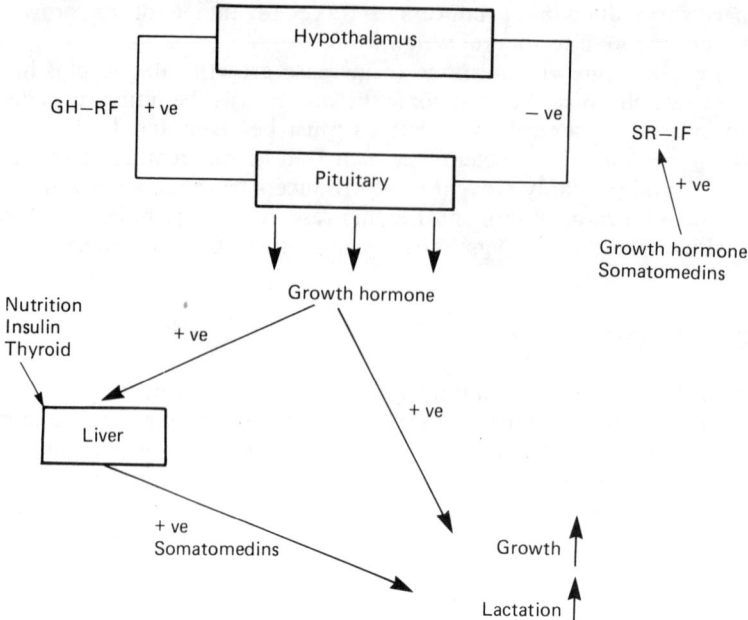

Figure 5.1 Hormonal control of growth hormone secretion

response. Further research is required in order to determine the optimum immunization conditions required to elicit maximum growth stimulation. Areas currently being researched include the effect of different carrier proteins (ovalbumin, human alpha-globulin or bovine serum albumin) as well as various adjuvants (Havlogen, Freund's Complete Adjuvant, Freund's Incomplete Adjuvant, Ribi Adjuvant or Regressin).

Repartitioning agents

The recent discovery of synthetic agents which have a repartioning effect on nutrient utilization in adipose tissue and skeletal muscle has caused a considerable amount of interest in the livestock industry (Asato *et al.*, 1984). These repartitioning agents, known as beta-agonists, have the ability to stimulate the production of lean muscle

Table 5.2 EFFECT OF SOMATOSTATIN IMMUNIZATION ON THE PERFORMANCE OF WEANER PIGS (4–9 WEEKS)

	Dose of somatostatin injected (µg)			
	0	5	25	100
Liveweight gain (g/day)	488	500	558	546
Feed intake (g/day)	869	877	939	987
Feed conversion ratio (kg feed/kg liveweight gain)	1.8	1.7	1.7	1.8

Thacker and Laarveld (1988)

while limiting the synthesis and deposition of subcutaneous and internal fat (Jones *et al.*, 1985; Dalrymple and Ingle, 1986).

Beta-agonists derive their name from the way that they act on individual cells in the body. Cells have receptors on their outer surfaces that bind or latch onto blood-bound messengers (Stiles, Caron and Lefkowitz, 1984). These receptors are divided into alpha and beta types and are very specific as to what messengers they will accept. However, when the right one is present, its action on the receptor causes the whole cell to alter its metabolism.

A beta-agonist is a chemical messenger that activates a beta-adrenergic receptor. These beta-agonists stimulate the breakdown of fat in the cell and increase the rate at which the released fatty acids are oxidized (Mersmann, 1979). Under normal conditions, a large part of the energy obtained from the oxidation of these fatty acids would be lost as heat (Stock and Rothwell, 1981). However, under the influence of beta-agonists, more of the energy obtained from fatty acid oxidation is made available to the body for the protein synthesis (Baker *et al.*, 1983). The end result is a decrease in the amount of fat and an increase in the amount of protein in the body.

The beta-agonists which have been most widely tested are clenbuterol and cimaterol. Some typical research results (Jones *et al.*, 1985) are presented in *Table 5.3*. These results indicate that pigs fed diets containing cimaterol gain at approximately the same rate as control pigs but consume less feed. As a consequence, pigs treated with cimaterol exhibit a trend towards an improved feed conversion efficiency. In addition, there was a 13.8% reduction in carcass fat and a 10.6% increase in loin eye area as a result of dietary inclusion of the beta-agonist. Similar results have been reported by other workers (Moser *et al.*, 1986; Cromwell, Kemp and Stahly, 1987; Hanrahan *et al.*, 1986).

Unfortunately, the withdrawal of cimaterol from the diet for as short a period as seven days has been shown to result in compensatory accumulation of fat in subcutaneous and internal depots (Jones *et al.*, 1985). Therefore, if a withdrawal period is prescribed, it is unlikely that there would be any benefit from including beta-agonists in pig diets. In addition, several experiments have indicated a greater incidence of hoof lesions when beta-agonists were included in the diet (Jones *et al.*, 1985; Cromwell, Kemp and Stahly, 1987). The aetiology of these hoof lesions has not been determined. More research would appear to be warranted before repartitioning agents can be recommended for routine inclusion in commercial swine diets.

Probiotics

Probiotics have been widely promoted as an alternative to the use of antibiotics in swine rations (Hale and Newton, 1979; Pollman, Danielson and Peo, 1980). Probiotics have the opposite effect to antibiotics on the micro-organisms in the digestive tract. Whereas antibiotics control the microbial population in the intestine by inhibiting or destroying micro-organisms, probiotics actually introduce live bacteria into the intestinal tract (Pollmann, 1986).

Both beneficial and potentially harmful bacteria can normally be found in the digestive tract of pigs. Examples of harmful bacteria are *Salmonella, Escherichia coli, Clostridium perfringens* and *Campylobacter sputorum*. Not only can these bacteria produce specific diseases known to be detrimental to the host but through competition for essential nutrients they can also decrease animal performance. In contrast to the effects of these disease causing micro-organisms, bacteria such as *Lactobacillus*

Table 5.3 PERFORMANCE OF GROWING PIGS FED REPARTITIONING AGENTS (CIMATEROL)

	Dietary level (ppm)			
	0	0.25	0.50	1.00
Liveweight gain (g/day)	760	800	770	790
Feed intake (kg/day)	2.98	2.83	2.76	2.72
Feed conversion ratio (kg feed/kg liveweight gain)	3.92	3.54	3.58	3.44
Leaf fat (kg)	1.27	1.16	1.13	1.09
Eye muscle area (cm^2)	29.85	31.96	33.96	32.09

After Jones et al. (1985)

and the vitamin B-complex producing bacteria can be beneficial to the host. By encouraging the proliferation of these bacteria in the intestinal tract, it may be possible to improve animal performance.

The ideal situation would be always to have specific numbers of beneficial bacteria present in the intestinal tract. However, physiological and environmental stress can create an imbalance in the flora of the intestinal tract allowing pathogenic bacteria to multiply. When this occurs, disease and poor performance may result. Probiotics increase numbers of the desirable microflora in the gut thereby swinging the balance towards a more favourable microflora.

The mode of action of probiotics has not been clearly defined. It has been suggested that probiotics increase the synthesis of lactic acid in the gastrointestinal tract of the pig (White et al., 1969 ; Thomlinson, 1981). This increased production of lactic acid is postulated to lower the pH in the intestine thereby preventing the proliferation of harmful bacteria such as E. coli (Mitchell and Kenworthy, 1976). The decrease in the number of E. coli may also reduce the amount of toxic amines and ammonia produced in the gastrointestinal tract (Hill, Kenworthy and Porter, 1970). In addition, there are reports which suggest that probiotics may produce an antibiotic-like substance (Shahani, Vakil and Kilara, 1976) and also stimulate the early development of the immune system of the pig.

The research conducted to determine the value of probiotics in pig diets has been inconclusive. The results of one experiment conducted to determine the effect of probiotics in starter diets are shown in Table 5.4 (Fralick and Cline, 1982). The results of this experiment are typical of most of the research conducted with starter pigs with most workers reporting slight improvements in daily gain and feed conversion efficiency as a result of probiotic inclusion. However, this is not always the case and other workers have reported the opposite effect (Bebiak, 1979; Combs and Copelin, 1981; Pollmann et al., 1982).

Table 5.4 PERFORMANCE OF STARTER PIGS FED DIETS CONTAINING A PROBIOTIC[a]

	Control	Probiotic
Liveweight gain (g/day)	304	322
Average daily intake (g)	843	889
Feed conversion ratio (kg feed/kg liveweight gain)	2.8	2.8

After Fralick and Cline (1982)
[a]Bio-T (Ag-Mark, Inc., Frankfort, IN, USA)

Some of the reasons for the variability of results include the fact that the viability of microbial cultures may be dependent on storage method, strain differences, dose level, frequency of feeding, species specificity problems and drug interactions (Pollmann, 1985). The difficulty in maintaining a viable *Lactobacillus* culture in pig feeds may also partially explain the inconsistency in research results (Pollmann and Bandyk, 1984). It is well documented that temperature, change in pH and various antibiotics will decrease the viability of *Lactobacillus* cultures.

The value of adding probiotics to growing-finishing rations would appear to be questionable based on experimental data such as that shown in *Table 5.5* (Fralick and Cline, 1982). Several other experiments conducted using probiotics during the growing-finishing phase have also shown little benefit (e.g. Pollmann, Danielson and Peo, 1980). Several researchers have speculated that probiotics may actually have some negative effects on pig performance during the growing-finishing phase by competing for nutrients with indigenous organisms of the digestive tract, decreasing carbohydrate utilization and increasing the intestinal transit rate of digesta (Pollmann, 1985). Therefore, although the theoretical concept of probiotics appears promising, the documented evidence of their therapeutic value suggests that the search must continue for a workable alternative to antibiotics.

Enzyme supplementation

Endogenous enzymes are required to break down the carbohydrates, proteins and fats in the diet into a form that can be utilized by the animal. Enzymes are also involved in activating and hastening the many chemical reactions which take place in the animal's body. Therefore, it is possible that pig performance could be improved by the addition of supplementary enzymes to the diet.

A considerable amount of research has been conducted in order to determine the value of supplementing pig feeds with some of the enzymes normally secreted in the digestive tract (Lewis *et al.*, 1955; Cunningham and Brisson, 1957; Combs *et al.*, 1960). Most of this work has been conducted with piglets weaned shortly after birth and has involved enzymes such as amylase, sucrase, pepsin, trypsin and pancreatin. Although of academic interest, this practice would appear to have limited practical application.

Recently, enzymes have been discovered which have the potential to break down deleterious compounds commonly found in pig rations such as the beta-glucans contained in barley and the soluble pentosans found in rye. This discovery has rekindled interest in the use of enzyme supplementation as a means of growth promotion.

Beta-glucans are water soluble polysaccharides found in the aleurone layer and

Table 5.5 PERFORMANCE OF GROWING-FINISHING PIGS FED DIETS CONTAINING A PROBIOTIC[a]

	Control	Probiotic
Liveweight gain (g/day)	710	700
Feed intake (kg/day)	2.33	2.37
Feed conversion ratio (kg feed/kg liveweight gain)	3.28	3.38

After Fralick and Cline (1982)
[a]Bio-T (Ag-Mark, Inc., Frankfort, IN)

endosperm of barley kernels (Prentice and Faber, 1981). They consist of glucose units linked together by beta-1,4 and beta-1,3 linkages (Fleming and Kawakami, 1977; *Figure 5.2*). The beta-1,3 linkages confer upon the molecule a step-like structure that interferes with hydrogen bonding between adjacent chains resulting in increased water solubility.

Beta-glucans greatly lower the nutritional value of barley. They restrict weight gain through an increase in the viscosity of the intestinal fluid (Burnett, 1966) which interferes with the digestive process by impeding enzyme–substrate association as well as affecting the rate at which released nutrients approach the mucosal surface for absorption (Campbell *et al.*, 1986). It has also been suggested that beta-glucans allow microbial populations to assimilate a greater proportion of the nutrients contained in the feed into their own system thereby reducing the availability of these nutrients to the host.

The level of beta-glucan in a barley sample can vary from 1.5 to 8% depending on the cultivar of barley and the environmental conditions under which it was grown (Willingham *et al.*, 1960; Gohl and Thomke, 1976). Barley grown in areas with low rainfall will have higher levels of beta-glucans than that grown under conditions of adequate moisture (Aastrup, 1979). In addition, the hull-less varieties of barley contain higher levels of beta-glucans than the hulled varieties (Fox, 1981) while the malting varieties of barley may be lower in beta-glucans than the so-called feed varieties. Therefore, variations in beta-glucan content may explain some of the differences in feeding value often seen among barley cultivars.

Enzymes capable of breaking down beta-glucans are termed beta-glucanases. Treatment of barley with these enzymes has been shown to improve the nutritive

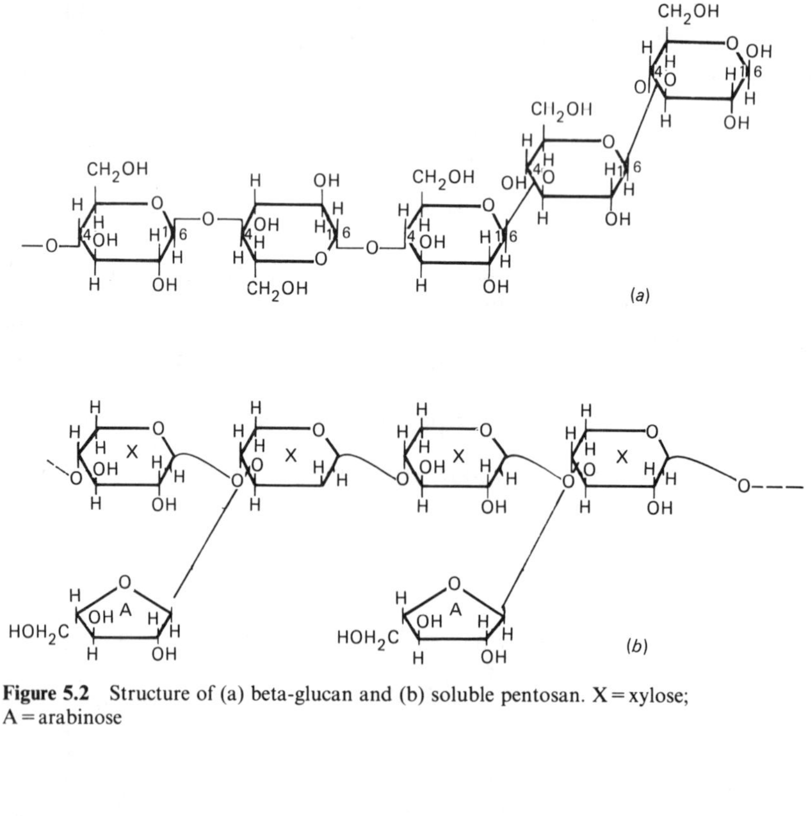

Figure 5.2 Structure of (a) beta-glucan and (b) soluble pentosan. X = xylose; A = arabinose

value of barley for poultry (Burnett, 1966; White *et al.*, 1983). In one study, bodyweight gains were improved 9.4% and feed conversion efficiency 5.8% by the addition of a beta-glucanase enzyme to barley-based poultry rations (Campbell *et al.*, 1984; *Table 5.6*).

There has been little research conducted to determine the effects of beta-glucans on the performance of pigs. However, the few research reports available indicate that supplementation of barley-based pig diets with beta-glucanase may be beneficial (Newman *et al.*, 1980; Newman, Eslick and El-Negoumy, 1983).

The results of a recent trial conducted at the University of Saskatchewan are presented in *Table 5.7* (Thacker *et al.*, 1988). In this study, there were no significant differences in growth rate, feed intake or feed conversion efficiency between pigs fed hull-less barley diets supplemented or unsupplemented with beta-glucanase. However, digestibility coefficients for dry matter, crude protein and energy were marginally improved by beta-glucanase supplementation (*Table 5.8*).

Enzyme supplementation may also have potential as a means of improving the nutritive value of rye. At the present time, rye is not widely used in pig diets. The classic explanation for this relates to its high ergot content. However, recent research with poultry indicates that high levels of soluble pentosans in rye (*Figure 5.2*) may pose an even greater problem than does ergot. These pentosans are solubilized during digestion and result in a highly viscous intestinal fluid that interferes with digestion in a manner similar to the beta-glucans found in barley.

From an agronomic standpoint, rye is an attractive crop. It is high yielding, makes more effective use of water during spring run-off than cereals planted in the spring and allows for a more equitable distribution of a farmer's workload due to its early harvest. If the detrimental effects of the soluble pentosans could be overcome by enzyme treatment, then the market potential for rye would increase providing an alternative feed resource for use in pig feeding.

The results of an experiment conducted to determine the effects of enzyme supplementation on the performance of grower pigs fed diets containing rye are shown in *Table 5.9* (Thacker *et al.*, 1988). Although these are preliminary data, the results indicate that the performance of pigs fed rye-based diets supplemented with an

Table 5.6 EFFECT OF BETA-GLUCANASE SUPPLEMENTATION OF BARLEY DIETS ON THE PERFORMANCE OF BROILER CHICKENS FROM 0–6 WEEKS OF AGE

	Feed intake (g)	Body weight (g)	Feed conversion efficiency	Cost of gain (cents)
Wheat-corn	3484	1996	1.75	0.81
Barley	3530	1718	2.08	0.73
Barley + enzyme	3738	1891	1.96	0.70

After Campbell *et al.* (1984)

Table 5.7 EFFECT OF BETA-GLUCANASE ON THE PERFORMANCE OF GROWING PIGS

	Control	Enzyme
Liveweight gain (kg/day)	740	760
Feed intake (kg/day)	2.32	2.35
Feed conversion ratio (kg feed/kg liveweight gain)	3.13	3.11

Thacker *et al.* (1988)

Table 5.8 EFFECT OF BETA-GLUCANASE SUPPLEMENTATION ON THE DIGESTIBILITY OF DRY MATTER, PROTEIN AND ENERGY BY PIGS (55 kg LIVEWEIGHT)

	Control	Beta-glucanase
Dry matter digestibility (%)	80.51	82.69
Protein digestibility (%)	75.06	77.72
Energy digestibility (%)	77.97	80.82

Thacker *et al.* (1988)

Table 5.9 EFFECT OF ENZYME SUPPLEMENTATION OF RYE ON PIG PERFORMANCE

	Barley	Rye	Rye + enzyme
Liveweight gain (g/day)	750	700	750
Feed intake (kg/day)	2.20	1.96	1.88
Feed conversion ratio (kg feed/kg liveweight gain)	2.92	2.72	2.50

Thacker *et al.* (1988)

enzyme capable of breaking down soluble pentosans can equal that of pigs fed barley-based diets.

Conclusion

Pig production continues to be an industry concerned with converting feed ingredients into meat with a degree of technical efficiency and economic management that allows a profit. New technology must be developed in order to allow the continued development of a viable pig industry. Several of the methods discussed in this chapter may find application as the growth promoters of the future.

References

AASTRUP, S. (1979). *Carlsberg Research Communications*, **44**, 381–393
ALTHEN, T.B. and GERRITS, R.J. (1976). *Endocrinology*, **99**, 511–515
ASATO, G., BAKER, P.K., BASS, R.T., BENTLEY, J., CHARI, M., DALRYMPLE, R.H., FRANCE, R.J., GINGHER, P.E., LENCES, B.L., PASCAVAGE, J.J., PENSACK, M. and RICKS, C.A. (1984). *Agriculture and Biochemical Chemistry*, **48**, 2883–2888
BAILE, C.A., DELLA-FERA, M.A. and MCLAUGHLIN, C.L. (1983). *Growth*, **47**, 225–236
BAKER, P.K., AUST, T., DALRYMPLE, R.H., INGLE, D.L. and RICKS, C.A. (1983). *Symposium on Novel Approaches and Drugs for Obesity*. The 4th International Congress on Obesity, New York
BEBIAK, D.M. (1979). *Report of Swine Research*, Michigan State University, pp. 8–10.
BOYD, R.D., WRAY-CAHEN, D., BAUMAN, D., BEERMANN, D., DENEERGARD, A. and SOUZA, L. (1987). *Proceedings of the Maryland Nutrition Conference*, pp. 58–66.
BRAZEAU, P., VALE, W., BURGUS, R., LING, N., BUTCHEN, M., RIVIER, J. and GUILEMIN, R. (1973). *Science*, **179**, 77–79

BURNETT, G.S. (1966). *British Poultry Science*, **7**, 55–75
CAMPBELL, G.L., CLASSEN, H.L. and SALMON, R.E. (1984). *Feedstuffs*. 1984, May 7, 26–27
CAMPBELL, G.L., CLASSEN, H.L., THACKER, P.A., ROSSNAGEL, B.G., GROOTWASSNIK, J.W. and SALMON, R.E. (1986). *Proceedings of the 7th Western Nutrition Conference, Saskatoon*, pp. 277–250
CHUNG, C.S., ETHERTON, T.D. and WIGGINS, J.P. (1985). *Journal of Animal Science*, **60**, 118–129
COMBS, G.E. and COPELIN, J.L. (1981). Department of Animal Science Research Report AL-1981-6, Florida Agricultural Experimental Station, pp. 7–9.
COMBS, G.E., ALSMEYER, W.L., WALLACE, H.D. and KOGER, M. (1960). *Journal of Animal Science*, **19**, 932–937
CROMWELL, G.L., KEMP, J.D. and STAHLY, T.S. (1987). *Feed Management*, **38**, 8–11
CUNNINGHAM, H.M. and BRISSON, G.J. (1957). *Journal of Animal Science*, **16**, 370–376
DALRYMPLE, R.H. and INGLE, D.L. (1986). *Proceedings of the 47th Minnesota Nutrition Conference*, pp. 102–114
DAUGHADAY, W.H. (1982). *Proceedings of the Society of Experimental Biology and Medicine*, **170**, 257–263
ETHERTON, T.D., WIGGINS, J.P., CHUNG., C.S., EVOCK, C.M., REBHUN, J.F., WALTON, P.E. (1986). *Journal of Animal Science*, **63**, 1389–1399
ETHERTON, T.D. and KENSINGER, R.S. (1984). *Journal of Animal Science*, **59**, 511–528
ETHERTON, T.D., WIGGINS, J.P., CHUNG., C.S., EVOCK, C.M., REBHUN, J.F. and WALTON, P.E. and STEELE, N.C. (1987). *Journal of Animal Science*, **64**, 433–443
FLEMING, M. and KAWAKAMI, K. (1977). *Carbohydrate Research*, **57**, 15–23
FOX, G.J. (1981). PhD Dissertation. Montana State University
FRALICK, C. and CLINE, T.R. (1982). *Proceedings of the Purdue University Swine Day*, pp. 7–10
GOEDDEL, D.V., HEYNEKER, H.L., HOZUMI, T., ARENTZEN, R., ITAKURA, K., YANSURA, D.G., ROSS, M.J., MIOZZARI, G., CREA, A. and SEEBERG, R.H. (1979). *Nature*, **281**, 544–548
GOHL, B. and THOMKE, S. (1976). *Poultry Science*, **55**, 2369–2374
GUEST, G.B. (1976). *Journal of Animal Science*, **42**, 1052–1057
HALE, O.M. and NEWTON, G.L. (1979). *Journal of Animal Science*, **48**, 770–775
HANRAHAN, J.P., QUIRKE, J.F., BOMANN, W., ALLEN, P., MCEWAN, J.C., FITZSIMONS, J.M., KOTZIAN, J. and ROCHE, J.F. (1986). *Recent Advances in Animal Nutrition—1986*, pp. 125–138. Ed. Haresign, W. and Cole, D.J.A. Butterworths, London
HAYS, V.W. and MUIR, V.M. (1979). *Canadian Journal of Animal Science*, **59**, 447–456
HENRICSON, B. and ULLBERG, S. (1960). *Journal of Animal Science*, **19**, 1002–1008
HILL, I.R., KENWORTHY, R. and PORTER, P. (1970). *Research in Veterinary Science*, **11**, 320–326
JONES, R.W., EASTER, R.A., MCKEITH, R.K., DALRYMPLE, R.H., MADDOCK, H.M. and BECHTEL, P.J. (1985). *Journal of Animal Science*, **61**, 905–913
LANCE, V.A., MURPHY, W.A., SUEIRAS-DIAZ, J. and COY, B.H. (1984). *Biochemistry Biophysics Research Communication*, **119**, 265–272
LEWIS, C.J., CATRON, D.V., LIU, C.H., SPEER, V.C. and ASHTON, G.C. (1955). *Agriculture and Food Chemistry*, **12**, 1047–1050
LIND, K.D., HOWARD, R.D., KROPF, D.H. and KOCH, B.A. (1968). *Journal of Animal Science*, **27**, 1763 (Abstract)
LINDSAY, D.G. (1984). *Pig News and Information*, **5**, 219–222
MACHLIN, L.J. (1972). *Journal of Animal Science*, **35**, 794–800
MERSMANN, H.J. (1979). *Proceedings of the Reciprocal Meat Conference*, **32**, 93–107
MITCHELL, I. and KENWORTHY, R. (1976). *Journal of Applied Bacteriology*, **4**, 163–174

MOSER, R.L., DALRYMPLE, R.H., CORNELIUS, S.G., PETTIGREW, J.E. and ALLEN, C.E. (1986). *Journal of Animal Science*, **62**, 21–26
NEWMAN, C.W., ESLICK, R.F., PEPPER, J.W. and EL-NEGOUMY, A.M. (1980). *Nutrition Reports International*, **22**, 833–837
NEWMAN, C.W., ESLICK, R.F. and EL-NEGOUMY, A.M. (1983). *Nutrition Reports International*, **28**, 139–145
PHILLIPS, L.S. and VASSILOPOULOU-SELLIN, R. (1979). *American Journal of Clinical Nutrition*, **32**, 1082–1096
POLLMANN, D.S., DANIELSON, D.M. and PEO, E.R. (1980). *Journal of Animal Science*, **51**, 577–581
POLLMANN, D.S., KENNEDY, G.A., KOCH, B.A. and ALLEE, G.L. (1982). Agricultural Experimental Station Report of Progress No. 422, Kansas State University, pp. 86–91
POLLMANN, D.S. and BANDYK, C.A. (1984). *Animal Feed Science and Technology*, **11**, 261–267
POLLMANN, D.S. (1985). *Guelph Pork Symposium*, pp. 59–74
POLLMANN, D.S. (1986). *Recent Advances in Animal Nutrition—1986*, pp. 193–205. Ed. Haresign, W. and Cole, D.J.A. Butterworths, London
PRENTICE, N. and FABER, S. (1981). *Cereal Chemistry*, **58**, 77–79
QUIRKE, J.F. (1985). *Livestock Production Science*, **13**, 1–2
SCHANBACHER, B.D. (1984). *Journal of Animal Science*, **59**, 1621–1630
SHAHANI, K.M., UAKIL, J.R. and KILARA, A. (1976). *Journal of Cultured Dairy Produce*, **11**, 14–20
SOLOMONS, I.A. (1978). *Journal of Animal Science*, **46**, 1360–1368
SPENCER, G.S. (1985). *Livestock Production Science*, **12**, 31–46
SPENCER, G.S. (1986). *Domestic Animal Endocrinology*, **3**, 55–68
STILES, G.L., CARON, M.G. and LEFKOWITZ, R.J. (1984). *Physiological Reviews*, **64**, 661–743
STOCK, M.J. and ROTHWELL, N.J. (1981). *Biochemical Society Transactions*, **9**, 525–527
THACKER, P.A., CAMPBELL, G.L. and GROOTWASSINK, J.W. (1988). *Department of Animal and Poultry Science Research Reports*, University of Saskatchewan, Saskatoon, Saskatchewan, pp. 140–145
THACKER, P.A. and LAARVELD, B. (1988). Unpublished data
THACKER, P.A., CAMPBELL, G.L. and GROOTWASSINK, J.W. (1988). *Nutrition Reports International* (in press)
THOMLINSON, J.R. (1981). *Veterinary Record*, **109**, 120–122
TURMAN, E.J. and ANDREWS, F.N. (1955). *Journal of Animal Science*, **14**, 7–18
WELSH, T.H. (1985). *Animal Nutrition and Health*, **40**, 14–19
WHITE, F., WENHAM, G., SHARMAN, G.A., JONES, A.S., RATTRAY, E.A. and MCDONALD, I. (1969). *British Journal of Nutrition*, **23**, 847–858
WHITE, W.B., BIRD, H.R., SUNDE, M.L. and MARLETT, J.A. (1983). *Poultry Science*, **62**, 853–862
WILLINGHAM, H.E., LEONG, K.C., JENSEN, L.S. and MCGINNIS, J. (1960). *Poultry Science*, **39**, 103–108

III

Poultry Nutrition

6

THE NUTRITIONAL REQUIREMENTS OF TURKEYS TO MEET CURRENT MARKET DEMANDS

C. NIXEY
British United Turkeys Ltd, Tarvin, Chester, UK

Market opportunities

Many of the feed mills in Britain produce little if any turkey food. They may be ignoring a market opportunity that already exists and certainly ignoring a potentially very significant market in the future. Of the various feed sectors, turkey feed had by far the largest percentage increase in 1986 over the previous year (*Table 6.1*). It is an area which could be exploited by general farmers but without encouragement from outside forces such as feed companies, they will miss out on the opportunity. Very few hotels or restaurants have turkey meat regularly on the menu, yet it is now an inexpensive meat, with a broad popular appeal. The reason it is not on menus is that restaurants cannot be assured of a regular supply of turkeys delivered to their door. It is not an area that the larger operations are interested in because of transport costs. However, a very nice income can be obtained by building up a local trade in large turkeys. The interest from the viewpoint of the feed industry is that it is a growth area offering all the year round business.

The weakness of empirical experimentation

A review of the published literature reveals that not much work has been published on turkey nutrition and what has been is open to criticism. The situation is further complicated by the tremendous genetic progress achieved in the last 20 years (*Table 6.2*). The little work that has been carried out on females in the past must be viewed in the light of the fact that they now grow as fast as large type males of 1969.

The initial comments on nutritional requirements relate to turkeys in general and are followed by comments on more specific requirements.

The turkey is a high protein, low fat containing animal. It is also a rapidly growing animal with the males multiplying their hatching weight by 300 times in 140 days. It is not surprising that, in a variety of situations all over the world, a growth response to increasing protein intake, or more specifically the first limiting amino acid, can normally be seen. Unfortunately very little work on the growing turkey's requirements for specific amino acids has been carried out. Most of that has been open to criticism for faulty design or interpretation. Most of the papers are on the lysine requirements with a lesser number on the sulphur amino acids and very few on the others.

Table 6.1 UK COMPOUND FEED PRODUCTION (MILLION TONNES)

Type of food	1985	1986	% change 1985/86
Calf	0.41	0.40	− 2.5
Dairy	3.33	3.63	+ 9.0
Other cattle	0.66	0.74	+ 12.1
Pig starters	0.18	0.20	+ 5.9
Pig breeding	0.67	0.67	+ 0.2
Other pig	1.22	1.27	+ 4.3
Broiler chicken	1.27	1.36	+ 6.8
Layer	1.09	1.09	—
Turkey	0.40	0.51	+ 27.5

Source: MAFF

Table 6.2 GENETIC PROGRESS OF TURKEYS

| | | 18-week liveweight (kg) | |
Year	Brand name	Males	Females
1966	Triple 6	9.45	6.75
1969	Triple 6	9.77	6.95
1972	BUT 6	10.27	7.27
1974	BUT 6	10.68	7.56
1977	BUT 6	10.96	7.80
1981	BIG 6	12.67	8.78
1982	BIG 6	12.80	8.84
1984	BIG 6	13.40	9.13
1986	BIG 6	13.96	9.88

(Based on the published performance goals of British United Turkeys Ltd)

LYSINE EXPERIMENTS

There are very few papers on lysine requirements of the female turkey. *Table 6.3* shows work on male turkeys with the author's conclusions expressed as g amino acid/ MJ ME. A decreasing requirement with age is seen but there is considerable range in the indicated lysine requirements. However, if the highest indicated requirements are plotted against age, the correlation is very good, decreasing by 5.7 mg lysine/MJ/day.

SULPHUR AMINO ACID EXPERIMENTS

It will be seen in *Table 6.4* that even fewer papers were found on total sulphur amino acids (TSAA). There is a further complication in that there is a requirement for methionine in addition to its contribution to the TSAA. However, it was not found feasible to separate out the methionine data. A contribution can be made to the total sulphur amino acids by cystine. Behrends and Waibel (1980) estimated that between 55 and 58% of the total sulphur amino acids, depending on age, could be cystine. There has been very little work on other amino acids but these are referred to later in this chapter.

Table 6.3 LYSINE REQUIREMENTS OF MALE TURKEYS

Approximate age range (days)	No. of data sets	Mean indicated lysine requirement (g/MJ ME)	Range of indicated requirement (g/MJ ME)	
			Low	High
0–28	6	1.236	1.135	1.328
28–56	3	1.254	1.219	1.278
56–84	10	0.896	0.729	1.110
84–112	3	0.836	0.646	0.978
112–140	9	0.557	0.478	0.717
140–168	2	0.512	0.470	0.569

Table 6.4 TOTAL SULPHUR AMINO ACID (TSAA) REQUIREMENTS OF TURKEYS

Approximate age range (days)	No. of data sets	Mean indicated TSAA requirement (g/MJ ME)	Range of indicated requirement (g/MJ ME)	
			Low	High
0–28	5	0.783	0.664	0.876
0–56	2	0.865	0.814	0.915
28–56	2	0.740	0.696	0.785
56–84	4	0.568	0.492	0.647
84–112	0	—	—	—
112–140	2	0.312	0.282	0.341
140–168	0	—	—	—

The Reading model

The Reading model produced by Fisher, Morris and Jennings (1973) to analyse and describe the response curve of a laying flock of hens to essential amino acids is well known. It has since been proposed for use with growing birds by Clark et al. (1982) and Fisher and Emmans (1982). If the broken line method is used to fit data, the indicated requirement for maximum bodyweight gain occurs at a lower level than that indicated by the Reading model. In the Reading model, the amino acid requirement for maximum bodyweight gain is calculated as $AA = a.\Delta W + b\bar{W}$ where ΔW is the potential bodyweight gain per day and $\bar{W}$ is the mean bodyweight. The constant a is defined as the amino acid intake (mg/day) per unit (g/day) bodyweight gain. The constant b is defined as the amino acid intake (mg/day) per unit (g) of bodyweight maintained. These constants are derived from fitting a curve as described by Fisher, Morris and Jennings (1973).

Very few of the papers reviewed gave sufficient data for use in the Reading model for lysine response and even fewer for TSAA. Those that did gave the a and b values as shown in *Table 6.5*. These confirm that more lysine than TSAA is required to produce 1 g of bodyweight gain. On the other hand, more TSAA than lysine is required to maintain 1 g of bodyweight for one day. The relatively greater significance of the maintenance requirement as the birds get older is reflected in the increased TSAA requirement.

Table 6.5 READING MODEL a AND b VALUES[1] DERIVED FROM SUITABLE DATA IN PUBLISHED PAPERS

Age (days)	a		$b\ (\times 10^3)$	
	Lysine	TSAA	Lysine	TSAA
0–28	22.32	13.75	22.8	64.1
28–56	19.91	ND[2]	17.2	ND
56–84	21.16	11.281	19.3	98.0
84–112	ND	ND	ND	ND
112–140	22.96	10.941	10.5	118.6

[1] a is the amino acid intake (mg/day) per unit (g/day) bodyweight gain.
b is the amino acid intake (mg/day) per unit (g) of bodyweight maintained
[2] ND, no data

Input and output procedures

The advantage of the Reading model is that the constants a and b can be used to predict the requirements of birds with different genetic potentials for bodyweight gain. An example of the input and output predictions that can be obtained is shown in Tables 6.6, 6.7 and 6.8.

The model predicts the bodyweight gain to be expected from amino acid intake or—vice versa—amino acid required to produce the bodyweight gain. This type of information has a very practical value provided that the bodyweight gain and food intake are measured to an age. If one of these amino acids is the first limiting amino acid, it will pinpoint which. It will indicate the likely growth response to be expected

Table 6.6 READING MODEL: AMINO ACID INPUT AND BODYWEIGHT GAIN OUTPUT PREDICTIONS FOR BIG 6 MALES, 0–4 WEEKS OF AGE

Bodyweight gain (g/bird day)	Lysine required (g/bird day)	TSAA required (g/bird day)	TSAA as % of lysine
14	0.328	0.246	75.0
16	0.372	0.263	70.7
18	0.418	0.290	69.4
20	0.462	0.318	68.8
22	0.508	0.346	68.1
24	0.550	0.373	67.8
26	0.596	0.400	67.1
28	0.640	0.428	66.9
30	0.685	0.456	66.6
32	0.730	0.483	66.2
34	0.775	0.512	66.1
36	0.820	0.540	65.9
38	0.870	0.570	65.5
40	0.920	0.602	65.4
42	0.973	0.634	65.2
44	1.023	0.665	65.0
46	1.080	0.703	65.1
48	1.320	0.860	65.1
48.2	1.390	0.900	64.7

Based on a and b values shown in Table 6.5, a 55 g day old poult and a potential daily weight gain for the period of 48.2 g

Table 6.7 READING MODEL: AMINO ACID INPUT AND BODYWEIGHT GAIN OUTPUT PREDICTIONS FOR BIG 6 MALES, 8–12 WEEKS OF AGE

Bodyweight gain (g/bird day)	Lysine required (g/bird day)	TSAA required (g/bird day)	TSAA as % of lysine
80	1.812	1.506	83.1
84	1.892	1.552	82.0
88	1.980	1.596	80.6
92	2.066	1.642	79.5
96	2.150	1.688	78.5
100	2.236	1.735	77.6
104	2.322	1.784	76.8
108	2.410	1.834	76.1
112	2.498	1.886	75.5
116	2.592	1.942	74.9
120	2.690	1.992	74.1
124	2.790	2.054	73.6
128	2.890	2.105	72.8
132	2.985	2.154	72.2
136	3.083	2.206	71.6
140	3.200	2.290	71.5
144	3.620	2.585	71.4
146	3.950	2.770	70.1
146.4	4.210	2.950	70.1

Based on *a* and *b* values shown in *Table 6.5*, 8 week liveweight of 4.1 kg and a potential daily weight gain for the period of 146.4 g

Table 6.8 READING MODEL: AMINO ACID INPUT AND BODYWEIGHT GAIN OUTPUT PREDICTIONS FOR BIG 6 MALES, 16–20 WEEKS OF AGE

Bodyweight gain (g/bird day)	Lysine required (g/bird day)	TSAA required (g/bird day)	TSAA as % of lysine
74	1.845	2.465	133.6
78	1.938	2.508	129.4
82	2.030	2.554	125.8
84	2.075	2.576	124.1
88	2.168	2.630	121.3
92	2.260	2.667	118.1
96	2.353	2.713	115.3
100	2.445	2.760	112.9
104	2.542	2.808	110.5
108	2.640	2.858	108.3
112	2.743	2.912	106.2
116	2.850	2.966	104.1
120	2.960	3.027	102.3
124	3.065	3.094	100.9
128	3.163	3.172	100.3
132	3.284	3.270	99.6
136	3.590	3.536	98.5
139	4.130	3.980	96.4
139.3	4.370	4.170	95.4

Based on *a* and *b* values shown in *Table 6.5*, 16 week liveweight of 12 kg and a potential daily weight gain for the period of 139.3 g

LYSINE AND TOTAL SULPHUR AMINO ACID RELATIONSHIPS

The relationship between TSAA and lysine is of importance. At maximum growth rate it is 64.7% for 0–4 weeks. However, from 8–12 weeks, the TSAA requirement relative to lysine has increased to 70.1%. This is even more pronounced from 16 to 20 weeks (*Table 6.6*). If these values are correct, it means that there is concern about TSAA levels because no one appears to be using similar TSAA levels to lysine levels at these ages.

A further reason for looking at TSAA levels, is the effect seen if the turkeys have a setback and do not grow to their potential for a period. This has the effect of increasing the relative influence of maintenance as opposed to growth. This increases the TSAA relative to the lysine requirement. It was also seen at the other ages. In situations of high stocking density where the effect becomes progressively more stressful, it would seem logical to use a higher TSAA level relative to lysine than in situations where good growth rate is expected. To make the best use of our knowledge, more information is required during the growing cycle, doing check weighings and measuring feed consumption. A lot of information is collected on breeding birds and it is time to do the same with growing birds.

Amino acid profiles

It is important to become conscious of the amino acid profiles of turkey diets using lysine value as the marker at 100 as the intake of one amino acid cannot be viewed in isolation. Interrelationships exist between amino acids.

Two groups of workers, one in Israel (Hurwitz *et al.*, 1983) the other in Edinburgh (Fisher and Emmans, 1983) have tried to overcome the problems of empirical experimentation by producing calculated models for the growing turkey. Their predictions (*Table 6.9*) serve as a useful introduction to the subject of amino acid profiles. The obvious amino acid on which to base the profile would seem to be lysine, being often first limiting and lacking the complications of total sulphur amino acid utilization.

The significance of these profiles is that they may indicate the likely first limiting amino acids with the use of particular ingredients. The profile of common ingredients is shown in *Table 6.10*. It will be seen that in high protein diets used early in life which

Table 6.9 A COMPARISON OF THE AMINO ACID PROFILES PREDICTED BY THE EDINBURGH (EDIN) AND ISRAELI (ISR) MODELS FOR LARGE MALE TURKEYS

Age (weeks)	Methionine		Total SAA		Tryptophan		Threonine	
	Isr	*Edin*	*Isr*	*Edin*	*Isr*	*Edin*	*Isr*	*Edin*
4	36.0	36.0	70.8	64.0	14.7	17.3	78.9	70.3
8	36.5	36.5	79.3	64.5	15.1	17.2	82.5	70.0
12	37.2	37.4	90.3	65.4	16.0	17.1	87.0	69.5
16	39.1	38.6	101.0	66.7	16.9	17.0	91.3	68.8
20	39.3	40.4	94.2	68.6	16.2	16.7	89.0	67.9

Table 6.10 THE AMINO ACID PROFILES OF INGREDIENTS COMMONLY USED IN TURKEY DIETS

Ingredient	Lysine	Methionine	TSAA	Tryptophan	Threonine
Maize	100	83.3	145.8	37.5	162.5
Wheat	100	48.3	119.4	38.7	103.2
Soya	100	22.2	45.7	22.2	61.7
Fish (white)	100	37.1	53.6	14.8	56.7
Meatmeal	100	25.0	47.0	12.0	58.0
Sunflower	100	50.0	100.0	45.0	105.0
Indicated profiles for turkey diets					
4 weeks					
Israeli	100	36.0	70.8	14.7	78.9
Edinburgh	100	36.0	64.0	17.3	70.3
20 weeks					
Israeli	100	39.3	94.2	16.2	89.0
Edinburgh	100	40.4	68.6	16.7	67.9

NRC (1984)

contain large quantities of soya, fishmeal and meatmeal, there is a very real possibility of the threonine level being well below even the Edinburgh predictions. Tryptophan is unlikely ever to be the first limiting amino acid.

TYPES OF DELETERIOUS AMINO ACID PROFILES

There are three types of deleterious amino acid profiles. The most common causes a decrease in food intake and growth rate which can be prevented by a supplement of the limiting amino acid.

The second problem which is amino acid toxicity may be common in a mild form. In this problem an excess of an amino acid reduces food intake and hence growth rate. The effects of toxicity wear off in that food intake gradually comes back to normal. Each time the food is changed, the amino acid profile will change and risk the effects of a different amino acid being in excess. It is an argument for not changing diets frequently. The third problem which is amino acid antagonisms is more complicated. Sometimes the adverse effects of an excess of an amino acid can be rectified by the addition of another amino acid. Of most practical significance in the UK is an excess of lysine which can be alleviated by increasing the arginine level. This relationship has been shown to exist in other animals and was illustrated in young turkeys by D'Mello and Emmans (1975). Their results are shown in *Table 6.11*. It will be seen that as the dietary lysine level increases so the optimum arginine level increases. There are insufficient data points to assess the precise relationship required between arginine and lysine but the ratio would appear to be around 1.1 arginine to 1.0 lysine. There is a poverty of other work on the subject so it is an area which needs investigation because practical starter diets are usually below this ratio.

Commercial recommendations

The commercial nutritionist must take decisions about his formulations and *Table*

Table 6.11 WEIGHT GAIN (g/BIRD DAY) OF TURKEY POULTS FED DIETS OF DIFFERENT LYSINE AND ARGININE CONTENTS FROM 7 TO 21 DAYS

Dietary arginine content (%)	Dietary lysine content (%)		
	1.05	1.30	1.55
1.00	15.0	18.3	20.8
1.25	14.1	23.3	24.2
1.50	15.1	24.1	26.6
1.75	13.7	21.5	29.4

D'Mello and Emmans (1975)

Table 6.12 SUGGESTED DIETS FOR BIG 6 MALES WHERE THE CONDITIONS AND PELLET QUALITY ARE GOOD

Age fed (weeks)	0–4	4–8	8–12	12–16	16–20	20–24
Nutrient (g/MJ ME)						
Lysine	1.57	1.34	1.10	0.89	0.75	0.65
Methionine	0.57	0.53	0.47	0.44	0.43	0.42
Methionine + cystine	1.02	0.94	0.82	0.76	0.75	0.71
Tryptophan	0.27	0.23	0.19	0.15	0.13	0.11
Threonine	1.00	0.86	0.75	0.58	0.48	0.42
Arginine	1.69	1.46	1.21	1.02	0.88	0.80
Calcium	1.10	1.05	0.95	0.85	0.80	0.70
Available phosphorus	0.62	0.60	0.55	0.50	0.45	0.40
Sodium	0.13	0.13	0.13	0.13	0.13	0.13
NACL	0.30	0.30	0.30	0.30	0.30	0.30
Essential fatty acids	1.27	1.09	—	—	—	—
ME (MJ/kg)	12.0	12.1	12.2	12.4	12.6	12.8

6.12 shows the basis of the author's formulations for use with Big 6. These recommendations however are not written in stone. They must be adjusted according to the situation. This is best assessed by comparing the liveweight for age against the breed's target liveweights.

Food intake factors

The most common formulation fault is the use of too wide ME to amino acid ratio. Even the best formulation cannot give optimum results if it is not consumed. Liveweights are often reduced because of adverse feed intake factors.

Turkey growers are most sensitive to pellet quality. Recent work in the USA (Hassibi, personal communication) illustrates the influence the form of the feed can have on results (*Table 6.13*).

High stocking densities will reduce liveweights. The main effect is probably via reduced food intake. This may occur because the temperature in the microclimate surrounding individual birds is high or because of the competition to get to feeders and water points.

FACTORS AFFECTING FOOD INTAKE OF YOUNG TURKEYS

Recent work (Nixey, 1987) shows the influence of various factors on four week

Table 6.13 THE INFLUENCE OF THE FORM OF THE FEED ON GROWTH PERFORMANCE

	Liveweight at 20 weeks (kg)	Food conversion ratio (kg feed/kg liveweight gain)
Mash	12.58	3.31
Poor pellet	12.77	3.22
Good pellet	13.08	3.18

liveweights and also the influence of four week liveweights on subsequent growth rates. In this experiment, all birds received the same formulation. Three types of influence were investigated (a) crumb versus mash feeding, (b) day-old versus seven-day debeaking, and (c) supplementary food in floor trays for seven days versus plastic hanging feeders alone. The crumb-fed birds weighed 23% heavier than the mash-fed birds at 28 days. A 16% difference was present even at seven days. Day-old debeaking had an adverse effect of 6% with mash-fed birds and no effect on crumb-fed birds. Most surprisingly supplementary food in floor trays gave an 8% improvement in liveweight. There was a 30% difference in 28-day liveweight between the best and worst combinations which illustrates the influence of food intake factors.

Early bodyweight and subsequent growth rate

The logical question to ask was 'what effect does 28-day liveweight have on subsequent growth rate?' To investigate this, at 28 days each treatment was split into either a high or low protein feed programme. The results are shown in *Table 6.14*.

It will be seen that there is no evidence of compensatory growth even on the high protein programme. The explanation for the difference compared with the classic work by Auckland, Morris and Jennings (1969) probably lies in the fact that the Big 6 males grow more than twice as fast as the birds used in their work and are also later maturing.

Table 6.14 INFLUENCE OF 4 WEEK WEIGHT ON SUBSEQUENT GROWTH RATE OF BUT BIG 6 MALE TURKEYS

	Liveweights (kg)					
	4 weeks	8 weeks 1 day	12 weeks 2 days	16 weeks 1 day	20 weeks	24 weeks 1 day
(a) High protein programme						
Protein % 0–28 days	30	27	23	18	18	18
Crumbs	1.02	4.16	8.76	12.73	16.54	20.00
Mash	0.78	3.67	8.21	11.85	15.63	18.79
Diff.	0.24	0.49	0.55	0.88	0.91	1.21
(b) Low protein programme						
Protein % 0–28 days	30	23	18	18	14	14
Crumbs	1.02	3.96	8.12	10.95	12.62	16.74
Mash	0.78	3.42	7.34	10.06	11.48	14.79
Diff.	0.24	0.54	0.78	0.89	1.14	1.77

Nixey (1987)

Leg problems

If however a strain of turkey or a particular farm has a history of leg problems, an improvement in the leg problems and hence subsequent growth will be seen if the growth rate is slowed down prior to 12 weeks, particularly from six to 12 weeks. The results however will be inferior to those achieved if the cause of the leg problems can be removed and the birds grown fast throughout.

One of the keys to the problem would appear to be to prevent diarrhoea particularly between six and 12 weeks. The skeleton of the turkey, particularly the male is growing very fast at that time. The adverse effect may be malabsorption of nutrients or it may be that the weakness of the birds, illustrated by the shaky legs and reluctance to walk far, is causing abnormal bone growth. Martland (1984) created leg problems just by damping down the litter so that diarrhoea may also be having an influence by causing wet litter. The suggested causes of diarrhoea are several. It may be of infective origin (Saif, 1987). Nutritional aspects can also be the cause. Among the nutritional aspects under suspicion are (1) marked changes in the ingredients used between formulations, (2) the use of new season cereals, (3) the use of manioc, (4) poor quality fats, (5) an adverse balance between K, Na and Cl, (6) excess protein. Doubtless there are other suggestions for causes of diarrhoea.

Nutrition and meat yields

Increasingly the turkey is being used as a source of meat and not sold as a whole bird. In this situation, the yield of breastmeat is very significant as it is usually worth at least twice the value of dark meat. The nutritional programme will have a marked influence on the breastmeat yield. In general the better the growth rate achieved the better will be the breastmeat yield. However the yield of breastmeat is very sensitive to the rate of growth in the latter period. This is illustrated in *Table 6.15*. The small breastmeat yield of the males grown poorly between 16 and 20 weeks was not unexpected because at this age the breast muscles develop rapidly under a normal growth regime. A common problem occurs when the males are moved to a finisher type diet too early. The definition of too early will of course depend upon the nutrient content of the finisher diet. However a 15% protein finisher type diet is bad for a large type turkey before 20 weeks of age.

Traditional farm-fresh turkeys

GROWTH RATE

Of most interest is probably the Christmas turkey market. The most common complaint that the feed industry receives from this type of customer is the turkeys have grown too big. Customers order a turkey of a particular weight, most commonly around 14 lb (6–6.5 kg) plucked weight. They do not take kindly to being offered a bird 25% bigger than they ordered which may be the case if the farmer only has one flock being grown for Christmas. Often he has done his planning wrong, or was unable to obtain poults on the date required. Also, almost invariably, turkeys grown on the general farm for Christmas will exceed their breed target liveweights by a considerable margin. They will have received a lot of attention, been given ample food

Table 6.15 GROWTH PATTERN AND BREASTMEAT YIELD IN LARGE TYPE MALES

Age (weeks)	Growth pattern[a]		
0–16	Good	Good	Poor
16–20	Good	Poor	Poor
Liveweight (kg)	15.9	14.7	13.0
Breastmeat as percentage of liveweight	27.1	24.3	23.9

Nixey (1983)
[a] Good and poor growth rates were achieved by different protein intakes

and floor space and been grown at a favourable time of the year with cool temperatures. Their feed representative can help them with planning, encouragement or even help to do sample weighing at various ages and adjust the feed programme accordingly. Sample weighing is essential to forewarn of problems. Too often the first time the farmer realizes his birds are very heavy is on the day he comes to kill them.

SUBCUTANEOUS FAT COVER

The main nutritional problem we face with this market is to achieve a good finish on the birds. Finish is a term used to describe the subcutaneous fat cover which should be particularly evident on the breast and thighs of the bird giving a creamy white appearance. A poorly finished bird will appear bluish and red and look unattractive. The older the bird gets, the greater the fat cover that would be laid down naturally. The problem is occurring because the modern industrial strains of turkey are growing extremely fast and so to achieve the required weights they must often be killed around 15 weeks of age instead of 20 weeks as in the past. It is more difficult to induce the younger bird to lay down subcutaneous fat by nutritional manipulation.

Moran (1979) working with chickens found that finish could be improved by widening the energy to protein ratio but liveweights and fleshing were worsened. Salmon (1974) found that carcass finish was more strongly influenced by the level of added dietary fat than by varying the energy to protein ratio. Subsequently Salmon (1986) found that progressively increased nutrient density with age was effective in improving carcass finish.

Rose and Michie (1985) found it difficult to alter the amount of subcutaneous fat by nutritional means even though abdominal fat amounts were altered. It was hypothesized that any changes in total fatness of the bird would be mostly reflected in the fat depots with a better blood supply, which does not include those in the feather tract of the breast. Their work, however, did find the suggestion that high energy starter feed could subsequently have a long-term influence to make it easier to lay down subcutaneous fat.

In the author's experience the most predictable change that can be accomplished is to alter the type of fat laid down. It is directly influenced by the fatty acid profile of the diet (Neudoerffer and Lea, 1967; Salmon, 1967). Unsaturated fatty acids as found in vegetable oils produce an oily yellow fat whereas animal fats or fat metabolized from carbohydrates will result in a solid white pork-type fat. The latter gives a much more attractive finish and will also be visually more apparent. As a result even though the amount of subcutaneous fat has not been changed, the visual appearance, which is the point of the exercise, has been influenced beneficially.

References

AUCKLAND, J.N., MORRIS, T.R. and JENNINGS, R.C. (1969). Compensatory growth after under nutrition in market turkeys. *British Poultry Science*, **10**, 293–302

BEHRENDS, B.R. and WAIBEL, P.E. (1980). Methionine and cystine requirements of growing turkeys. *Poultry Science*, **59**, 849–859

CLARK, F.A., GOUS, R.M. and MORRIS, T.R. (1982). Response of broiler chicken to well balanced protein mixtures. *British Poultry Science*, **23**, 433–446

D'MELLO, J.P.F. and EMMANS, G.C. (1975). Amino acid requirements of the young turkey: lysine and arginine. *British Poultry Science*, **16**, 297–306

FISHER, C. and EMMANS, G.C. (1983). Calculated amino acid requirements for growing turkeys. *Turkeys*, **31** (No. 1), 39–43

FISHER, C., MORRIS, T.R. and JENNINGS, R.C. (1973). A model for the description and prediction of the response of laying hens to amino acid intake. *British Poultry Science*, **14**, 469–484

HURWITZ, S., FRISCH, Y., BAR, A., EISNER, U., BENGAL. I. and PINES, M. (1983). The amino acid requirements of growing turkeys. 1. Model construction and parameter estimation. *Poultry Science*, **62**, 2208–2217

MARTLAND, M.F. (1984). Wet litter as a cause of plantar pododermatitis, leading to foot ulceration and lameness in fattening turkey. *Avian Pathology*, **13**, 241–252

MORAN, E.T.J. (1979). Carcass quality changes with the broiler chicken after dietary protein restriction during the growing phase and finishing period compensatory growth. *Poultry Science*, **58**, 1257–1270

MORRIS, T.R. (1988). The place of the turkey in the animal industry of the future. In *Proceeedings of 21st Poultry Science Symposium on Recent Advances in Turkey Science*. Ed. Nixey, C. and Grey, T.C. Butterworths, London (in press)

NEUDOERFFER, T.S. and LEA, C.H. (1967). Effects of dietary polyunsaturated fatty acids on the composition of the individual lipids of turkey breast and leg muscle. *British Journal of Nutrition*, **21**, 691–714

NIXEY, C. (1983). Feeding turkeys on the farm. In *Proceedings of 7th Colborns Feed Industry Conference*, Colborn Dawes Ltd, Eastbourne

NIXEY, C. (1988). Nutritional responses of growing turkeys. In *Proceedings of 21st Poultry Science Symposium on Recent Advances in Turkey Science*. Ed. Nixey, C. and Grey, T.C. Butterworths, London (in press)

NRC (1984). National Research Council (US) Sub-Committee on Poultry Nutrition. *Nutrient Requirements of Poultry*, 8th Revised Edition. National Academy Press, Washington

ROSE, S.P. and MICHIE, W. (1985). Improving the fatness of turkeys by diet. Research and Development note No. 29 September 1985. The Scottish Agricultural Colleges

SAIF, Y.M. (1988). Enteric conditions of turkeys. In *Proceedings of 21st Poultry Science Symposium on Recent Advances of Turkey Science*. Ed. Nixey, C. and Grey, T.C. Butterworths, London (in press)

SALMON, R.E. (1974). Effect of dietary fat concentration and energy to protein ratios on the performance, yield of carcass components and composition of skin and meat of turkeys as related to age. *British Poultry Science*, **15**, 543–560

SALMON, R.E. (1976). The effect of age and sex on the rate of change of fatty acid composition of turkeys following a change of dietary fat source. *Poultry Science*, **55**, 201–208

SALMON, R.E. (1986). Effect of nutrient density and energy to protein ratio on performance and carcass quality of small white turkeys. *British Poultry Science*, **27**, 629–638

7

MINERAL AND TRACE ELEMENT REQUIREMENTS OF POULTRY

R. HILL
Royal Veterinary College, University of London, Potters Bar, Hertfordshire, UK

Introduction

There are several sets of values of the nutrient requirements of poultry, each including sections on minerals. Those most commonly quoted in the UK are given by the Agricultural Research Council (1975) and the National Research Council of the USA (1984). No attempt will be made to compare, in detail, all the values available for mineral requirements, though attention is drawn to an interesting set of tables prepared by the Subcommittee on Mineral Requirements of Poultry (1981), showing a large number of values used by European countries and including those of the ARC (1975) and NRC (1977).

Data presented and discussed here will be largely those assembled over recent years by the AFRC Working Party on Nutrient Requirements of Poultry, to be used in revisions of values given previously (ARC, 1975). The requirement values listed here as ARFC, may be modified slightly before their publication with the technical reviews.

Some sources give requirement values while others give recommended allowances. Allowances include a margin of safety which is intended to recognize the presence of a number of uncontrolled variables, such as nutrient values of different batches of feeds, genetic make-up of the birds and subclinical infections. Values of ARC (1975) and NRC (1984), and those to be published by the AFRC Working Party, are requirements, and for practical purposes a margin of safety should be added. This is commonly about + 0.05–0.10, but for some minerals, particularly the trace elements, larger margins are often safely adopted.

Some elements are present in conventional diets in quantities that are adequate or more than adequate to meet requirements and in consequence no supplements are needed. These elements are included in the tables of requirements but, except for potassium, are not considered further.

The assessment of requirements

The factorial method of determining nutrient requirements is accepted as superior to the alternative, dose-response method. This is because as new data become available for one or more of the factors involved (e.g. net requirement for maintenance) this can be incorporated into the requirement value. The dose-response method on the other hand, although giving results more quickly, provides requirement values that apply only to the particular set of circumstances in which the values were obtained, the particular environment, the type of bird, diet and level of production.

The Subcommittee on Mineral Requirements of Poultry (1984) has published values for Ca in adult birds calculated using the factorial method, and Weigand and Kirchgessner (1979) have given values using this method for Cu, Fe, Mn and Zn. However, although the case for the use of the factorial method is clearly established, it cannot be used satisfactorily at present for minerals in poultry, there being incomplete data on the factors involved, for example, net requirements for maintenance and different levels of production.

Thus, it remains necessary to use dose-response data to deduce requirement values. The choice of the appropriate response to be measured poses a problem for some elements. In growing birds, rate of gain is an obvious choice, but some aspects of tissue composition may also be important. This was considered in detail for Zn by Dewar (1986) who took account of data for net retention and tibial content of Zn, as well as rate of gain. For each element, the most appropriate criteria for the assessment of requirements need to be determined.

The value of data available in the literature from dose-response experiments in providing a firm requirement value depends primarily on the number of dietary concentrations included in the experiment, but also on a number of other items such as the type of bird and diet used. For some nutrients there is a number of sets of satisfactory data on which reliable requirement values can be based, but still, for most elements the data are few and the requirement value deduced is regarded as 'tentative' and this is indicated in the tables. Having accepted this distinction, there is, nevertheless, no sharp line between firmly based requirement values and tentative values, and there remains a subjective element in providing most requirement values for minerals in poultry.

Requirement values given in ARC and AFRC tables have been adjusted wherever possible to an ME content of 12.6 MJ (3.0 Mcal)/kg diet.

Calcium and phosphorus

The influence of phytate and vitamin D on the absorption of calcium and phosphorus, though reasonably well understood, is of continuing interest. Mohammed (1987) showed that intestinal phytase activity in broilers was stimulated by adding a moderately large supplement of vitamin D to the diet (50 000 i.u./kg): this increased the proportion of phytate hydrolysed and reduced the quantity of phosphate supplement needed for satisfactory growth and bone mineralization. In other studies, also with broilers, Lee *et al.* (1986) found that in certain diets phytic acid was a useful source of phosphorus. However, in most diets composed of common ingredients, with sufficient calcium to meet requirements and moderate quantities of Vitamin D, phytate phosphorus is sufficiently poorly utilized to justify quoting requirements in terms of non-phytate, or available phosphorus: this is particularly true for chicks and layers. In young chicks the effect of high phytate on calcium absorption justifies an adjustment of calcium requirement when phytate phosphorus in the diet exceeds 2 g/kg. In diets containing high proportions of soyabean meal this value commonly may be exceeded. The adjustment suggested, based on the composition of calcium phytate, is an increase of 1.3 g Ca/kg for each 1 g phytate P/kg in excess of 2 g/kg.

In most experiments with broilers in which phosphorus requirements have been determined, the source of supplementary calcium in the diets was precipitated chalk or ground limestone of an unspecified particle size, but it has now been shown that the available phosphorus requirements of chicks vary according to particle size of the calcium supplement. McNaughton (1981) found that at the dietary concentrations of available phosphorus studied (1.5–4.5 g/kg) growth and bone ash of broiler chicks at 21 days of age was greater when a limestone supplement of medium particle size (20–60 mesh) was used than with either coarse (12–20 mesh) or fine (100–200 mesh) limestone. Requirements for available phosphorus were reduced from 4.5 to 3.5 g/kg with medium 20–60 mesh limestone.

The introduction of a high-calcium diet (30 g/kg) to pullets during the growing period at 8 weeks of age, caused severe damage to the renal system (Watkins, Dilworth and Day, 1977). When this change in diet was made at 18 or 19 weeks of age, no problem of this kind occurred (Leeson, Julian and Summers, 1986), but there is no experimental evidence to indicate the minimum age at which pullets can safely be given a high-calcium layers diet.

The calcium requirement for maximum shell thickness is greater than for maximum egg production and as shell thickness is related to strength, the requirement quoted is for maximum or near maximum thickness. Most of the results available show a continuing increase in thickness with all increases in intake, even at 5 g or more daily (ARC, 1975), but there is some evidence that voluntary feed intake may be depressed when diets containing high levels of calcium are given (Scott, Hull and Mullenhoff, 1971). The optimum daily intake is about 4.0 g but Simons (1986) suggested amounts related to the weight of egg produced daily per bird, that is 3.8 g Ca for 50 g of egg and 4.5 g for 60 g.

The translation of the daily intake of calcium to a proportion of calcium in the diet clearly requires knowledge of feed intake: the values given in *Table 7.1*, 35.0–36.5 g/kg, are based on an intake of 110 g/day when production is high, but when the daily intake of the flock differs fairly widely from 110 g it will be desirable to maintain an overall intake close to 4.0 g/day by adjusting the proportion of calcium in the diet. Differences in requirement within a flock will be met in some measure by comparable differences in intake of feed.

The influence of source of calcium carbonate and of particle size on shell quality has been studied in numerous experiments. Results of early studies cannot be readily interpreted because comparisons were made between limestone and oyster shell of different particle size: when particle size was controlled most experiments showed no difference between limestone and oyster shell (Roland, 1986). In very few studies has the effect of particle size with a single source been determined, but the evidence on particle size shows that shell strength is improved, without affecting production or feed efficiency when grits contribute 67–75% of the total supplementary calcium (Scott, Hull and Mullenhoff, 1971; Kuhl, Holder and Sullivan, 1977; Watkins, Dilworth and Day, 1977; Bradley and Krueger, 1982). There is little evidence at present to indicate the precise particle size or mix of particle sizes that will give maximum shell quality.

The requirement of phosphorus for the laying hen has been studied closely in recent years, stimulated by the rising cost of phosphate and the possibility that earlier values were unnecessarily high, 350 mg/day non-phytate phosphorus. Low phosphate intakes may also have a beneficial effect on shell quality since increases in plasma phosphate cause a decline in egg specific gravity (Boorman, Volynchook and Belyavin, 1985). In some experiments intake of non-phytate phosphorus below 200 mg/day throughout the laying period gave as good shell quality as larger amounts

(Owings, Sell and Balloun, 1977; Mikaelian and Sell, 1981), but a value around 150 mg/day decreased egg production in an experiment of Rodriguez, Owings and Sell (1984). From these reports a value of 200–230 mg/day is suggested, corresponding to about 2 g non-phytate phosphorus/kg diet at an intake of 110 g/day. The requirement for phosphorus evidently decreases during the first part of the egg-laying year and to be sure of avoiding any ill-effects from low phosphorus intakes it has been suggested that the changes from a pre-laying high-phosphorus diet be made in phases: with diets decreasing from 3.5 to 2.5 and 1.5 g non-phytate phosphorus/kg diet fed during the first six months of production, egg numbers and shell quality were equal to those from diets containing 3.5 g/kg throughout the laying period (Mikaelian and Sell, 1981; Rodriguez, Owings and Sell, 1984).

The calcium supplement in these experiments on low-phosphorus diets for laying birds was ground limestone or limestone flour. The use of larger particles of limestone could reduce further the requirement of available phosphorus. By providing calcium for absorption over a greater part of each 24 h period, the need for bone resorption would be decreased. This would decrease the phosphorus required to repair bone resorbed to supply calcium for shell formation. However, the magnitude of a further decrease in phosphorus requirement is likely to be relatively small, bearing in mind the quantity leaving the bird as egg at high levels of production will be around 100 mg/day (Subcommittee for Mineral Requirements of Poultry, 1984): the new requirement value suggested is 200–230 mg/day or 1.8–2.1 g/kg (*Table 7.1*).

Sodium, potassium and chloride

Requirements of sodium have been studied very little in recent years and in consequence although some adjustments of earlier amounts are made in the proposed ARFC values (*Table 7.2*), almost all have still to be regarded as 'tentative'. The exception is for turkey breeders: Harms, Buresh and Wilson (1985) reported experiments with Large White hens given maize–soyabean diets containing 0.04, 0.34, 0.63, 0.93 and 1.22 g Na/kg. Maximum egg production was achieved with the diet containing 0.63 g Na/kg but for maximum fertility and hatchability 0.93 g Na/kg was required, and 1.0 g/kg is the preferred value.

Requirements of potassium also have been studied very little recently and all those given in *Table 7.2* are regarded as tentative. With conventional diets, the concentrations of potassium present always exceed the requirement value, for all classes of poultry.

Chloride requirement values for several classes of bird are more firmly based by the results of recent studies. Results of Gardiner and Dewar (1976) for chicks supported earlier estimates of 1.1 g/kg, and a study with broiler breeder hens from Harms and Wilson (1984) provides more data on egg production. Supplements of chloride ranging from 0.45 to 1.80 g Cl/kg were added to a basal maize–soyabean diet containing 0.3 g Cl/kg. With a chloride supplement of 0.9 g/kg, giving a total concentration of 1.2 g/kg, maximum egg production, egg weight, feed conversion efficiency and hatchability were recorded. This value, 1.2 g/kg, was also obtained by Vogt (1977) for laying birds, when the data were adjusted to an intake of 110 g/day, and is used in *Table 7.2*.

Turkey poults evidently need more chloride than chicks: Harms (1982) fed young turkey poults on maize–soyabean diets containing chloride concentrations of from 0.40–2.56 g Cl/kg and observed maximum growth with a diet providing 0.47 g/Mcal,

Table 7.1 THE REQUIREMENTS OF Ca AND P (g/kg DIET)[a] OF POULTRY

Element	Class of poultry		ARC (1975)	NRC (1984)	AFRC[e]
Ca	Fowl	0–4 week	11.5–12.0[b]	8.0–9.5	11.5–12.0[b]
		4–8 week	7.0–8.0[c]	7.0–8.5	7.0–8.0
		8–16 week	4.0	6.0	4.0
		Layers and breeders[d]	35.0–36.5[b]	35.0	35.0–36.5[b]
	Turkeys	0–8 week	7.5–8.5[b]	11.5	7.5–9.4[b]
		8–16 week	4.0–6.0	8.0	6.0
		17 + week	—	5.5	5.0
		Breeders	22.5–25.0[b]	23.5	22.5–25.0[b]
P non-phytate	Fowl	0–4 week	4.6–4.8	4.2	4.6–4.8
		4–8 week	—	3.5	3.7–4.3
		8–16 week	—	3.3	1.0
		Layers and breeders[d]	3.2	3.2	1.8–2.1
	Turkeys	0–8 week	4.5–5.5	5.8	4.6–5.5
		8–16 week	—	4.0	4.0
		17 + week	—	3.0	3.0
		Breeders	3.0	3.6	4.0
P total	Fowl	0–4 week	—	—	6.6–6.8
		4–8 week	5.7–6.3	—	5.7–6.3
		8–16 week	3.0	—	3.0
		Layers and breeders[d]	—	—	4.0–4.2
	Turkeys	0–8 week	6.0–7.0	—	6.6–7.5
		8–16 week	4.0–6.0	—	6.0
		17 + week	—	—	5.0
		Breeders	5.0	—	6.0

[a] Adjusted to 12.6 MJ ME (3.0 Mcal)/kg
[b] Provided diet contains not more than 2 g phytate P/kg. There is no direct evidence for layers and breeders
[c] Values in italics are regarded as 'tentative'
[d] Assuming a feed intake of 110 g/day
[e] AFRC Working Party on Nutritive Requirements of Poultry (to be published)

corresponding to 1.4 g Cl/kg for a diet containing 12.6 MJ (3 Mcal) ME/kg, and this is the value proposed. For breeding turkey hens, Harms, Junqueira and Wilson (1983) found that for maximum egg weight and specific gravity, as well as egg production, fertility and hatchability, 1.2 g Cl/kg diet was needed, the same concentration as that required for egg production in the fowl.

Under practical conditions the acid-base status of birds is determined primarily by the amounts of sodium and potassium in the diet in addition to chloride and therefore the proportions of these elements in the diet may be important as well as the requirement values. An excessive intake of sodium and/or potassium intake of chloride leads to an alkalosis and an excessive intake of chloride an acidosis, and ill-effects have been observed from each of these conditions. Alkalosis reduced the rate of weight gain of chicks (Mongin and Sauveur, 1977) and depressed production in laying hens (Junqueira, Miles and Harms, 1984), while acidosis reduced growth in chicks (Mongin and Sauveur, 1977) and impaired egg shell secretion in hens (Hunt and Aitken, 1962; Mongin, 1968). Attempts have been made to define acceptable values or ranges of values for the acid-base balance of poultry diets using the difference, in mEq/kg, between the strongly basic cations and acidic anion, Na + K − Cl.

Table 7.2 THE REQUIREMENTS OF Na, K, Cl AND Mg (g/kg DIET)[a] OF POULTRY

Element	Class of poultry		ARC (1975)	NRC (1984)	AFRC[c]
Na	Fowl	0–4 week	1.5	1.4–1.5	1.3
		4–8 week	1.2[b]	1.4–1.5	1.2
		8–16 week	0.7	1.5	0.7
		Layers and breeders	1.0	1.5	1.5
	Turkeys	0–8 week	1.5–2.0	1.6	1.5–2.0
		8–16 week	1.5–2.0	1.2	1.5–2.0
		17 + week	1.5–2.0	1.2	1.5–2.0
		Breeders	1.5–2.0	1.5	1.0
K	Fowl	0–4 week	2.5	4.0	2.6
		4–8 week	2.5	3.0	2.3
		8–16 week	2.5	3.0	1.4
		Layers and breeders	2.5	1.5	1.4
	Turkeys	0–8 week	4.4	7.0	6
		8–16 week	4.4	5.0	6
		17 + week	4.4	5.0	6
		Breeders	4.4	6.2	6
Cl	Fowl	0–4 week	1.4	1.5	1.1
		4–8 week	1.1	1.5	1.0
		8–16 week	0.6	1.5	0.7
		Layers and breeders	0.9	1.5	1.2
	Turkeys	0–8 week	—	1.5	1.4
		8–16 week	—	1.3	1.2
		17 + week	—	1.3	1.2
		Breeders	—	1.2	1.2
Mg	Fowl	0–4 week	0.4	0.6	0.4
		4–8 week	0.4	0.5	0.4
		8–16 week	0.4	0.4	0.4
		Layers and breeders	0.4	0.5	0.4
	Turkeys	0–8 week	0.45	0.6	0.45
		8–16 week	0.45	0.6	0.45
		17 + week	0.45	0.6	0.45
		Breeders	0.45	0.6	0.45

[a]Adjusted to 12.6 MJ ME (3.0 Mcal)/kg
[b]Values in italics are regarded as 'tentative'
[c]AFRC Working Party on Nutritive Requirements of Poultry (to be published)

For chicks, Mongin and Sauveur (1977) found that optimal growth occurred with an acid-base or electrolyte balance of 250 mEq/kg. However, Karunajeewa, Barr and Fox (1986) gave to broilers, diets with electrolyte balances of 150, 200, 250 and 300 mEq/kg and found no effects of these on weight gain; neither was there any effect on moisture content of droppings. It seems, therefore, there are other factors influencing the effects of electrolyte balance calculated from Na + K − Cl. The results of Mongin and Sauveur (1977) were obtained largely using semipurified diets, while those of Karunajeewa, Barr and Fox (1986) were from conventional diets, and there were other differences in the concentrations of the elements and the ratios of sodium to potassium: any of these may have influenced the results. Furthermore, in the experiment of Karunajeewa, Barr and Fox (1986) a depression of bodyweight caused by excess phosphate was alleviated by increasing the electrolyte balance to 250 or 300 mEq/kg showing that phosphorus has also to be considered in this situation.

The electrolyte balance value suggested for laying hens by Sauveur and Mongin

(1978) was about 200 mEq/kg. However, Cohen and Hurwitz (1974) fed laying hens on diets with (Na + K − Cl) values as high as 338 mEq/kg without any reduction in egg production and a diet used by Hunt and Aitken (1962) containing ammonium chloride, with an electrolyte balance of approximately 60 mEq/kg had no ill-effect on production or shell thickness. It is clear, therefore, that the value of 200 mEq/kg suggested by Sauveur and Mongin (1978) carries with it a large margin of safety.

As indicated for broilers, phosphorus should be considered along with sodium, chloride, and possibly potassium, in layer diets. The interrelations of phosphorus and sodium, as chloride or bicarbonate, studied by Junqueira, Miles and Harms (1984), showed that a supplement of sodium bicarbonate increased egg specific gravity but only when the phosphorus content was either low or high. It is clear that at least phosphorus should be set alongside sodium, potassium and chloride when studying the interactions among the cations and anions in poultry diets.

Manganese

The availability of manganese in the diet is influenced by the quantities of calcium carbonate and calcium phosphate present and it is well established that for similar amounts of calcium, calcium phosphate decreases availability to a greater degree than calcium carbonate (Schaible and Bandemer, 1942; Longstaff and Hill, 1971). However, the addition of supplements of calcium and phosphorus separately also evidently influence the availability of manganese, judged by the incidence of hock disorder (Smith and Kabaija, 1985). With a diet containing 6 g P/kg, increasing the calcium content from 10 g/kg to 20 and 30 g increased hock disorders markedly and the circumference of the tibia. On increasing the phosphorus content from 6 to 12 g/kg, the low incidence of hock problems with the diet containing 10 g Ca increased, and the high incidence of hock problems with the diet containing 30 g Ca decreased. The addition of manganese, the basal content was 77 mg Mn/kg, to diets that caused a high incidence of hock problems reduced them substantially, but did not prevent them entirely. It has been suggested that manganese availability is reduced by its adsorption on to finely divided calcium phosphate in the digestive tract, and this could explain some of the effects occurring in this study of Smith and Kabaija (1985), but not all, notably the decrease in hock disorders observed on increasing the phosphorus content from 6 to 12 g/kg of the diet containing 30 g Ca/kg. The absorption of manganese is, in general, very low: the processes causing variation in availability and absorption have received very little attention.

The quantity of manganese required for maximum shell quality in laying hens is still not readily assessed from the data available. The variability among results is partly associated with the range of measurements carried out on the shell to assess quality. In addition to thickness, deformation, and static breaking force, Whisenhurst and Maurice (1985) determined cracking caused by dynamic loading, an assessment that is relevant to mechanical handling and transport of eggs. In that study, a basal diet containing 25 mg Mn/kg was compared with the same diet supplemented to bring the manganese content to 200 mg/kg. The additional manganese did not increase shell thickness, static breaking strength nor uronic acid content of the shell, but it reduced cracking from dynamic loading. These results suggest that the ideal commercial shell is produced by diets containing more than 25 mg Mn/kg. Panic, Apostolov and Knezevie (1978) compared the effects of diets containing 20, 40, 50, 60 and 100 mg Mn/kg on several measurements of shell quality and found that in each case the highest concentration of manganese, 100 mg/kg, gave the highest value. From other experiments smaller amounts of manganese gave maximum shell strength, 20 mg/kg (Cox and Balloun, 1969) and 76 mg/kg (Gatowska and Parkhurst, 1942). It seems possible, from these results, that there may remain further aspects of the diet or the bird that influence the efficiency of utilization of dietary manganese. Meanwhile, the 'tentative' requirement value suggested for egg production is 100 mg Mn/kg (*Table 7.3*).

Zinc

The availability of zinc is much lower in conventional than semipurified diets, a difference caused, at least partly, by phytate in cereals and extracted oil seed meals, and in consequence, requirement values are greater for conventional than semipurified diets. Requirement values obtained for broilers with conventional diets, at least 85 mg Zn/kg (Tortuero and Brenes, 1976; Brenes and Tortuero, 1978), were more than fourfold greater than those with semipurified diets, 18 mg Zn/kg (Dewar and Downie, 1984). However, the relation between phytate and zinc availability has not been described sufficiently precisely to allow a recommended increase in requirement to be made according to the phytate content of the diet. Furthermore, in experiments described by Leese and Williams (1967) it was evident that factors in addition to phytate influenced the availability of zinc: these remain to be identified.

Table 7.3 THE REQUIREMENTS OF TRACE ELEMENTS (mg/kg DIET)[a] OF POULTRY

Element	Class of poultry		ARC (1975)	NRC (1984)	AFRC[d]
Cu	Fowl,	chick	*4*[b]	8	8
		breeder	—	8	10
	Turkey,	poult	—	8	8
Fe	Fowl,	chick	75	80	80
		breeder	—	60	55
	Turkey,	poult	75	80	80
Mn	Fowl,	chick	50	60	60
		layer	30	30	*100*
		breeder	50	60	50
	Turkey,	poult	50	60	60
		breeder	100	60	*100*
	Pheasant,	chick	75	—	75
Zn	Fowl,	chick	40	40	*85*
		breeder	60	65	*80*
	Turkey,	poult	70	75	*85*
	Pheasant,	chick	60	—	*85*
I	Fowl,	chick	0.4	0.35	*0.3*
		breeder	0.2	0.30	*0.2*
Se	Fowl,	chick	—	0.15(10)[c]	*0.10(5)* or *0.05(10)*
		layer	—	0.10(5)	*0.05(15)*
		breeder	—	0.10(10)	*0.15(15)* or *0.05(25)*
	Turkey,	poult	—	0.20(12)	*0.20(10)*
		breeder	—	0.20(25)	*0.20(15)*

[a] Adjusted to 12.6 MJ ME (3 Mcal)/kg
[b] Values in italics are regarded as 'tentative'
[c] Vitamin E value in parentheses
[d] AFRC Working Party on Nutritive Requirements of Poultry (to be published)

Requirement values for zinc are particularly difficult to propose, the only studies providing satisfactory dose-response data having been made with semipurified diets. In this situation the values given in *Table 7.3* are all regarded as tentative.

Selenium

The availabilities of selenium in a range of feeds have been reported but the determinations are indirect, and include some aspect of metabolism in addition to absorption. Cantor, Scott and Noguchi (1975) used the incidence of exudative diathesis as the method of assessment, while Gabrielson and Opstvedt (1980a,b) and Ikumo and Yoslinden (1981) used blood glutathione peroxidase, and in each case the values are given as proportions of the effect of the same quantity of selenium as sodium selenite, which is known to be readily absorbed. Gabrielsen and Opstvedt (1980a,b) obtained availabilities of 0.34 and 0.48 for two batches of fish meal, 0.18 for soyabean meal and 0.26 for maize gluten. There is evidently no information on factors limiting the availability of selenium in feeds.

Selenium is known to be associated with a number of other dietary substances— vitamin E, unsaturated fats, ascorbic acid, vitamin A, zinc, protein and cystine. Each of these nutrients is likely to influence the requirements of selenium but, with the exception of vitamin E, there are no data from which quantitative estimates of their effects could be derived.

Athough the importance of the interaction of vitamin E with selenium has been recognized for some time, the determination of vitamin E was difficult and unsatisfactory until relatively recently. An improved method has been described (Manz and Philipp, 1981), but no studies on selenium and vitamin E requirements using this method have so far been published. The most useful data available at present were obtained from experiments in which the basal diet was treated to reduce the vitamin E content to zero, or as near zero as possible. Using diets prepared in this way, Thompson and Scott (1969) obtained maximum growth in clinically healthy chicks when the concentrations of selenium and vitamin E (mg/kg) were:

Se 0.015 and E 100
   0.025   10
   0.055   0

When no selenium was added to the basal diet containing 0.005 mg Se/kg, even a very large amount of vitamin E, 1000 mg/kg, did not give quite maximum weight. The basal diet contained 125 mg antioxidant/kg. Similar results were obtained by Seier and Bragg (1973), but much larger estimates of selenium and vitamin E requirements were derived from *in vivo* measurements of peroxidation (Combs and Scott, 1974). Zero peroxidation in liver microsomal preparations was obtained from birds given diets containing 0.15 mg Se and 40 mg E/kg, 0.06 mg Se with 100 mg E/kg, or suitable intermediate combinations. These values seem unnecessarily high: it is probable that *in vivo*, maximum weight gain and clinical health occur even with some degree of peroxidation. The values proposed for chicks (*Table 7.3*) are slightly higher than those that gave maximum gain in healthy birds in the studies of Thompson and Scott (1969) and Seier and Bragg (1973).

Other values given in *Table 7.3* are based on useful evidence, but adequate dose-response data are not available. For example, Combs and Scott (1979) using a practical-type diet selected for its low selenium content, found requirements were greater for breeding than laying fowl. The basal diet containing 0.04 mg Se/kg, an estimated 13.6 mg vitamin E, and 125 mg antioxidant, gave maximum egg production, but the addition of vitamin E or selenium was needed for maximum hatchability. In these experiments, a single large supplement of either vitamin E or selenium was added, thus the proposed values (*Table 7.3*) are tentative.

Requirement values of ARC (1975), NRC (1984) and those proposed from AFRC (1988) are shown in *Tables 7.1–7.3*. There are few marked differences between the most recent NRC and AFRC values. Where differences are large, background information related to these was considered earlier in the appropriate sections; examples are the lower available phosphorus value proposed for laying and breeding hens, and the higher manganese suggested for egg production.

References

AGRICULTURAL RESEARCH COUNCIL (1975). *The Nutrient Requirements of Farm Livestock, No. 1: Poultry.* Agricultural Research Council, London
BRADLEY, J.W. and KRUEGER, W.F. (1982). *Poultry Science,* **61,** 1423
BRENES, A. and TORTUERO, F. (1978). In *Third World Congress on Animal Feeding.* International Veterinary Association for Animal Production **8,** 149
BOORMAN, K.N., VOLYNCHOOK, J.G. and BALYAVIN, C.G. (1985). In *Recent Advances in Animal Nutrition—1985,* pp. 181–195. Ed. Haresign, W. and Cole, D.J.A. Butterworths, London
CANTOR, A.H., SCOTT, M.L. and NOGUCHI, T. (1975). *Journal of Nutrition,* **105,** 96–105
COHEN, I. and HURWITZ, S. (1974). *Poultry Science,* **53,** 378–383
COMBS, G.F. JR. and SCOTT, M.L. (1974). *Journal of Nutrition,* **104,** 1297–1303
COMBS, G.F. JR. and SCOTT, M.L. (1979). *Poultry Science,* **58,** 871–884
COX, A.C. and BALLOUN, S.L. (1969). *Poultry Science,* **48,** 745–747
DEWAR, W.A. (1986). In *Nutrient Requirements of Poultry and Nutritional Research,* pp.155–171. Ed. Fisher, C. and Boorman, K.N. Butterworths, London
DEWAR, W.A. and DOWNIE, J.N. (1984). *British Journal of Nutrition,* **51,** 467–477
GABRIELSON, B.O. and OPSTEVDT, J. (1980a). *Journal of Nutrition,* **110,** 1089–1095
GABRIELSON, B.O. and OPSTEVDT, J. (1980b). *Journal of Nutrition,* **110,** 1096–1100
GARDINER, E.E. and DEWAR, W.A. (1976). *British Poultry Science,* **17,** 337–340
GATOWSKA, M.S. and PARKHURST, R.T. (1942). *Poultry Science,* **21,** 227–287
HARMS, R.H. (1982a). *Poultry Science,* **61,** 2447–2449
HARMS, R.H., JUNQUEIRA, O.M. and WILSON, H.R. (1983). *Poultry Science,* **62,** 242–244
HARMS, R.H. and WILSON, H.R. (1984). *Poultry Science,* **63,** 835–837
HARMS, R.H., BURESH, R.E. and WILSON, H.R. (1985). *British Poultry Science,* **26,** 217–220
HUNT, J.R. and AITKEN, J.R. (1962). *Poultry Science,* **41,** 434–438
IKUMO, H. and YOSLINDEN, M. (1981). *Japanese Poultry Science,* **18,** 307–311
JUNQUEIRA, O.M., MILES, R.D. and HARMS, R.H. (1984). *Poultry Science,* **63,** 1229–1236
KARUNAJEEWA, H., BARR, D.A. and FOX, M. (1986). *British Poultry Science,* **27,** 601–612
KUHL, H.J. JR., HOLDER, D.P. and SULLIVAN, T.W. (1977). *Poultry Science,* **56,** 605
LEE, K.U., HA, J.K., HAN, I.K. and KIM, M.K. (1986). *Korean Journal of Animal Sciences,* **28,** 146–152
LEESE, J.G. and WILLIAMS, W.P. (1967). *Poultry Science,* **46,** 233–241

LEESON, S., JULIAN, R.J. and SUMMERS, J.D. (1986). *Canadian Journal of Animal Sciences*, **66**, 1087–1095
LONGSTAFF, M. and HILL, R. (1971). *British Poultry Science*, **12**, 401–411
MCNAUGHTON, J.L. (1981). *Poultry Science*, **60**, 179–185, 197–203
MANZ, U. and PHILIPP, K. (1981). *International Journal for Vitamin and Nutritional Research*, **51**, 342–348
MIKAELIAN, K.S. and SELL, J.L. (1981). *Poultry Science*, **60**, 1916–1924
MOHAMMED, A.A. (1987). The utilisation of phytate phosphorus by chicks. PhD Thesis. University of Southampton
MONGIN, P. (1968). *World's Poultry Science Journal*, **24**, 200–230
MONGIN, P. and SAUVEUR, B. (1977). In *Growth and Poultry Meat Production*, pp. 235–247. Ed. Boorman, K.N. and Wilson, B.J. British Poultry Science Ltd, Edinburgh
NATIONAL RESEARCH COUNCIL (1977). *Nutritional Requirements of Poultry*, 7th edition. National Academy of Sciences, Washington DC
NATIONAL RESEARCH COUNCIL (1984). *Nutritional Requirements of Poultry*, 8th edition. National Academy of Sciences, Washington DC
OWINGS, W.J., SELL, J.L. and BALLOUN, S.L. (1977). *Poultry Science*, **56**, 2056–2060
PANIC, B., APOSTOLOV, N. and KNEZEVIE, J. (1978). In *Trace Element Metabolism in Man and Animals 3*, pp. 511–514. Ed. Kirchgessner, M. Arbeitskreis fur Trernahrungs forschung Weihenstephen, German Federal Republic
RODRIGUEZ, M., OWINGS, W.J. and SELL, J.L. (1984). *Poultry Science*, **63**, 1553–1562
ROLAND, D.A. (1986). *World's Poultry Science Journal*, **42**, 166–171
SAUVEUR, B. and MONGIN, P. (1978). *British Poultry Science*, **19**, 475–485
SCHAIBLE, P.J. and BANDEMER, S.L. (1942). *Poultry Science*, **21**, 8–14
SCOTT, M.L., HULL, S.J. and MULLENHOFF, P.A. (1971). *Poultry Science*, **50**, 1055–1063
SEIER, L. and BRAGG, D.B. (1973). *Canadian Journal of Animal Science*, **53**, 371–375
SIMONS, P.C.M. (1986). In *Nutrient Requirements of Poultry and Nutritional Research*, pp. 141–154. Ed. Fisher, C. and Boorman, K.N. Butterworths, London
SMITH, O.B. and KABAIJA, E. (1985). *Poultry Science*, **64**, 1713–1720
SUBCOMMITTEE ON MINERAL REQUIREMENTS FOR POULTRY (1981). Working Group No. 2—Nutrition—of the European Federation of Branches of the WPSA. *World's Poultry Science Journal*, **37**, 127–138
SUBCOMMITTEE ON MINERAL REQUIREMENTS FOR POULTRY (1984). Working Group No. 2—Nutrition—of the European Federation of Branches of the WPSA. *World's Poultry Science Journal*, **40**, 183–187
THOMPSON, J.N. and SCOTT, M.L. (1969). *Journal of Nutrition*, **97**, 335–342
TORTUERO, F. and BRENES, A. (1976). *Advances in Alimentacion y Mejora Animal*, **18**, 139–143
VOGT, H. (1977). *Archiv fur Geflugelkunde*, **41**, 125–129
WATKINS, R.M., DILWORTH, B.C. and DAY, E.J. (1977). *Poultry Science*, **56**, 1641–1647
WEIGAND, E. and KIRCHGESSNER, M. (1979). *Proceedings of the Second European Symposium on Poultry Nutrition*, pp. 64–71. Ed. Kan, C.A. and Simons, P., Beekbergen, Netherlands.
WHISENHURST, J.E. and MAURICE, D.V. (1985). *Nutrition Reports International*, **31**, 757–764

8

RESPONSE OF LAYING HENS TO ENERGY AND AMINO ACIDS

R. M. GOUS and F. J. KLEYN
Department of Animal Science and Poultry Science, University of Natal, Pietermaritzburg, South Africa

Introduction

Twenty years ago Morris (1968) presented a review concerning the response of laying hens to nutrient density from which it is possible to optimize this characteristic of the feed for a flock of laying hens. Very little has been added to our knowledge of this subject since then. Significant advances have, however, been made during the past two decades in the characterization of responses among laying hens to the more important amino acids. In an elegant paper on this subject, Fisher, Morris and Jennings (1973) produced a model by which such responses could be used to determine the optimum intakes of amino acids in a population of laying hens. Responses are measured as changes in egg weight, rate of lay, body size and composition, and food intake. Information published on these responses is reviewed briefly here; some recent research is discussed pertaining to food intake regulation; and information is presented regarding the relative changes in egg weight and rate of lay that take place with changes in protein supply.

In spite of the financial and biological benefits that result from the use of the above research reports, there is a reluctance among many feed compounders to make use of such methods to optimize feeding strategies in laying flocks.

There are two particular reasons why all nutritionists do not make use of biological responses in feed formulation. Firstly, compounders tend to 'play safe' with feeds whose composition has been found to produce satisfactory results in the field, i.e. they do not wish to move outside the narrow bounds defined in their linear programming (LP) matrix. Secondly, the calculation of marginal costs of amino acids and the determination of optimum nutrient densities is both tedious and time-consuming, as least-cost solutions have to be obtained over a range of concentrations of amino acids and nutrient densities in order to obtain such values. It is easier to ignore changes in ingredient prices which may (but may not) make only small differences to profitability. More emphasis has been directed towards the development of linear programs to optimize bulk handling and usage of raw materials in the feed mill than in optimizing the feeding strategy of laying hens. Milling costs have certainly been reduced in this way, with a consequent increase in millers' margin, but further benefits should be possible on behalf of the egg producer. Such an approach would be especially beneficial in large integrated operations. An efficient and rapid computer program has been developed recently as a means of determining the optimum strategy

Effect of amino acid and energy concentrations on egg output

AMINO ACIDS

It is common knowledge that dietary amino acid supply will affect egg production. The requirement for these amino acids should, however, not be seen as a fixed concentration in the feed but should take account of the relationship between amino acid intake and egg output, which is dependent on both feed intake and level of production. A series of response experiments, mainly at the University of Reading (Fisher, Morris and Jennings, 1973; Pilbrow and Morris, 1974; Morris and Wethli, 1978; Wethli and Morris, 1978; Morris and Blackburn, 1982) has improved our knowledge of efficiencies of utilization of amino acids for egg production, and culminated in a review by McDonald and Morris (1985) in which best estimates of the coefficients of response by laying hens to different amino acids were presented (*Table 8.1*).

With such information the Reading Model (Fisher, Morris and Jennings, 1973) can be used effectively to determine the optimum economic intake of each of the amino acids for existing stocks of laying hens as well as future stocks, which may differ in body weight, in potential egg output, or in both. Because this optimum is dependent on such variables as flock uniformity, marginal costs of amino acids and marginal revenue for eggs, the most economical intake of each amino acid will differ under the variable conditions experienced between farms, between flocks and between different countries. No single recommendation could be expected to be applicable under all circumstances.

The optimum intake of the first-limiting amino acid will influence the optimum intakes of all other amino acids. Because the marginal costs of essential amino acids are likely to vary one from the other, one, or possibly two, amino acids with the highest marginal costs will limit egg output of the flock. Under most circumstances it would, therefore, be unnecessary to feed the remaining amino acids at their optimum intakes, some economic advantage being derived by reducing the intakes of these amino acids to the level governed by the egg output sustainable by the first-limiting amino acid.

A major difficulty confronting feed formulators is the conversion of these optimum intakes of amino acids into dietary concentrations. This difficulty would be resolved if

Table 8.1 COEFFICIENTS OF RESPONSE (mg) FOR SOME ESSENTIAL AMINO ACIDS FOR INDIVIDUAL PULLETS

Lysine	10.0 E	+	73 W
Methionine	4.8 E	+	31 W
Tryptophan	2.6 E	+	11 W
Isoleucine	8.0 E	+	67 W
Valine	8.9 E	+	76 W

After McDonald and Morris (1985)
Where E = Egg output (g/bird day) and W = bodyweight (kg)

food intake could be predicted accurately. Further discussion on this subject is presented later.

Partitioning the response to amino acids between egg weight and egg number

Egg size can be manipulated only to a very limited extent by nutrition. A high energy diet containing supplementary fat will increase egg weight by a maximum of 1 g (Morris, 1985), provided that the protein:energy ratio is maintained. Dietary amino acid supply also has a small effect on egg weight, mainly on albumen content (Fisher, 1969) but because egg numbers are influenced also, there is little scope for manipulating egg size by adjusting amino acid supply.

For a detailed financial analysis of the consequences of varying protein supply it is necessary to predict separately the expected changes in rate of lay and egg size, since in most markets eggs of different sizes have different values per unit weight. Morris and Gous (1988) surveyed 42 sets of data in which responses in egg number and egg size to dietary amino acids were measured. Contrary to the common belief that only egg size is reduced when protein supply is marginally reduced (i.e. that slightly more protein is needed to maximize egg size than to maximize egg number) the evidence indicated that small increments in protein (or amino acids), close to the optimum, resulted in equal proportional responses in rate of lay and egg size. When protein supply is below the optimum (egg output below 0.95 of the maximum potential of the flock), the expected reduction in rate of lay is greater than the expected reduction in egg size. In the above analysis it was found that reduction in egg size did not fall below 0.9 of the maximum value, until amino acid supply was well below 0.5 of its optimum value, whereas rate of lay was reduced below 0.7 of its potential value when amino acid intake is half the optimum (*Figure 8.1*).

The proportion of eggs falling into different grades will differ with protein supply, and hence influence the marginal revenue for eggs. The equation relating egg size to protein supply (*Figure 8.1*) can therefore be used to determine marginal revenue. This is a prerequisite if the Reading Model is to be used accurately. However, the optimum intakes of amino acids appear to be relatively insensitive to changes in the marginal revenue for eggs (see later).

It can be concluded that, because under most commercial conditions the amino acid supply that maximizes profit in a layer operation will be that which allows the flock to produce at a near-maximal rate, any effort to use nutritional means to reduce egg size will result in a concomitant decrease in rate of lay, which would be likely to reduce profits. Egg size can be manipulated to a greater extent by adjusting lighting patterns (e.g. ahemeral cycles longer than 24 h, or 6 h repeating light–dark cycles (Sauveur and Mongin, 1983) than by altering nutrient levels (Morris, 1985) and the effects of such cycles on egg size can be switched on and off when necessary, a further advantage over the use of nutritional methods of altering egg size.

ENERGY

A certain amount of confusion exists regarding the terms 'energy concentration' and 'nutrient density'. It has been recognized for a long time that when the energy content of a poultry feed is increased, the concentration of most of the other nutrients in the

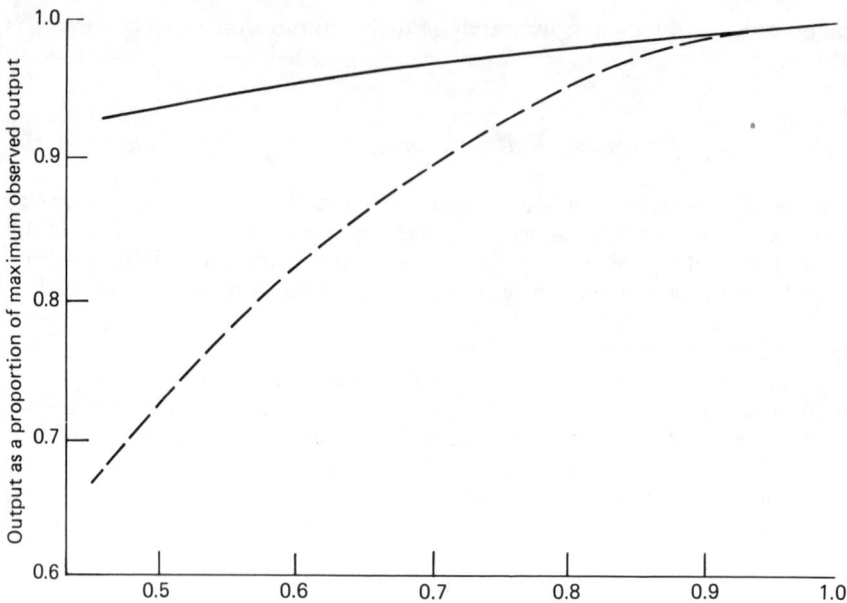

Figure 8.1 The relationship between intake of a limiting amino acid and rate of lay (---) or egg weight (——). Equations for the two responses are: relative egg weight = $1 - 0.07353x - 0.10424x^2$; relative rate of lay = $1 - 0.03734x - 1.02927x^2$. (After Morris and Gous, 1988)

feed should be increased proportionately (Combs, 1962). Thus the more useful term 'nutrient density' should be used to describe the concentrations of dietary energy, assuming that nutrients have been adjusted accordingly. Some of the confusion has arisen as a result of the use of ME as the unit measure of nutrient density.

Although much of the following discussion concerns nutrient density, some interesting experiments involving changes in energy at constant amino acid concentrations will be mentioned also.

Effect of nutrient density on egg production

The production of eggs by laying hens is not influenced by nutrient density, except at the extremes (Morris, 1968). Evidence regarding the effect of nutrient density on egg weight is less well defined, recent reports suggesting that feeds high in nutrient density produce slightly larger eggs. McDonald (1984) found the regression of egg weight on nutrient density to be 0.2487 g/MJ kg. This effect is so small as to have virtually no influence on the optimum density of feeds.

The positive effect of nutrient density on egg weight could probably be ascribed to either a linoleic acid deficiency at very low nutrient densities, which is overcome with the use of high energy ingredients, or to the use of oil in the feeds of high nutrient density, this latter being known to increase egg size (Edwards and Morris, 1967).

Effect of dietary energy concentration on egg output

In three experiments, designed to measure the responses to lysine, methionine and

R.M. Gous and F.J. Kleyn 115

isoleucine at different energy concentrations (Gous, Griessel and Morris, 1987) the response in egg output was such that a common curve for each amino acid adequately described the response at each of the dietary energy concentrations (*Figure 8.2*). Egg output was, therefore, not influenced by energy concentration, other than through its effects on food intake, with a consequent change in intake of the first-limiting amino acid. It was demonstrated that, although the main effect of energy concentration on egg weight and rate of lay in the methionine experiment was significant in both cases, these differences were brought about by the particularly low egg output associated with the low energy, low methionine treatment. When the response in egg weight and rate of lay were compared with the actual intakes of methionine it was evident that at

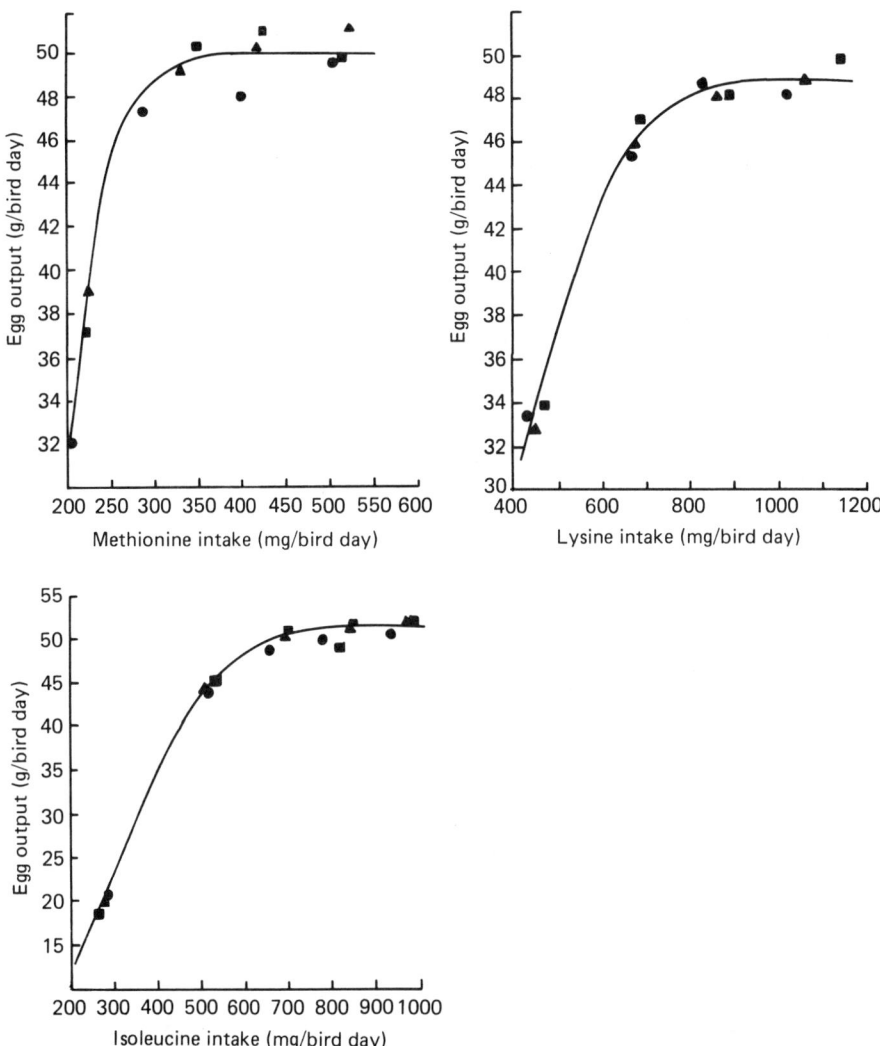

Figure 8.2 Response of laying hens to intakes of methionine, lysine and isoleucine at three energy concentrations (low ●, medium ◁ and high ■). (After Gous, Griessel and Morris, 1987)

the lowest methionine concentration the difference in productivity between energy levels had not been caused by the low energy concentrations *per se*, but rather by the low intake of methionine (*Figure 8.3*). The apparent effect of energy concentration on egg output can therefore be explained by expressing output as a function of intake of the limiting nutrient.

Effect of environmental temperature on egg output

The principle described above applies to environmental temperatures also. High temperatures will reduce intake of all nutrients, and output, which is a function of intake of the first-limiting nutrient, will be reduced by a corresponding amount. Bray and Gesell (1961) showed that output at high temperatures (30°C) could be sustained by increasing the concentration of nutrients in the feed. Subsequently it was suggested (Smith and Oliver, 1972) that at temperatures above 27°C it is likely that egg output is restricted by an inadequate intake of energy, although another factor which probably plays a role in reducing egg output is a reduction in blood supply to the ovary (Wolfenson *et al.*, 1979). At temperatures in excess of 30°C, if energy intake is the limiting factor in determining egg output, it would be unnecessary and counterproductive to increase amino acid supply still further as the excess protein, by the process of digestion and deamination, will increase heat production by the hen thereby further aggravating her heat load. Under such circumstances the use of dietary fat would serve the purpose of increasing energy intake and, because of its lower heat increment compared with other feed materials, would reduce the heat load on layers at extremely high temperatures (Marsden and Morris, 1981). Egg size, which is reduced at high temperatures, can be increased in this manner, although the effect would be expected to be very small (Marsden, Morris and Cromarty, 1987).

The obvious conclusion regarding the effect of constant high temperatures on egg production is that the hen becomes too hot and no nutritional manipulations will enable her to overcome this problem. In the short term, utilization of lipid reserves may enable the hen to produce eggs at a higher rate than might be expected considering her low energy intake, but in the long term such a buffering capacity could not persist and production would decline.

Effect of amino acid and energy concentrations on food intake

Egg output of laying hens is determined largely by the intake of the first-limiting nutrient in the feed (Almquist, 1952; Emmans, 1981). Some accurate method of predicting food intake under varying environmental and nutritional circumstances would therefore be a prerequisite in calculating optimum dietary concentrations of nutrients in feeds for laying hens. Despite this, very little effort has been made to predict food intake accurately, yet this is central to the issue of maximizing profit in a layer enterprise.

Because of the vast body of publications stating that 'birds eat to satisfy their energy requirement', a major effort has been made to determine the energy requirements of laying hens, usually by multiple regression analysis (Byerly, 1941; 1979) but scant attention has been paid to the effects on intake of, for example, low dietary amino acid or calcium concentrations. Some observations and suggestions regarding this important issue are reported below, with the effects of energy being discussed first.

ENERGY

When fed *ad libitum*, birds will consume different amounts of energy depending on the nature of the diet presented to them (Morris, 1968). Most experiments in which dietary energy has been partitioned between body maintenance, egg output and growth have involved one feed only and under such circumstances the 'goodness of fit' of the resultant empirical equation is usually statistically highly significant (e.g. Leeson, Lewis and Shrimpton, 1973; McDonald, 1977; Gous *et al.*, 1978). The outliers are regarded as 'experimental errors' and usually no attempt is made to explain these anomalies. Prediction equations of this nature, then, are valuable only as a means of roughly estimating the average food intake of a population. Some sophistication can be introduced by considering the effects of temperature and feather cover (Emmans, 1974), but there are basic inaccuracies in estimating energy intake in this manner.

The relationship between environmental temperature and energy intake is curvilinear, with food intake declining more steeply as ambient temperature approaches body temperature (Marsden and Morris, 1987). The partition equation of Emmans (1974) expresses energy required for maintenance under different environmental temperatures but does not predict how much energy will be consumed by layers under such circumstances. A quadratic term would have to be introduced into the equation if the effect of temperature on energy intake is to be predicted. Alternatively, by expressing energy intake as a function of metabolic body size it can be represented as a linear function of temperature within the range 15–30°C, with a slope of -141.8 kJ/$kg^{0.75}$ day (Marsden and Morris, 1987).

A further source of error in partition equations is that the energy required for maintenance is expressed in terms of body weight. A bird with large lipid reserves probably does not need more energy than a bird of similar protein weight with no lipid reserves. Maintenance heat should thus be scaled according to protein weight (preferably feather-free) and possibly also to the degree of maturity (Emmans, personal communication). This would have a significant effect on the recommended energy requirement of broiler breeder hens and would reduce the discrepancy in maintenance energy requirements between brown and white strains of laying hens, although this will not be eliminated entirely because of behavioural (activity) and anatomical (comb and wattle size; rate of feather loss) differences between these strains. Bodyweight gain (or loss) could be a result of protein retention (or depletion) with the associated change in water, lipid retention (or loss) or both and the amount of energy required (or yielded) will differ in each case. A single coefficient for change in body weight is a best-estimate for a flock under certain circumstances, but the nutritionist should be aware that there are many individuals in the flock that will not conform to this estimate, nor would this coefficient be the same for all feeds or for all environmental conditions.

AMINO ACIDS

In the published reports of many trials designed to measure responses of laying hens to amino acid concentrations (e.g. Pilbrow and Morris, 1974; Wethli and Morris, 1978; Morris and Wethli, 1978; Gous, Griessel and Morris, 1987) food intake increased in almost all cases as the concentration of dietary amino acids decreased, the birds clearly attempting to eat more food to compensate for marginal deficiency of the first-limiting amino acid. As the deficiency became more severe food intake declined in virtually all cases. In the three experiments reported by Gous, Griessel and Morris (1987) in which reponses to lysine, methionine and isoleucine were measured at different energy concentrations, food intake was affected to a greater extent by the dietary amino acid concentrations than by the energy concentrations (*Figure 8.3*).

The reason for the decline in food intake at low amino acid concentrations is not known. Emmans (1981) lists three reasons for the failure of birds to consume sufficient feed to overcome the deficiency in the first-limiting nutrient—bulk, environmental heat demand and the presence of toxins.

The first and last of these seem unlikely to be involved here, as the feeds in the dilution series do not differ markedly in these respects. Environmental heat demand is likely to be responsible for the observed decline in food intake. A consequence of consuming the excessively large amounts of a food, limiting in an amino acid, necessary to maintain maximum egg output, would be that the bird would need to lose more heat than it is capable of losing, this maximum heat loss being some function of body size and enironmental temperature. The hen would thus have no alternative but to reduce food intake. Egg production could be sustained immediately after the switch to a feed of low amino acid content if the hen had protein reserves from which to draw. These would differ between birds, hence egg production would drop more rapidly in some hens than in others.

Differences in initial lipid stores and maximum storage capacity would also influence the extent to which food intake and egg output were affected immediately after a change in dietary amino acid concentration. A feed intake simulation model, based on these concepts, is in the process of being developed by Emmans and Gous (as yet unpublished).

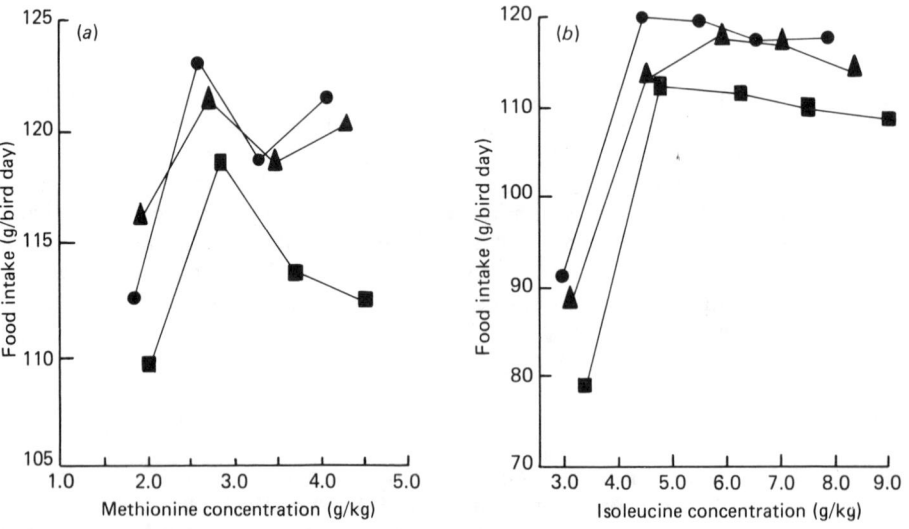

Figure 8.3 Food intakes of laying hens fed different concentrations of methionine (a) and isoleucine (b) at three energy concentrations (low ●, medium ▲ and high ■). (From Gous, Griessel and Morris, 1987)

R.M. Gous and F.J. Kleyn   119

Optimizing amino acid intakes and nutrient density

AMINO ACID INTAKES

The Reading model (Fisher, Morris and Jennings, 1973) is a population response model that optimizes the intake of amino acids for flocks varying in body weight and potential egg output, and for different relative values of amino acids and eggs.

The calculation of optimum intake of an amino acid is illustrated in *Tables 8.2* and *8.3*. At a dietary energy concentration the amino acid intake necessary to produce a specified mean egg output by the flock can be converted to a concentration in the feed, from a knowledge of the characteristic intake of food by the flock at that energy concentration. Feeds of increasing amino acid supply are formulated and the cost of feeding the bird is calculated as food cost, cents/g × food intake, g. Similarly the revenue derived can be calculated as value of egg, cents/g × mean egg output, g, the difference between these being the margin over food cost. The optimum intake is that which results in the highest margin.

A more rapid method of calculating intake of each amino acid in turn is described by Fisher, Morris and Jennings (1973), but the method described above, using basic

Table 8.2 AMINO ACID INTAKES REQUIRED TO SUSTAIN A RANGE OF EGG OUTPUTS IN A FLOCK OF LAYING HENS AND THE CONCENTRATIONS (g/kg) REQUIRED AT DIFFERENT DIETARY ENERGY CONCENTRATIONS

Mean egg output (g/bird day)	Amino acid intakes required[a] (mg/bird day)		Energy concentration (MJ/kg)						
			10.0	10.5	11.0	11.5	12.0	12.5	13.0
					Food intake (g/bird day)[b]				
			124	113	118	108	104	100	96
46	Lys	625	5.12	5.37	5.62	5.87	6.12	6.37	6.62
	Met	290	2.34	2.45	2.57	2.68	2.80	2.91	3.02
	Trp	145	1.17	1.23	1.28	1.34	1.40	1.46	1.51
48	Lys	652	5.26	5.52	5.77	6.03	6.29	6.54	6.80
	Met	303	2.44	2.56	2.68	2.80	2.92	2.04	3.16
	Trp	153	1.23	1.29	1.35	1.41	1.47	1.53	1.59
50	Lys	682	5.50	5.77	6.04	6.31	6.58	6.84	7.11
	Met	318	2.56	2.69	2.81	2.94	3.06	3.19	3.31
	Trp	161	1.29	1.36	1.42	1.48	1.55	1.61	1.67
52	Lys	720	5.81	6.09	6.38	6.66	6.94	7.22	7.51
	Met	336	2.71	2.84	2.98	3.11	3.24	3.37	3.50
	Trp	170	1.37	1.43	1.50	1.57	1.63	1.70	1.77
53	Lys	742	5.98	6.28	6.57	6.86	7.15	7.45	7.74
	Met	347	2.80	2.94	3.07	3.21	3.35	3.48	3.62
	Trp	175	1.41	1.48	1.55	1.62	1.69	1.76	1.82
54	Lys	772	6.23	6.53	6.84	7.14	7.44	7.75	8.05
	Met	360	2.90	3.05	3.19	3.33	3.47	3.61	3.75
	Trp	183	1.47	1.54	1.62	1.69	1.76	1.83	1.90
55	Lys	802	6.47	6.78	7.10	7.42	7.73	8.05	8.36
	Met	378	3.05	3.20	3.35	3.50	3.64	3.79	3.94
	Trp	192	1.54	1.62	1.70	1.77	1.85	1.92	2.00
56	Lys	875	7.06	7.40	7.75	8.09	8.44	8.78	9.12
	Met	413	3.33	3.49	3.65	3.81	3.98	4.14	4.30
	Trp	207	1.67	1.75	1.83	1.91	2.00	2.08	2.16

[a]Calculated from the coefficients published by McDonald and Morris (1985)
[b]Assuming a characteristic intake of 1243 kJ ME at a dietary energy concentration of 11.3 MJ/kg, using the equation of Morris (1968)

Table 8.3 MARGIN OVER FOOD COST[a] FOR A RANGE OF AMINO ACID CONCENTRATIONS, REQUIRED TO SUSTAIN A GIVEN MEAN EGG OUTPUT IN A LAYING FLOCK, AT DIFFERENT DIETARY ENERGY CONCENTRATIONS

Mean egg output[b] (g/bird day)	Energy concentration (MJ/kg)						
	10.0	10.5	11.0	11.5	12.0	12.5	13.0
46	91.4	91.7	91.9	92.1	92.0	91.9	91.7
48	96.3	96.6	96.9	96.9	96.8	96.7	96.2
50	98.5	98.8	99.0	99.1	98.9	98.7	98.1
52	99.2	99.6	99.8	99.7	99.5	99.3	98.5
53	99.4	99.7	100.0	99.8	99.6	99.2	98.5
54	99.3	99.6	99.8	99.6	99.4	98.9	98.2
55	99.1	99.5	99.5	99.3	99.1	98.4	97.6
56	98.3	98.6	98.6	98.3	97.9	97.1	96.3

[a]Calculated as (mean egg output, g × egg revenue, cents/g) − (food intake, g/bird × food cost cents/g) and expressed relative to the highest margin, which is given the value 100
[b]Mean egg output of a flock with E_{max} = 56 g/bird day, W = 2.0 kg and characteristic intake of energy at 11.3 MJ ME/kg of 1243 kJ/day. Amino acid concentrations required to sustain these egg outputs are given in *Table 8.2*

principles, will ensure a better understanding of the optimization procedure which follows.

The necessity of determining which amino acid is first-limiting and then to calculate all other amino acid intakes in terms of the limiting amino acid, is obviated if all amino acids are considered simultaneously in the above exercise. It is necessary to know the concentration of each amino acid required to sustain each level of egg output, and although the feeds formulated on this basis will not necessarily consist of an ideal balance of amino acids, nevertheless the feed resulting in the highest margin will be at the optimum economic amino acid balance, which will be dependent on the supply of protein-containing ingredients.

NUTRIENT DENSITY

From a knowledge of the characteristic intake of the strain or breed being considered, energy intakes, and hence food intakes, over a range of dietary energy concentrations can be calculated (Morris, 1968). If the amino acid supply is considered simultaneously, as described above, a matrix of energy and amino acid concentrations will result (*Table 8.2*). By calculating margin over food cost for each cell of the matrix (*Table 8.3*) the optimum nutrient density can be found which will maximize profitability under the prevailing economic conditions. The optimum feeding strategy under the conditions outlined in *Table 8.3* would be to supply sufficient amino acids to sustain a mean flock egg output of 53 g/bird day at an energy concentration of 11.0 MJ ME/kg. By adding a surcharge of 100 rand/ton to the cost of the feed the optimum strategy was shifted to an egg output of 52 g/bird day and an energy concentration of 12.0 MJ ME/kg, corresponding to a food intake for the flock of 105 g/bird.

It is interesting to note that profitability decreases only marginally in the immediate vicinity of the optimum combination of amino acid (egg output) and energy concentrations, with different combinations of these two dietary characteristics yielding similar profits.

USE OF MIXED INTEGER PROGRAMMING TO OPTIMIZE FEEDING STRATEGY

The method outlined above is time-consuming and repetitive. A rapid method of achieving the same goal has been described by Kleyn and Gous (1988) involving mixed integer programming, an extension of the usual linear programming method of feed formulation.

This technique 'offers' the computer a range of egg outputs, specified by the user, together with both the nutrient intakes required to sustain that level of output and the value of the output. This is referred to as an integer block of egg outputs, equivalent to the rows in *Table 8.2*. Because each level of output is assigned a maximum value of 1 and a minimum value of 0 the computer can be made to choose only one output under the conditions prevailing, the output chosen being that which maximizes profit. Similarly, in an integer block of energy concentrations (equivalent to the columns of the matrix in *Table 8.2*), the amount of each amino acid that would be consumed at each energy concentration is specified by the user. In the mixed integer program algorithm the computer to choose both the energy concentration and the egg output (amino acid supply) that will maximize profit, and will produce the same result as that in *Table 8.3*, which was the result of 56 least-cost feed formulations.

Because of the speed of such a method of optimizing the feeding strategy in a flock of laying hens, simulations are more readily carried out, and the remainder of this section will deal with examples of factors that may influence feeding strategy and the extent to which each factor is of importance. The standard situation from which changes were measured involved a flock of laying hens with a maximum egg output of 56 g/bird day and a mean body weight of 2.0 kg, with a characteristic intake of 110 g/bird day of a feed containing 11.3 MJ ME/kg. A surcharge of 100 rand/ton was added to food cost.

Changes in egg revenue

Changing the marginal revenue of eggs to values of 0.5 above and 0.5 below current revenue (*Table 8.4*) illustrate the relative inflexibility of optimum amino acid intake to this variable. Little financial benefit is derived by reducing feed quality when marginal revenue decreases, but when the ratio of feed cost to egg revenue is substantially

Table 8.4 THE EFFECT OF CHANGES IN EGG PRICE ON THE OPTIMUM COMBINATION OF FOOD INTAKE AND EGG OUTPUT FOR A FLOCK OF LAYING HENS

Egg price	Feed intake (g/bird day)	Mean flock egg output (g)	Margin over food cost[a]
Less 50%	104	51.5	15
Standard price	105	52.0	100
Plus 50%	105	54.0	271

[a]Relative to standard conditions described in the text

122 *Response of laying hens to energy and amino acids*

reduced, profit would be maximized by increasing the supply of amino acids thereby increasing egg output.

Changes in cost of availability of bulky ingredients

Ingredients such as wheat bran and middlings, lucerne and other bulky, low energy materials fluctuate in price and availability. By increasing the amount of these ingredients available in the daily food allocation from zero to 10 g/bird day the optimum nutrient density changed (*Table 8.5*) but the amino acid supply that maximized profit remained the same. Margin over food cost increased with wheat bran inclusion, implying that low density feeds are more cost-effective than feeds of high nutrient density. However, when surcharges are added to the cost of each ton of feed, a higher nutrient density is chosen as the surcharge increases (*Table 8.6*).

Change in characteristic food intake

Morris (1968) illustrated the differences in optimum food intake between a light, a medium and a heavy strain of laying hen which was the result of differences in their response to changes in dietary energy content. Such scenarios are modelled here to demonstrate how the optimum feeding strategy would differ for birds with characteristic intakes of 100, 110 and 120 g/day at an energy level of 11.3 MJ/kg (*Table 8.7*). The egg output sustainable at the optimum remained the same in all cases, with margin over food cost being highest for the bird with the lowest appetite. No

Table 8.5 EFFECT OF LOW DENSITY INGREDIENT (WHEAT BRAN) AVAILABILITY ON THE OPTIMUM COMBINATION OF FOOD INTAKE AND EGG OUTPUT FOR A FLOCK OF LAYING HENS

Availability of wheat bran (g/bird day)	Feed intake (g/bird day)	Mean flock egg output (g)	Margin over food cost[a]
0	100	52	99.4
2.5	102	52	99.7
5.0	103	52	99.8
7.5	104	52	99.9
10.0	105	52	100

[a] Relative to standard conditions described in the text

Table 8.6 EFFECT OF DIFFERENT SURCHARGES ON THE OPTIMUM COMBINATION OF FOOD INTAKE AND EGG OUTPUT FOR A FLOCK OF LAYING HENS

Surcharge (rand/ton)	Food intake (g/bird day)	Mean flock egg output	Margin over food cost[a]
0	113	53.0	116
+50	109	52.7	109
+100	105	52.0	100

[a] Relative to standard conditions described in the text

Table 8.7 EFFECT OF STRAIN OF BIRD WITH DIFFERENT CHARACTERISTIC FOOD INTAKES ON THE OPTIMUM COMBINATION OF FOOD INTAKE AND EGG OUTPUT FOR A FLOCK OF LAYING HENS

Characteristic intake of energy[a] (kJ/bird day)	Food intake (g/bird day)	Mean flock egg output (g)	Margin over food cost[b]
1130	100	52	103
1243	105	52	100
1356	117	52	97

[a]When given a feed containing 11.3 MJ ME/kg
[b]Relative to standard conditions described in the text, and assuming the same egg output

allowance was made for differences in egg size or potential output in the three strains, although this is possible, which would have influenced the relative margins.

Summary

Responses among laying hens to dietary energy and amino acids have been well characterized by quantitative nutritionists, who have simultaneously developed sophisticated techniques for optimizing the feeding strategy in flocks of laying hens. Although feed compounders are aware of these techniques there has been a reluctance on their part to make full use of them. Most feed formulators have however been made aware of certain feeding principles emanating from this research, such as the need to feed higher concentrations of amino acids to birds in hot climates or whose characteristic intake is low. It is possible, too, to ascertain the optimum feeding strategy of future laying flocks that might differ from those of today in potential egg output, in body weight, or both.

One of the most controversial aspects discussed in this chapter relates to the partitioning of the response to amino acids between egg weight and rate of lay, in which it is shown that egg weight cannot be manipulated by altering the dietary amino acid concentrations without simultaneously affecting rate of lay, the effect on rate of lay being more severe than that on egg weight.

A rapid technique for determining the feeding strategy that will maximize the profitability of a laying flock has been developed recently and this method is outlined, together with some practical examples of factors that may influence this strategy. This method provides a rapid and simple means of making use of the developments that have taken place in this area of quantitative nutrition.

References

ALMQUIST, H.J. (1952). *Archives of Biochemistry and Biophysics*, **59**, 197–202
BRAY, D.J. and GESELL, J.A. (1961). *Poultry Science*, **40**, 1328–1335
BYERLY, T.C. (1941). *Bulletin Maryland Agricultural Experimental Station*, A–1, 1–29
BYERLY, T.C. (1979). In *Food Intake Regulation in Poultry*, pp. 327–363. Ed. Boorman, K.N. and Freeman, B.M. British Poultry Science Ltd, Edinburgh
COMBS, G.F. (1962) In *Nutrition of Pigs and Poultry*, pp. 127–147. Ed. Morgan, J.T. and Lewis, D. Butterworths, London

EDWARDS, D.G. and MORRIS, T.R. (1967). *British Poultry Science*, **8,** 163–168
EMMANS, G.C. (1974). In *Energy Requirements of Poultry*, pp. 79–90. Ed. Morris, T.R. and Freeman, B.M. British Poultry Science Ltd, Edinburgh
EMMANS, G.C. (1981) In *Computers in Animal Production*, pp. 103–110. Ed. Hillyer, G.M., Wittemore, C.T. and Gunn, R.G. British Society of Animal Production
FISHER, C. (1969). *British Poultry Science*, **10,** 149–154
FISHER, C., MORRIS, T.R. and JENNINGS, R.C. (1973). *British Poultry Science*, **14,** 469–484
GOUS, R.M., BYERLY, T.C., THOMAS, O.P. and KESSLER, J.W. (1978). *16th Worlds Poultry Congress, Rio de Janeiro*, **2,** 1–8
GOUS, R.M., GRIESSEL, M.J. and MORRIS, T.R. (1987). *British Poultry Science*, **28,** 427–436
KLEYN, F.J. and GOUS, R.M. (1988). *Agricultural Systems*, **26,** (in press)
LEESON, S., LEWIS, D. and SHRIMPTON, D.H. (1973). *British Poultry Science*, **14,** 595–602
MARSDEN, A. and MORRIS, T.R. (1981). In *Intensive Animal Production in Developing Countries*, pp. 299–309, Ed. Smith, A.J. and Gunn, R.G. British Society of Animal Production
MARSDEN, A. and MORRIS, T.R. (1987). *British Poultry Science*, **28,** 693–699
MARSDEN, A., MORRIS, T.R. and CROMARTY, A.S. (1987). *British Poultry Science*, **28,** 361–380
MCDONALD, M.W. (1977). *Research Bulletin* 1/77, School of Environmental Studies, Queensland Agricultural College, Lawes, Australia
MCDONALD, M.W. (1984). *British Poultry Science*, **25,** 139–144
MCDONALD, M.W. and MORRIS, T.R. (1985). *British Poultry Science*, **26,** 253–264
MORRIS, T.R. (1968). *British Poultry Science*, **9,** 285–295
MORRIS, T.R. (1985). *South African Journal of Animal Science*, **15,** 120–122
MORRIS, T.R. and BLACKBURN, H.A. (1982). *British Poultry Science*, **23,** 405–424
MORRIS. T.R. and GOUS, R.M. (1988). *British Poultry Science*, (in press)
MORRIS, T.R. and WETHLI, E. (1978). *British Poultry Science*, **19,** 455–464
PILBROW, P.J. and MORRIS, T.R. (1974). *British Poultry Science*, **15,** 51–73
SAUVEUR, B. and MONGIN, P. (1982). *British Poultry Science*, **24,** 405–416
SMITH, A.J. and OLIVER, J. (1972). *Rhodesian Journal of Agricultural Research*, **10,** 43–60
WETHLI, E. and MORRIS, T.R. (1978). *British Poultry Science*, **19,** 559–565
WOLFENSON, D., FREI, Y.F., SNAPIR, N. and BERMAN, A. (1979). *British Poultry Science*, **20,** 167–174

IV

Ruminant Nutrition

9

PREDICTING THE METABOLIZABLE ENERGY (ME) CONTENT OF COMPOUNDED FEEDS FOR RUMINANTS

P. C. THOMAS, S. ROBERTSON and D. G. CHAMBERLAIN
Hannah Research Institute, Ayr, Scotland

and

R. M. LIVINGSTONE, P. H. GARTHWAITE, P. J. S. DEWEY, R. SMART and C. WHYTE
The Rowett Research Institute, Aberdeen, Scotland

Introduction

In the late 1970s, in the UK and in other member states of the European Economic Community, concern began to be expressed over the lack of information about the nutritional value of retailed compounded feeds for ruminants. By that time the Metabolizable Energy (ME) System had become well accepted in the UK as the preferred feeding system for ruminant livestock and, through the work of the Rowett Research Institute's Feed Evaluation Unit, contemporary directly-determined ME values had been provided for a range of the most common 'straight' feeds. However, representatives of farmers and of the feed supply industry recognized that for compounded feeds, which typically contain a wide variety of ingredients including by-product materials, it was essential to have methods to predict feed ME content from laboratory analysis. Only in that way could feed manufacturers implement fully adequate systems of quality control and provide the quality assurances and product specifications necessary to satisfy their customers.

After discussions between the United Kingdom Agricultural Supply Trades Association (UKASTA), the National Farmers Union (NFU), the Ministry of Agriculture, Fisheries and Food (MAFF) and the Department of Agriculture and Fisheries for Scotland (DAFS) a programme of research on compounded feeds was commissioned at the Rowett Research Institute in 1979. This work provided a basis for the development of equations to calculate feed ME content from chemical analysis (Wainman, Dewey and Boyne, 1981). Appropriate equations were subsequently derived and recommended by a joint Working Party of UKASTA, the Agricultural Advisory and Development Service and the Committee of Scottish Agricultural Colleges (UKASTA/ADAS/COSAC, 1985); these equations have since been taken up and used by the feed compounding industry.

The studies conducted at the Rowett Research Institute left some unanswered questions, however, and continuing concern over the adequacy of description of compound feeds has led over the past two years to a further programme of research. This work, which was conducted at the Rowett Research Institute and the Hannah

Research Institute and which, like the initial study, was commissioned and sponsored jointly by farmers, feed manufacturers and suppliers, and government agencies, forms the main focus of the present chapter. However, to place the new experiments in context some features of the initial Rowett study and of the UKASTA/ADAS/COSAC (1985) Report are also outlined and discussed.

The initial Rowett study

The primary objective of the work begun at the Rowett Institute in 1979 was to provide a framework of data from which to construct predictive equations to relate feed ME content to chemical composition. In the experiments, mature wether sheep given a standard, maintenance level of feeding were used to determine the amounts of dietary gross energy lost in faeces, urine and methane during the consumption of a selected series of test diets. Faecal and urine energy losses were determined by conventional methods of total collection over a ten-day 'balance' period and methane losses were measured over a two-day period when the sheep were confined to a closed-circuit respiration calorimeter. Full details of the methods used are given in the Report of the Feedingstuffs Evaluation Unit (Department of Agriculture and Fisheries for Scotland, 1975) and by Wainman, Dewey and Boyne (1981).

In the experiments, a total of 24 compounded feeds were investigated. These were selected to cover a range of compositions as specified by the following criteria:

(1) two ranges of ether extract, 20–39 g/kg DM and 50–70 g/kg DM;
(2) two ranges of crude fibre, 40–60 g/kg DM and 80–120 g/kg DM;
(3) three ranges of crude protein, 120–149 g/kg DM, 150–179 g/kg DM and 180–209 g/kg DM.

The feeds were formulated using selected ingredients from a list of 35 raw materials commonly used by the UK feed compounding industry (*Table 9.1*). Each of the feeds contained seven to 12 ingredients, the maximum level of inclusion of any starchy raw material being 400 g/kg and of any non-starchy material being 150 g/kg (*Table 9.1*). In terms of their chemical composition, the feeds ranged from 127 to 209 g crude protein/kg DM, 25 to 73 g ether extra/kg DM and 38 to 167 g crude fibre/kg DM with corresponding associated variations in other constituents.

The ME content of each of the feeds was determined using a 'by difference' procedure. Each was given along with hay or grass silage and with each forage type the compounded feed was tested at three levels of inclusion, 25, 50 and 75% of the total diet. Observations on each compounded feed and forage combination were made in duplicate using two sheep, and thus the experimental design provided 12 independent observations for each compound. These data were examined initially to test (1) for interaction effects related to forage type, and (2) for linearity with respect to the level of compounded feed included in the total diet; neither effect was found to be statistically significant. Thereafter, the 12 observations were fitted to a simple linear model which assumed that the ME content of the compounded feed, the hay and the silage were unique and that the ME content of any combined diet reflected the relative proportions of compound, hay or silage dry matter that it contained. The ME contents of the 24 compounded feeds determined in this way varied with feed composition from 9.8 to 13.7 MJ/kg DM, a range corresponding to that which might be encountered in commercial practice.

Table 9.1 FEED INGREDIENTS, NUMBER OF FEEDS IN WHICH INGREDIENTS WERE USED AND MAXIMUM LEVELS OF INCLUSION FOR THE 24 DIETS STUDIED BY WAINMAN, DEWEY AND BOYNE (1981)

Feed ingredient	No. of feeds	Maximum inclusion (g/kg)	Feed ingredient	No. of feeds	Maximum inclusion (g/kg)
Barley grain	17	400	Dried grass	5	56
Oat grain	6	140	Distillers' grains	8	100
Oat feed	11	150	Maize gluten feed	4	125
Wheat grain	14	400	Rapeseed meal	11	80
Flour	6	200	Palm kernel meal	4	50
Wheat feed	17	400	Sunflower meal	7	75
Sorghum grain	7	200	Cottonseed meal	4	54
Maize grain	6	300	Guar meal	5	36
Cassava	4	150	Groundnut cake	4	51
Fat blend	14	40	Soyabean meal	6	99
Dry fat premix	6	60	Denatured skim milk		
Sugar beet pulp	5	100	(with grass meal)	5	98
Straw	7	200	Herring meal	3	50
Improved straw	4	100	Meat and bone meal	3	40
Rice bran	7	150	Blood meal	3	28
Cocoa bean waste	9	50	Feather meal	5	30
Coffee waste	7	50	Urea	1	12
Grape follicle	9	84	Molasses	24	75

PREDICTION OF ME CONTENT

Using the data obtained with their 24 test diets, Wainman, Dewey and Boyne (1981) derived a total of 73 regression equations which allowed the observed ME contents to be predicted from laboratory analyses of the feeds with a residual standard deviation of less than 0.5 MJ/kg DM. However, it was pointed out that this gave a rather misleading impression of precision since, in practice, the error of prediction of ME would also reflect the between-laboratory variance in chemical analysis of the feed. This source of error in the predictions was considered further by the UKASTA/ADAS/COSAC Working Party (1985) who concluded that prediction equations should have:

(1) the lowest possible residual standard deviation (S'') calculated taking account of both regression and inter-laboratory variances,
(2) the minimum number of analytical determinations,
(3) the highest speed and lowest cost of analysis, appropriate regard being given to the purpose for which the ME value was required and the accuracy of prediction needed.

On the basis of these criteria, the Working Party recommended three prediction equations developed from the results of the Rowett study. These equations are given in *Table 9.2*. Equation U1 was proposed for legal purposes; it was based solely on analytical parameters which were a requirement of the existing UK Feedingstuffs Regulations, namely crude protein, ether extract, crude fibre and ash. Equation U2 was proposed for reference purposes; it had a very low S'' value but included the (then) relatively new analytical measurement of cellulose digestible organic matter in

Table 9.2 RECOMMENDED EQUATIONS FOR THE PREDICTION OF METABOLIZABLE ENERGY CONTENT FROM THE CHEMICAL ANALYSIS OF COMPOUNDED FEEDS

Equation designation	Recommended use	Equation[a]	Standard deviation (S'', MJ/kg DM)[b]
U1	Legal and voluntary	ME = 11.78 + 0.0654 CP + 0.0665 EE^2 − 0.0414 EE × CF − 0.118 TA	0.36
U2	Reference purposes	ME = 11.56 − 2.37 EE + 0.030 EE^2 + 0.030 EE × NCD − 0.034 TA	0.32
U3	Voluntary declaration	ME = 13.83 − 0.488 EE + 0.0394 EE^2 × CP − 0.0085 MADF × CP − 0.138 TA	0.35

[a]ME is MJ/kg DM, all other units are g/100 g DM. CP = crude protein; EE = ether extract; CF = crude fibre; TA = total ash; NCD = cellulase digestible organic matter in the dry matter following neutral detergent extraction; MADF = modified acid detergent fibre (for details see Wainman, Dewey and Brewer, 1984)
[b]Residual standard deviation taking account of between laboratory variances in chemical analysis of feed composition (see UKASTA/ADAS/COSAC, 1985)

the dry matter after neutral detergent extraction (NCD; Alderman, 1985). Finally, with the prospect of voluntary declaration in mind the Working Party proposed Equation U3, which was similar to U1 but offered the advantage that the determination of crude fibres was replaced by the more convenient determination of modified acid detergent fibre (MADF).

All three of the prediction equations recommended by UKASTA/ADAS/COSAC (1985) showed a good fit to the data for the 24 diets examined in the Rowett experiments (for example see *Figure 9.1*). However, reservations about the prediction of the ME content of compound feeds still remained, and there were concerns about a number of separate issues.

Firstly, the Rowett experiments had led to the exposure of an unresolvable inconsistency. Using the most up-to-date data on the ME contents of raw material feeds held on the UKASTA and ADAS databases, the UKASTA/ADAS/COSAC Working Party had calculated the ME contents of the feeds used in the Rowett study. The two databases gave rise to similar calculated values but both values showed a divergence from those estimated *in vivo* (*Figure 9.2*).

Secondly, the range of feeds selected for the Rowett study had been representative of those in common use in the late 1970s, but in the ensuing period there had been major changes in design of compounded feeds. The levels of inclusion and variety of types of fat and fibre sources being used had changed substantially, and it was consequently questionable whether the ME content of these new styles of feeds would be satisfactorily estimated by the existing prediction equations. The changes in diet design had already been acknowledged by the analytical chemists and to accommodate the increased use of fats and fatty acid sources the Feedingstuffs Regulations were amended in 1982 to add an acid extraction procedure for fats and oils (Method B) to the existing ether extract technique for neutral fats (Method A).

Finally, there was persisting uncertainty about whether the predicted values for feed ME content, based on measurements made in mature sheep at a maintenance level of feeding, were appropriate for dairy compounded feeds, which are given to lactating cows fed at three to four times maintenance.

It was against the background of these questions that further studies on the ME content of compounded feeds were initiated.

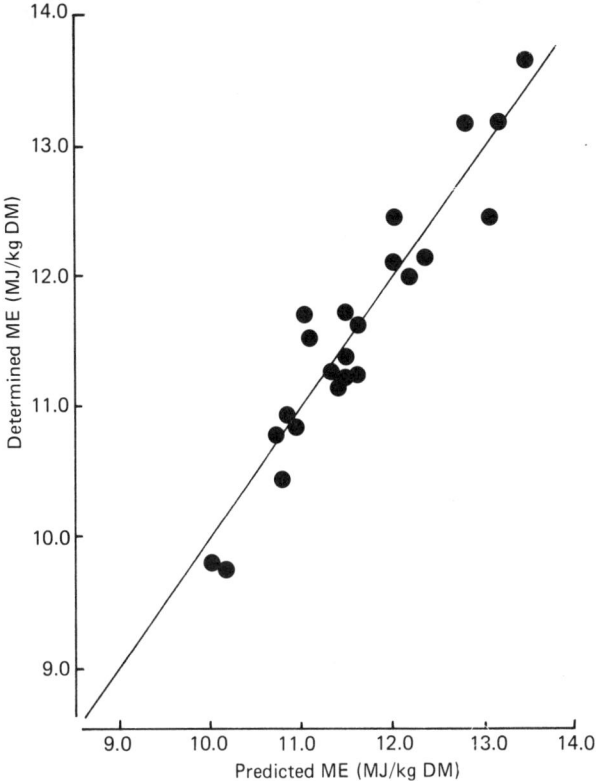

Figure 9.1 The relationships between *in vivo* estimates of feed ME content and those predicted from chemical analysis using Equation U1 (from UKASTA/ADAS/COSAC, 1985). The solid line shown is the line of equivalence

An outline of recent experiments

The study that has been completed recently at the Rowett Research Institute and Hannah Research Institute (RRI/HRI Study) consisted of a substantial new programme of work with sheep and a linked programme of work with dairy cows. The sheep experiments were undertaken at the Rowett Research Institute and the dairy cow experiments at the Hannah Research Institute. However, the experiments at the two centres were coordinated and designed so that they used precisely the same feedstuffs. Compounded concentrate feeds for both the sheep and the cows were formulated and manufactured at the Rowett Institute using single batches of raw materials. Correspondingly, the grass silages which were used as the basal forages were made as uniform clamps at the Hannah Institute and were transported in 1-tonne blocks to the Rowett Research Institute for the sheep experiments.

TECHNICAL DETAILS OF THE SHEEP EXPERIMENTS

In the sheep experiments, as in the earlier Rowett studies, measurements of gross energy intake and faecal, urine and methane energy losses were made in mature wether animals (liveweight 50–80 kg) given appropriate test diets at a maintenance

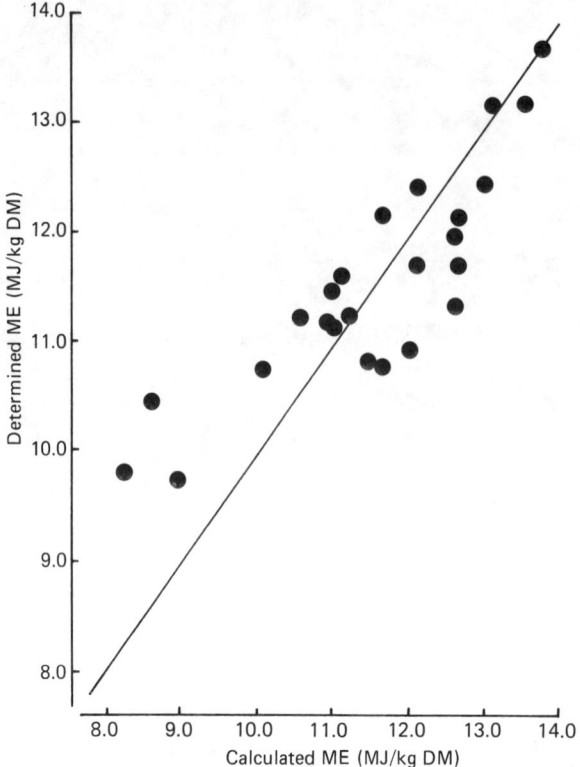

Figure 9.2 The relationship between *in vivo* estimates of feed ME content and those calculated from the ME content of the raw material ingredients derived from the UKASTA database (from UKASTA/ADAS/COSAC, 1985). The solid line shown is the line of equivalence. The line of best fit for the data was $y = 0.58x + 4.58$

level of feeding. After a 21-day establishment period on a diet, faecal and urine losses were determined over a nine-day collection period and methane loss was determined over a two-day period when the sheep were confined to an open-circuit respiration calorimeter.

The experiments were conducted in two parts. The main study was of an incomplete randomized design involving a total of 195 observations with 38 sheep and 78 dietary treatments. The treatments were a diet of highly digestible grass silage given alone and 77 diets in which the silage was given together with an equal proportion (50:50 on a dry matter basis) of one of a series of compounded feeds. These compounds were formulated from a restricted range of ingredients (*Table 9.3*) to provide feeds differing systematically in their content and source of 'added fat' and 'added fibre'. Five sources of fat were used, each being included at levels of 30 g/kg and 60 g/kg. The fats were palm fatty acid oil, maize/soya oil and three free flowing, 'dry' fat products, Megalac (Volac Ltd), Fat premix (FP1; BOCM Ltd), and fat prills (BP Ltd). Added fibre sources were ground straw, sodium hydroxide treated, nutritionally improved straw (NIS) or a mixture (50:50) of sugar beet pulp and citrus pulp, the levels of inclusion being 200 g/kg and 400 g/kg in each case (for summary of treatments see *Table 9.4*). Replications of the dietary treatments were designed to be unequal; for most treatments, observations were replicated with two or three sheep

Table 9.3 FEED INGREDIENTS, NUMBER OF FEEDS IN WHICH THE INGREDIENTS WERE USED AND THE MAXIMUM LEVELS OF INCLUSION FOR THE 86 DIETS USED IN THE MAIN AND SUBSIDIARY STUDY AND FOR THE SUPPLEMENTARY STUDY WITH 14 ADDITIONAL DIETS (SEE TEXT)

	Main and subsidiary study		Supplementary study	
Feed ingredient	No. of feeds	Maximum inclusion (g/kg)	No. of feeds	Maximum inclusion (g/kg)
Barley	86	150	11	270
Wheat	86	346	9	340
Wheat feed	86	199	8	420
Straw	22	400	—	—
NI straw	28	400	2	400
Citrus pulp	23	400	1	305
Sugar beet pulp	23	400	1	305
Rice bran	—	—	3	235
Molasses	86	80	14	80
Maize gluten	—	—	3	300
Full fat soya	—	—	3	315
Soyabean meal	86	232	12	248
Rapeseed meal	86	75	12	75
Blood meal	—	—	1	43
Tallow	—	—	3	60
Palm fatty acid oil	20	90	2	60
Maize/soya oil	17	90	—	—
Fat premix (FP1)	14	60	—	—
Fat prills	16	90	—	—
Megalac	14	60	—	—

but observations on the diet of silage alone and on the 'basal' diet containing the compound containing no added fat or fibre were repeated at intervals of approximately six weeks throughout the study, and for each of these treatments there were ten replications in total (*Table 9.4*).

Subsequent to the main study, 97 further metabolizable energy determinations were undertaken. Twelve of these were to provide additional replications of observations made in the main study (*Table 9.4*). A further 26 were for a subsidiary study with nine selected additional compounds similar in composition to those used in the main study but containing either 15 or 90 g/kg added fat (*Table 9.4*). These observations were designed to facilitate tests for non-linear relationships between ME content and fat inclusion. Additionally, 39 determinations were made in a supplementary study to measure the ME content of 14 additional compounded feeds containing ingredients or combinations of ingredients not represented by the range of compounds used in the main or subsidiary study (*see Table 9.3*). Finally, 20 determinations were made with nine of the compounds used in the main study and/or silage to test whether the quality of silage given had any influence on the ME value determined for the compound. For these determinations, a silage of lower digestibility than that given in the main experiment was used.

Statistical analysis and calculation of feed ME values

There was one missing observation in the experiments because a sheep had suffered a digestive disturbance during a collection period, otherwise the complete data were used for statistical analysis. Analysis of variance procedures were used, and, where appropriate, stepwise multiple regression models were adopted to test for variation

Table 9.4 A SUMMARY OF THE FAT AND FIBRE SOURCES AND THE LEVELS OF INCLUSION FOR THE COMPOUNDED FEEDS USED AND THE NUMBER OF EXPERIMENTAL OBSERVATIONS ON EACH FEED IN THE MAIN AND THE RELATED SUBSIDIARY SHEEP STUDY (SEE TEXT)

Fat sources	(g/kg)	Fibre source (g/kg)						
		None	Straw		NIS		Sugar beet and citrus pulp	
			200	400	200	400	200	400
None		(10)[a]	3	(3)	3	5[b]	3	(3)
	15	—	—	—	3	—	—	—
	30	2	2	2	2	2	2	2
Palm acid oil	60	(3)	2	(3)	4[b]	3	2	(3)
	90	3	—	—	3	3	—	—
Maize/soya oil	30	2	2	2	2	2	2	2
	60	3	2	3	2	3	2	3
	90	3	—	—	3	3	—	—
FP 1	30	2	4[b]	2	4[b]	2	2	2
	60	(3)	2	(3)	2	3	4[b]	(3)
Megalac	30	2	2	3[b]	2	2	2	2
	60	3	2	4[b]	2	3	2	3
	15	—	—	—	3	—	—	—
	30	2	2	2	2	2	2	2
Fat prills	60	3	2	3	2	3	2	3
	90	—	—	—	—	—	3	—

[a] Values in parenthesis indicate treatments which were common to the sheep study and the dairy cow study
[b] Indicates treatments where additional replicate observations were made (see text)

due to animals, diets and dietary ingredients; data were examined for both linear and non-linear effects and interactions. In contrast to the earlier studies with sheep (Wainman, Dewey and Boyne, 1981), direct determinations had been made of the ME content of the silage used as the forage component of the diets. Thus, the ME content of the compounded feeds could be determined by a simple proportionality and did not require the construction of a model (see above).

TECHNICAL DETAILS OF THE COW EXPERIMENTS

The cow experiments were conducted with three balanced groups of four multiparous Friesian animals (mean liveweight 492 kg) in the mid-phase of lactation. They were given diets of the high-digestibility silage alone or in combination (approximately 55:45 on a dry matter basis) with each of nine of the compounded feeds used in the main sheep experiment. The feeds contained no added fat, palm acid oil (60 g/kg) or FP1 (60 g/kg) in combination with no added fibre, straw (400 g/kg) or sugar beet and citrus pulp (400 g/kg) (see *Table 9.4*). The experiment was designed in the form of three 4 × 4 Latin Squares, each square having four animals, four four-week periods and four experimental treatments. In the last two weeks of each experimental period complete collections of faeces and urine were made over a seven-day period and methane production was measured over a two-day period when the cows were confined to an open-circuit respiration calorimeter. The diets for each square comprised one diet of silage alone and three diets containing a common level and sources of fibre but with a differing level and source of fat inclusion (see *Table 9.5*). The level of compounded feed allowance was fixed at a flat rate of 8 kg/day throughout the experiment, and for the mixed silage and compound diets the level of silage allowance was adjusted accordingly. For treatments in which cows were fed silage alone, forage was rationed to minimize feed refusals and allowances were adjusted to provide approximately 90% of the animals' voluntary feed intake. Using this approach broadly similar levels of feed intake and animal performance were maintained across the squares (*Table 9.5*).

Statistical analysis and calculation of feed ME values

In an initial analysis of results, uniformity between the Latin Squares was tested by comparison of the values determined in each square for the diet of silage alone. No significant differences between squares could be detected, and the coefficient of variation of the estimate of mean ME from the combined data was extremely low (CV 0.62%). The results for the three squares were therefore combined and treated as a single data set for analysis of variance, diet effects being separated from those due to animals and periods.

As in the corresponding sheep experiments, direct determinations of the ME content of the silage and of each of the silage plus compound diets allowed the ME content of the compounded feeds to be estimated by a simple 'by difference' procedure.

SUMMARY OF SELECTED RESULTS

The sheep experiments were designed with a view to regression analysis of the results

Table 9.5 FEED INTAKE AND MILK PRODUCTION OF DAIRY COWS GIVEN DIETS OF SILAGE ALONE OR WITH VARIOUS COMPOUNDED FEEDS IN THREE 4 × 4 LATIN SQUARE BLOCKS

Latin square	Type of compound Fibre source	Fat source	Feed intake (kg DM/day) Silage	Compound feed	Total	Milk production (kg/day)[c]
1	—	—	11.51	—	11.51	14.34
	None	None	8.41	6.81	15.22	22.48
	None	Palm acid oil	8.48	6.90	15.38	23.25
	None	FP1	8.09	6.90	14.99	22.38
	SEM[b]					1.24
2	—	—	11.37	—	11.37	16.63
	Straw	None	8.08	6.79	14.87	20.42
	Straw	Palm acid oil	8.20	6.86	15.07	24.60
	Straw	FP1	8.27	6.87	15.14	22.97
	SEM[b]					1.36
3	—	—	12.35	—	12.35	14.07
	Pulp[a]	None	9.03	6.76	15.79	19.50
	Pulp	Pulp acid oil	8.94	6.91	15.85	21.58
	Pulp	FP1	8.93	6.87	15.80	20.31
	SEM[b]					1.98

[a]Sugar beet and citrus pulp
[b]Standard error of a mean
[c]Fat corrected to 40 g/kg

Effects of fat and fibre sources on feed ME content

Table 9.6 summarizes results from the sheep experiments for compounded feeds in which there were systematic changes in the level and source of fat and fibre added to the diet. As can be seen, feed ME content varied with both the level and source of fat and fibre added. Statistical analysis of these data indicated that, in general, interaction effects between fat and fibre sources were non-significant and that the effects of 'fats' and 'fibres' were additive. There was some evidence that the effects of added fats were non-linear, but only in the case of the fat prills did this tendency reach statistical significance ($P < 0.001$). There was also clear evidence ($P < 0.001$) of non-linear effects on ME content arising from the addition of NIS to the diet.

To provide an overall guide to the effects of adding fat and fibre sources to the feeds, the results were fitted to a simplified model that included no interaction effects and treated the fats and the fibre sources as independent factors. The results of this analysis (*Table 9.7*) showed distinctive and substantial differences between fibre sources but the effects due to fats were more complex. Nonetheless, it should be noted that at the 60 g/kg level of addition, which probably represents the upper limit of use of the fats in commercial practice, differences between fat sources in their effects on ME were comparatively small.

Table 9.6 THE MEAN METABOLIZABLE ENERGY (ME) CONTENT (MJ/kg DM) IN SHEEP OF COMPOUNDED FEEDS CONTAINING VARIOUS SOURCES OF FAT AND FIBRE (FOR DETAILS OF TREATMENT REPLICATION SEE *Table 9.4*)

Fat sources	(g/kg)	Fibre source (g/kg)						
		None	Straw		NIS		Sugar beet and citrus pulp	
			200	400	200	400	200	400
None		13.08	10.63	10.78	11.55	11.32	12.30	12.23
	15	—	—	—	12.60	—	—	—
Palm acid oil	30	12.70	11.65	10.31	11.50	10.58	13.56	13.53
	60	15.07	12.12	11.25	12.56	13.26	14.58	14.64
	90	15.16	—	—	14.10	13.92	—	—
Maize/soya oil	30	14.06	12.51	10.89	12.13	12.21	14.07	13.79
	60	14.32	13.68	12.03	13.30	12.86	13.80	14.48
	90	14.57	—	—	14.78	13.45	—	—
Fat premix	30	13.63	11.33	10.16	11.26	11.74	13.25	13.76
(FP1)	60	13.87	12.80	10.99	13.13	12.95	13.89	14.31
Megalac	30	13.46	11.43	10.76	10.85	11.77	13.49	13.28
	60	14.47	14.10	10.33	13.61	13.34	13.81	13.79
Fat prills	15	—	—	—	11.77	—	—	—
	30	14.12	10.47	11.00	12.17	11.16	11.94	11.49
	60	14.23	12.69	10.38	13.29	13.16	14.83	14.98
	90	—	—	—	—	—	14.45	—

Table 9.7 THE ESTIMATED CHANGES IN THE METABOLIZABLE ENERGY (ME) CONTENT (MJ/kg DM) OF A COMPOUNDED FEED AS A CONSEQUENCE (a) OF THE INCLUSION OF VARIOUS FIBRE SOURCES AND (b) OF THE INCLUSION OF VARIOUS FAT SOURCES

(a) Fibre sources	Level of inclusion (g/kg)	
	200 g	400 g
Straw	−1.66[c]	−3.05[c]
NIS	−1.50[c]	−1.50[c]
Sugar beet and citrus pulp	−0.12	−0.06

(b) Fat sources		Level of inclusion (g/kg)		
	15	30	60	90
Palm acid oil	1.15[a]	0.22	1.57[c]	2.43[c]
Maize/soya oil	—	0.80[b]	1.65[c]	2.15[c]
Fat premix	—	0.31	1.30[c]	—
Megalac	—	0.28[a]	1.35[c]	—
Fat prills	0.62	0.17	1.49[c]	1.74[c]

[a] $P < 0.05$; [b] $P < 0.01$; [c] $P < 0.001$. Negative values indicated a reduction in ME content

ME values in sheep and cows

Nine of the compounded feeds used in the sheep experiments were also studied in the cows. The feeds were formulated from the same batches of ingredient materials, to the same specifications but showed small differences in composition reflecting random variations between replicated batches of the same feed mixture. The average ME content for the six diets estimated in the sheep was 12.91 ± 0.16 MJ/kg DM and did not differ significantly from the slightly lower value of 12.58 ± 0.16 MJ/kg DM determined in the cows. The mean difference of 0.33 MJ/kg DM did not reflect any systematic bias between diets, and values for corresponding diets were closely correlated between species ($r = 0.94$, $P < 0.001$). Nonetheless there was evidence of deviations for the fat-supplemented diets containing sugar beet and citrus pulp; for these diets values for cows were approximately 1 MJ/kg DM lower than those for sheep. It should be recognized that these comparisons are subject to the maximum influences of the combined experimental errors incurred in the sheep and cattle studies.

Prediction of the ME content of compound feeds

EXISTING PREDICTION EQUATIONS

Compared with the 24 feeds studied by Wainman, Dewey and Boyne (1981), the 100 diets used in the RRI/HRI study were on average higher in gross energy, crude protein, oil, fibre and sugar and lower in starch (*Table 9.8*). More importantly, the diets covered a very wide range in composition and thus provided an excellent data set against which to test the UKASTA/ADAS/COSAC ME prediction equations or from which to develop improved equations. Initial examination of the RRI/HRI sheep results indicated clearly that the exisiting equations were not accurate in their prediction of ME. For example, the average determined ME content for all the feeds

Table 9.8 THE MEAN CHEMICAL COMPOSITION AND THE RANGE IN CHEMICAL COMPOSITION FOR THE COMPOUNDED FEEDS USED IN THE STUDY OF WAINMAN, DEWEY AND BOYNE (1981) AND IN THE ROWETT RESEARCH INSTITUTE/HANNAH RESEARCH INSTITUTE (RRI/HRI) COLLABORATIVE STUDY

	Wainman et al. (1981)		RRI/HRI study	
	Mean ($n = 24$)	Range	Mean ($n = 100$)	Range
Gross energy (MJ/kg DM)	18.12	16.30–19.15	18.66	17.50–20.30
Crude protein (g/kg DM)	167	127–209	200	186–232
Oil (g/kg DM)[a]	44	27–73	67	19–127
Crude fibre (g/kg DM)	88	37–167	116	48–218
Neutral detergent fibre (g/kg DM)	224	116–363	252	116–428
Acid detergent fibre (g/kg DM)	134	57–239	160	50–288
Sugar (g/kg DM)	69	50–100	96	58–179
Starch (g/kg DM)	320	162–437	194	42–397
NCD g/kg DM)[b]	787	666–881	777	620–849

[a]Oil by method B
[b]Cellulase digestible organic matter after neutral detergent extraction

examined was 12.79 MJ/kg DM but the corresponding value estimated from equation U1 was 11.76 MJ/kg DM and from equation U2 was 12.21 MJ/kg DM. Moreover, it was apparent that the discrepancy between the observed and predicted values was not due to aberrant predictions for a few isolated feeds. Rather, there was a bias in the equations leading to a general tendency to underpredict the ME values (for example see *Figure 9.3*).

DEVELOPMENT OF IMPROVED EQUATIONS

In considering the development of an improved series of prediction equations, a number of statistical approaches was explored. Similarly, a wide variety of chemical components was examined as potential 'terms' in multi-component regression equations. The analyses included: gross energy, oil (method B), crude protein, neutral detergent fibre, acid detergent fibre, crude fibre, modified acid detergent fibre (MADF), digestible organic mater in dry matter (DOMD), cellulase digestible organic matter after neutral detergent extraction (NCD), starch, sugar, and ash. In the following sections, only the main features of the most relevant statistical analyses have been outlined and the most important regression equations have been selected for consideration.

STATISTICAL CONSIDERATIONS

As is discussed at length by UKASTA/ADAS/COSAC (1985), prediction equations for the routine estimation of ME from chemical composition of feeds must be judged on a basis that takes account of laboratory errors in analysis, as well as errors arising from inaccuracies in the experimentally-derived prediction equations. The laboratory variance in analysis was established by UKASTA/ADAS/COSAC by estimation of the coefficient of variation of individual analytical measurements between five collaborating laboratories (Wainman, Dewey and Boyne, 1981). A corresponding test between four laboratories was also undertaken as part of the RRI/HRI study. The results (*Table 9.9*) were in accord with previous findings and showed that there were

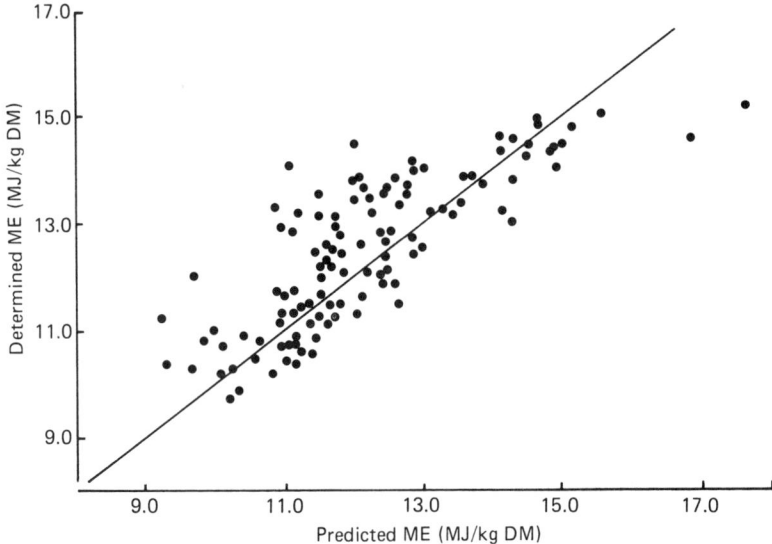

Figure 9.3 The relationship between the ME content of compound feeds determined *in vivo* in sheep and estimated from chemical composition using Equation U2. (Data are from the RRI/HRI study. The solid line is the line of equivalence)

marked differences between analytical procedures in between-laboratory reproducibility.

To comment on the errors of prediction of ME in a way which would take account of errors arising from both 'laboratory analysis' and 'prediction equations' UKASTA/ADAS/COSAC (1985) employed a combined residual standard deviation term S'''. A similar approach could have been adopted with respect to the RRI/HRI study but at an early stage of the data analysis it was decided that the preferred

Table 9.9 THE ESTIMATED REPRODUCIBILITY, STANDARD DEVIATION AND THE COEFFICIENT OF VARIATION FOR SELECTED CHEMICAL ANALYSES AS DETERMINED BETWEEN FOUR REPLICATE ANALYSES UNDERTAKEN IN DIFFERENT LABORATORIES

	Mean concentration for all diets (g/kg)	Reproducibility (s.d. g/kg)	Coefficient of variation (%)
Gross energy	18.66[a]	0.22[a]	1.2
Crude protein	200	3.5	1.8
Oil[b]	68	3.5	5.3
Crude fibre	116	4.5	3.9
MAD fibre	151	6.0	4.0
Neutral detergent fibre	253	7.4	2.9
Acid detergent fibre	160	6.7	4.1
Sugar	96	6.1	6.3
Starch	195	5.2	2.7
NCD[c]	777	14.2	1.8

[a] MJ/kg DM
[b] Oil by method B
[c] Cellulase digestible organic matter after neutral detergent extraction

Table 9.10 A SUMMARY OF EQUATIONS FOR THE PREDICTION OF METABOLIZABLE ENERGY (MJ/kg DM) FROM THE CHEMICAL COMPOSITION (g/100 g DM) (STUDY 1 REFERS TO DATA TAKEN FROM THE WORK OF WAINMAN, DEWEY AND BOYNE (1981); STUDY 2 REFERS TO DATA FROM THE RRI/HRI STUDY)

Equation	Study	Constant term	Model terms and coefficients[a]				S_A	S_B	RMSE
			CP	OiF	Oil.CF	TA			
U1	1	12.20	0.051	0.061	−0.037	−0.148			0.27
	2	13.34	−0.066	0.034	−0.018	0.046			0.42
	1+2	12.43	0.0558	0.0347	−0.018	−0.145	0.13	0.57	0.49
			Oil	OiF	Oil.NCD	TA			
U2	1	12.20	−2.75	0.055	0.031	−0.050			0.13
	2	11.72	−1.34	0.005	0.020	−0.035			0.29
	1+2	11.77	−1.491	0.0033	0.0216	−0.0753	0.23	0.38	0.28
			Oil	Oil.CP	MADF.CP	TA			
U3	1	13.55	−0.42	0.036	−0.0084	−0.13			0.20
	2	11.72	0.06	0.010	−0.0069	0.14			0.36
	1+2	13.52	−0.183	0.0230	−0.00620	−0.104	0.13	0.47	0.38
			Oil	Protein	Starch	Sugar			
E1	1	2.38	0.29	0.187	0.096	0.24			0.43
	2	7.02	0.32	0.023	0.066	0.20			0.30
	1+2	4.36	0.331	0.134	0.0749	0.198	0.18	0.44	0.33
			Oil	NCD					
E2	1	−1.48	0.24	0.152					0.28
	2	−0.15	0.25	0.145					0.23
	1+2	−0.62	0.249	0.1475			0.23	0.34	0.24
			Oil	NCD	DOMD				
E4	1	0.09	0.43	−0.002	0.138				0.20
	2	−0.44	0.27	0.135	0.014				0.24
	1+2	−0.99	0.298	0.1074	0.0468		0.23	0.33	0.23

			Oil	NDF					
E6	1	0.67	0.43	-0.007	DOMD				0.19
	2	6.35	0.33	-0.063	0.130				0.25
	1+2	3.87	0.365	-0.0420	0.085		0.34	0.41	0.25
			Oil	NDF	0.1071				
E7	1	19.1	0.12	-0.170	Ash	Starch			0.21
	2	16.3	0.20	-0.145	-0.23	0.067			0.25
	1+2	18.18	0.186	-0.1507	-0.03	0.044	0.14	0.31	0.25
			Oil	NDF	-0.193	-0.057			
E8	1	5.82	0.32	-0.059	DOMD	Ash·starch			0.19
	2	12.4	0.24	-0.124	0.090	0.0023			0.23
	1+2	11.15	0.249	-0.1117	0.043	-0.0050	0.19	0.29	0.22
			Oil	NDF	0.0493	-0.00421			
E9	1	16.62	0.11	-0.165	Ash·starch				0.23
	2	16.58	0.19	-0.154	-0.0065				0.23
	1+2	16.43	0.178	-0.1552	-0.0060		0.14	0.28	0.23
			GE	NDF	-0.00597				
E10	1	-4.43	0.58	-0.071	NCD	Starch			0.23
	2	-3.80	0.82	-0.115	0.100	-0.025			0.22
	1+2	-3.06	0.752	-0.109	0.064	-0.043	0.20	0.31	0.22
			GE	NDF	0.0650	-0.0372			
E12	1	4.00	0.06	-0.064	Starch	GE·NCD			0.21
	2	0.90	0.54	-0.110	-0.027	0.0061			0.22
	1+2	1.97	0.458	-0.105	-0.042	0.0037	0.21	0.31	0.22
			Oil	NCD	-0.0367	0.00372			
E14	1	23.1	0.1099	-0.0336	NDF	Ash	Starch		0.22
	2	6.90	0.2254	0.0855	-0.200	-0.272	-0.075		0.23
	1+2	8.02	0.221	0.0851	-0.065	-0.017	-0.022	0.16	0.30	0.23
					-0.0712	-0.106	-0.033		

[a]CP = crude protein; Oil = oil by method B; CF = crude fibre; TA = total ash; MADF = modified acid detergent fibre; NCD = cellulase digestible organic matter after neutral detergent extraction; DOMD = *in vitro* digestible organic matter in the dry matter; GE = gross energy; NDF = neutral detergent fibre

approach to the development of new prediction equations was to combine the data with those derived from the earlier Rowett experiments. This was possible since contemporary analysis had been carried out on the original 24 feeds to provide information on oil content determined by method B. However, combination of the data precluded the use of S''' since that would have implied that ME values in the two studies were determined by identical methodology. As has been pointed out earlier, this was not the case because the approach to 'by difference' calculations of the ME of compounded feeds was not the same in the two studies and the studies differed in their approach to the replication of treatments. An alternative statistical approach was therefore adopted. Errors associated with the between-laboratory variance were described through use of a standard deviation S_A, whilst errors applicable after the inaccuracy in the prediction equations had also been allowed for were described by a standard deviation S_B. Then overall predictive accuracy of the derived equations was summarized through the calculation of a root mean square error (RMSE).

EQUATIONS TO PREDICT ME

Stepwise weighted linear regression analysis was initially used to select and fit equations to data from Wainman, Dewey and Boyne (1981) and the RRI/HRI study. A number of equations were found to have a reasonably good predictive accuracy but no single multiple regression equation satisfactorily fitted both sets of data (*Table 9.10*). Rather, it was found that the data from the two studies could best be described by two parallel regression lines having different intercept values. This implied that the results of the two studies could be combined through the use of an appropriate weighting procedure. This approach was examined and combined equations were calculated by weighting the data from the RRI/HRI study and the initial Rowett study in the ratio of 2:1. This ratio was arbitrary but was selected with regard to the number of feeds examined and the number of observations made in each study. Using this approach, combined equations with a very low RMSE were obtained and are shown in *Table 9.10*.

Compared with other equations examined, Equations U1, U2, U3 and E1 had a relatively poor fit to the combined experimental data. These equations were therefore eliminated from further consideration. Similarly, whilst Equations E4, E6, E10 and E12 had low RMSE values they were not favoured since they included analyses of gross energy and *in vitro* DOMD, which depend on equipment and techniques that are not routinely available in all laboratories. Finally, it was argued that in selecting a prediction equation the aim should be to minimize the number of analytical terms and to avoid complex interactions, and this tended to rule out Equations E7, E8, E9 and E14. There remained Equation E2.

This last equation was particularly interesting since it contained only two terms, oil and NCD, and because in the separate analysis of the two data sets the regression coefficients in the derived equations were virtually identical. Moreover, for the combined equation the constant term was close to zero. With that in mind, further regression analysis was undertaken and each set of experimental results and the weighted combined results were fitted to regression lines which were constrained to pass through the origin. The equations so derived showed a very close similarity one to another, the coefficients were essentially the same and RMSE values were acceptably low (*Table 9.11*). The combined equation from this analysis was designated Equation E3.

On this basis therefore the most useful equations for the prediction of ME appeared

Table 9.11 EQUATIONS FOR THE PREDICTION OF METABOLIZABLE ENERGY (MJ/kg DM) BASED ON AN ANALYSIS OF OIL AND CELLULASE DIGESTIBLE ORGANIC MATTER AFTER NEUTRAL DETERGENT EXTRACTION (NCD; g/100 g DM). (EQUATIONS ARE DERIVED FROM DATA FROM THE INITIAL ROWETT STUDY (1), FROM THE RRI/HRI STUDY (2) AND FROM THE WEIGHTED COMBINATION OF THE TWO AND HAVE BEEN CONSTRAINED TO PASS THROUGH A ZERO INTERCEPT. THE COMBINED EQUATION WAS DESIGNATED E3)

Study	Equation	S_A	S_B	RMSE
1	ME = 0.252 oil + 0.133 NCD	—	—	0.30
2	ME = 0.251 oil + 0.143 NCD	—	—	0.23
1 + 2	ME = 0.250 oil + 0.140 NCD	0.22	0.34	0.24

to be Equations E2 and E3. These equations had a similar degree of fit to the combined experimental data (*Figure 9.4*) but E3 had the advantage that it was little affected by assumptions concerning the most appropriate ratio of weighting to be given to data from the initial Rowett study and the RRI/HRI study.

VALIDITY OF PREDICTED ME VALUES

Application of predicted ME values to dairy feeds

Nine of the feeds used in the RRI/HRI study were tested in both sheep and cows and this provided a basis for examining the validity of Equations E2 and E3 in relation to the ME values of dairy feeds. *Table 9.12* summarizes the results for determined ME values for the range of feeds studied in cows together with corresponding values calculated using Equations E2 and E3. As can be seen, there was on average no difference between determined and estimated values and the correspondence of values was exceptionally close. Only for feeds containing high levels (400 g/kg) of sugar beet and citrus pulp plus high levels (60 g/kg) of added fat was there any indication of divergence between predicted and determined values. However, whether this divergence is systematic or simply reflects random experimental errors cannot be ascertained. As has already been indicated, an exactly similar divergence was evident in the *in vivo* ME values for these diets between the tests undertaken in sheep and those undertaken in cows. Thus for the sugar beet and citrus pulp diet containing palm acid oil (*Table 9.12*), the ME content determined in sheep was 14.64 MJ/kg DM and for the corresponding feed containing FP1 the ME content was 14.31 MJ/kg DM. Further research on diets of this type is needed to confirm the apparent differences in feed ME as determined in sheep and cows.

Comparison with database values

The determined ME values for the feeds used in the RRI/HRI study agreed reasonably well with those calculated from the current ADAS and UKASTA feed databases, although on average the database values were approximately 0.25 MJ/kg DM lower than those determined in sheep. Thus, the mean (with SE) value determined for the 77 feeds used in the main sheep study was 12.66 ± 0.15 MJ/kg DM compared with 12.35 ± 0.11 MJ/kg DM calculated from the ADAS database and 12.40 ± 0.11 MJ/kg DM calculated from the UKASTA database. There was generally

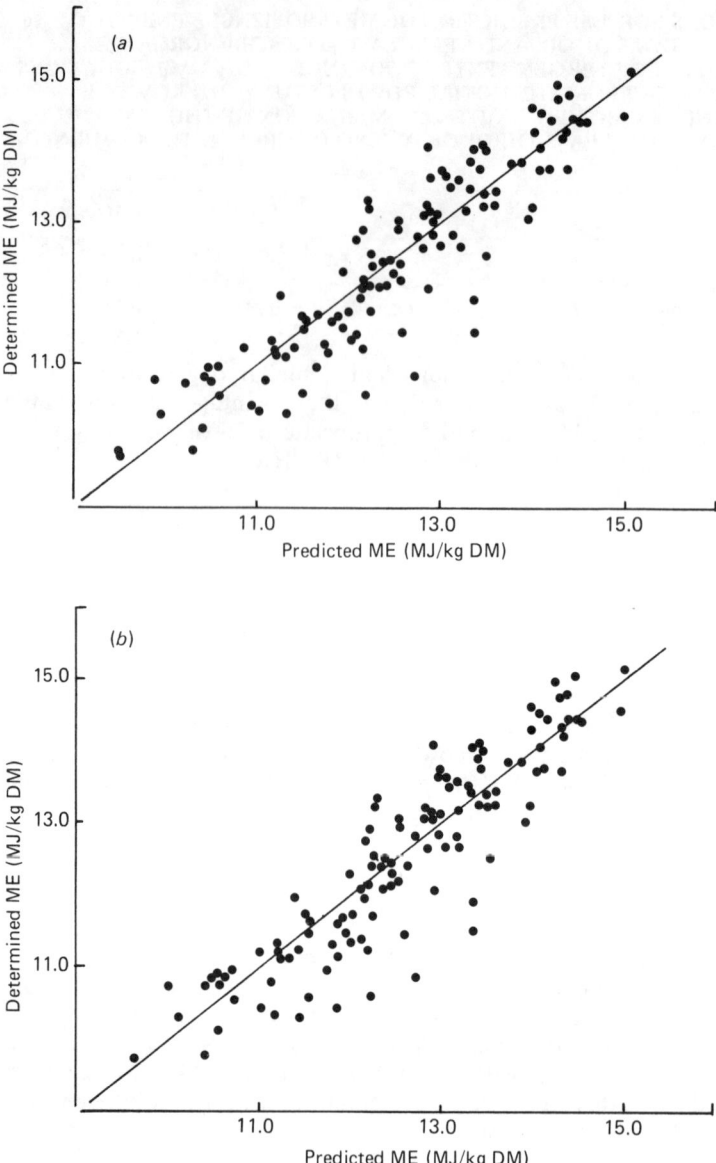

Figure 9.4 The relationship between *in vivo* estimates of ME content of compound feeds and those predicted from Equations (a) E2 and (b) E3 (Data are from Wainman, Dewey and Boyne (1981) and from the PRI/HRI study. The solid line is the line of equivalence)

very close accord between the database values and those derived from prediction equations; the mean value for the 77 sheep feeds calculated from E3 was 12.41 ± 0.13 MJ/kg DM, for example. Similarly the ME values calculated from the feed databases were in excellent agreement with the values determined *in vivo* in lactating cows (see *Table 9.12*).

Table 9.12 THE METABOLIZABLE ENERGY CONTENT (MJ/kg DM) OF VARIOUS COMPOUNDED FEEDS DETERMINED IN LACTATING COWS, ESTIMATED FROM CHEMICAL COMPOSITION USING EQUATIONS E2 AND E3 AND CALCULATED FROM ADAS DATABASE VALUES

Type of feed		Determined ME	Estimated ME from equations		ME from database values
Fibre source	Fat source		E2	E3	
None	None	12.43	12.53	12.47	12.43
None	Palm acid oil	14.42	14.30	14.24	13.73
None	FP1	13.49	13.70	13.65	13.81
Straw	None	10.35	10.02	10.10	9.94
Straw	Palm acid oil	11.87	11.58	11.65	11.39
Straw	FP1	11.55	11.26	11.34	11.48
Pulp[a]	None	12.43	12.44	12.39	12.01
Pulp	Palm acid oil	13.30	14.21	14.16	13.49
Pulp	FP1	13.40	14.05	13.99	13.79
Mean (with SEM)		12.58 ± 0.40	12.68 ± 0.50	12.67 ± 0.50	12.45 ± 0.46

[a]Sugar beet and citrus pulp

SIGNIFICANCE OF BASAL DIET

A feature of the regression analysis summarized in *Table 9.10* was that the ME values recorded in sheep in the RRI/HRI study were generally higher than those observed in the earlier study of Wainman, Dewey and Boyne (1981). There was no clear technical difference between the two studies which could explain this effect, although, as has been indicated above, there were some small modifications in approach. One difference between the studies which was considered to be of potential significance was that the silage used in the RRI/HRI study was of a much higher quality than the silage or hay used in the previous experiments. To examine the importance of this, calorimetric measurements on nine of the feeds from the main RRI/HRI study were repeated with the feeds given in conjunction with a silage which was similar to that used in the main study but of lower digestibility (ME was 8.75 MJ/kg DM versus 11.68 MJ/kg DM). The ME content of the compounded feeds determined in conjunction with the low-digestibility silage did not differ significantly ($P > 0.05$) from that with the high-digestibility silage. However, the mean values were 12.27 MJ/kg DM and 12.83 MJ/kg DM (s.e.m. 0.22 MJ/kg DM) respectively, indicating a trend which may account for the apparently systematic differences between the work of Wainman, Dewey and Boyne (1981) and the more recent studies.

Conclusions

On the basis of the evidence summarized above it appears that Equations E2 or E3 would provide an accurate and reliable basis for the routine prediction of the ME content of compound feeds from laboratory measurements of their oil and NCD content. The accuracy of the two equations is similar but, as has been discussed, there are other considerations which favour E3. The adoption of this equation for routine use is therefore recommended.

The results which have been described highlight two areas where further research may be justified. There is a need for more information about the ME value in sheep as compared with cows of feeds containing high levels of sugar beet pulp and citrus pulp together with high levels of added fat. Similarly, the interaction effects of forage quality on the determined ME content of supplementary compound feeds warrants further consideration.

Acknowledgements

The authors are grateful to the members of the UKASTA/ADAS/NFU/FAC Technical Committee on ME Content of Ruminant Feeds for helpful comments and advice and for information from UKASTA and ADAS databases. P.H. Garthwaite is a member of the Department of Mathematical Sciences, University of Aberdeen, whose contribution to the work is gratefully acknowledged.

References

ALDERMAN, G. (1985). In *Recent Advances in Animal Nutrition—1985*, pp 3–52. Eds W. Haresign and D.J.A. Cole. Butterworths, London

DEPARTMENT OF AGRICULTURE AND FISHERIES FOR SCOTLAND (1975). *First Report of the Feedingstuffs Evaluation Unit*. Department of Agriculture and Fisheries for Scotland, Edinburgh

WAINMAN, F.W., DEWEY, P.J.S. and BOYNE, A.W. (1981). Compound Feedingstuffs for Ruminants. *Third Report of the Feedingstuffs Evaluation Unit*. Department of Agriculture and Fisheries for Scotland, Edinburgh

WAINMAN, F.W., DEWEY, P.J.S. and BREWER, A.C. (1984). *Fourth Report of the Feedingstuffs Evaluation Unit*. Department of Agriculture and Fisheries for Scotland, Edinburgh

UKASTA/ADAS/COSAC (1985). *Prediction of the Energy Value of Compound Feeds*. Report of the UKASTA/ADAS/COSAC Working Party. United Kingdom Agricultural Suppliers and Traders Association, London

10

NUTRIENT ALLOWANCES FOR RUMINANTS

J. D. OLDHAM
Edinburgh School of Agriculture, West Mains Road, Edinburgh, UK

Introduction

Ruminants require nutrients and energy in certain amounts and proportions in order to achieve a level of performance within the limit fixed by genotype. Differences between animals in a group might arise through genetic differences, age, stage of lactation, previous nutritional history, social rank and so on, so that performance of the individuals within a group varies, although nominally the amounts of nutrients and energy consumed may be the same. Similarly, there is variation in performance and voluntary consumption of feed when all or part of a diet is available *ad libitum*. In order to enable average performance of a group to reach a preferred level, the variability between animals in their responses to available food must be taken into account—hence, the concept of an allowance, i.e. an allocation of food designed to be sufficient for a defined proportion of animals in a group to achieve performance target or not to be underfed relative to their potential.

If an allowance of food was changed from one proscribed level to another, then it is of interest to know if performance changes in a predictable manner, i.e. what would be the response to a change in allowance? This question is at the nub of current debate in nutritional science and the purpose of this chapter will be to discuss recent developments in the estimation of requirements and to frame them in relation to progress towards nutritional systems which allow prediction of responses. Within this context, the increasing emphasis on control of animal performance so as to yield animal products which relate to market and consumer requirements needs to be noted. While our current nutritional systems contain descriptions of energy (metabolizable energy; ME) and protein requirements as input, the user of this information is concerned with the influence of feeding on the growth of lean and fat for the production of saleable animal carcasses or of the yield of major milk constituents (fat, protein, lactose) because it is for these products that the producer is paid.

Growth of body protein and body fat and the secretion of milk solids cannot be predicted adequately using nutrient requirement systems such as ARC (1980) (Beever and Oldham, 1986; MacRae, Buttery and Beever, 1988; AFRC, TCORN, 1988) understandably, because the systems were not designed for this purpose. Future developments in nutrient allowance systems can, however, be expected to relate nutrients derived from food to the production of particular animal products. This approach represents a considerable change from conventional descriptions of requirements but is a logical development from that philosophy. The approach puts increased emphasis on the description of animal characteristics ('potential') and calls for a change in the description of feedstuffs so as to enable prediction of nutrient yields (*see* Chapter 11).

Historical note

The historical development of feeding standards for ruminants spans almost two centuries. Tyler (1959) and Blaxter (1986) have summarized these developments. Important publications which represent recent milestones in the British literature are the various issues of MAFF Bulletins on Rations for Livestock (Evans, 1960), the Proceedings of the Conference 'Scientific Principles of Feeding Farm Livestock' held at Brighton in 1958 and the subsequent establishment of the Agricultural Research Council (ARC) Technical Committee on the Nutrient Requirements of Farm Livestock with its publications on ruminant requirements for water, ME, protein, minerals, trace elements and vitamins in 1965 and 1980. The ARC Committee also published an updating supplement on estimates of protein requirements in 1984.

Following the ARC publications in 1965 and 1980, Inter-Departmental Working Parties (IDWPs) were appointed, with representatives of MAFF, DAFS, DANI, BVA and UKASTA, to consider the practical application of the ARC estimates of requirements. *Maff Technical Bulletin 33*, which was the report of the Working Party on energy allowances, published in 1975, was the effective 'launch vehicle' for the ARC ME system officially presented by the ARC in 1965.

Of the three IDWP groups established after publication of ARC 1980, only the mineral, trace element and vitamin group has published its report (MAFF, DAFS, DANI, UKASTA, BVA, 1983); the report of the Energy Working Party will be available early in 1988 and it is hoped that the Protein Working Party will also be able to report within 1988.

The ARC Technical Committee on Nutrient Requirements of Farm Livestock was disbanded in 1983 and shortly afterwards the Agricultural and Food Research Council (AFRC) Technical Committee on Responses to Nutrients (TCORN) was formed. One of its aims is to enable the prediction of animal responses to nutrients made available from food. The formation of TCORN represents a significant change in attitude in nutritional science. It indicates a high level of confidence in our current ability to measure the amount of energy and nutrients which are needed to promote unit rates of performance (for example, the amounts of energy and protein which are needed to deposit 1 kg/day of body tissue or milk of defined composition) such that we can now proceed to use our knowledge of requirements to help describe how animals respond to available food. Within the context of 'nutrient allowances' the movement from description of requirements to prediction of responses is logical and can only be made when requirements for unit processes have been established. The requirements which need to be known follow from the kinds of response to be predicted. In general these are:

(1) animal performance relevant to the generation of marketable animal products
(2) voluntary food consumption.

Energy and protein requirements

The IDWP Energy Working Party (IDWP, 1988) have concluded that the ME system as described in ARC (1980) is an appropriate basis on which to base practical application of the ME system. They found little difference in the ability of ARC (1980) and *MAFF Technical Bulletin 33* (MAFF, 1975) as systems to predict the performance of ruminants eating measured amounts of food of known or estimated ME content. But on the grounds that ARC (1980) is founded on more rigorous principles than MAFF (1975) and that the framework is capable of accepting future modifications to improve its precision ARC (1980) was preferred.

For growth, ARC (1980) give predictions of k_f which either vary with diet class or are generalized across all diets. The all-diets generalization has been adopted by IDWP (1988).

For formulating rations using ARC (1980) as a basis, IDWP has adopted the variable net energy approach described by Harkins *et al.* (1974). This is done to solve simultaneously the problems of estimating ME needs at a particular plane of feeding and of the metabolizability of a particular mixture of dietary ingredients offered to meet this need.

The IDWP Protein Working Party has yet to report officially but the major features of its recommendations have been given by Webster (1987) and Cottrill (1988). The principles outlined in ARC (1980 and 1984) have been adopted as a working basis from which to develop a practical system. The major developments likely to be introduced by the Working Party are:

(1) Partition of Rumen Degradable Nitrogen (RDN) into fractions which are quickly degraded (QDN) or slowly degraded (SDN). While under RDN limiting conditions, SDN can be incorporated into microbial N with an apparent conversion efficiency of 1, QDN can be converted to microbial N with an apparent conversion efficiency of only 0.8.

(2) Microbial requirements for RDN are considered to be variable in relation to the supply of fermentable ME such that, at high planes of feeding, microbial crude protein generated per MJ fermented ME is substantially greater than would be predicted from ARC (1980 or 1984).

(3) The true digestibility of Undegraded Dietary Nitrogen (UDN) in the small intestine is taken to be a variable depending on the acid detergent insoluble nitrogen (ADIN) content of the raw ingredients contributing UDN. Digestible UDN is taken to be:

0.9 (UDN − ADIN)

(4) The term Basal Endogenous Nitrogen (BEN) replaces Total Endogenous Nitrogen (TEN) employed by ARC (1984) but the numerical value of TEN (0.35 g N/kg$^{0.75}$) is adopted for BEN. The efficiency with which BEN is replaced by absorbed amino acid N is taken to be 1.

(5) The efficiency with which observed amino acids are used to meet production requirements for tissue accretion, milk protein secretion or wool production is taken to be a function both of k_{naai} (the efficiency with which an amino acid mixture ideally balanced for the purpose (e.g. milk protein production) is used for that purpose) and RV, the value of amino acids available for a productive

purpose relative to that of an ideal amino acid mixture. Observed efficiency of amino acid use (EAAU) is then:

$$\text{EAAU} = k_{naai} \cdot \text{RV}$$

This modification was required to enable a more structured approach to the estimation of efficiencies of amino acid use and to provide the opportunity for incorporating data on the influence of amino acid balance on cow performance when such data become available. In effect it merely puts the description of tissue amino acid needs on a conceptual par with that used in monogastric nutrition and allows a proper separation of an animal characteristic (k_{naai}) and an attribute of absorbed nutrients (RV). A working value of k_{naai} is 0.85 (Oldham, 1987).

Rations formulated according to these modified principles are given by Cottrill (1988). No comprehensive evaluation of the IDWP protein system has yet been published but, at least for dairy cows, the system shows promise of accommodating more realistically the types of response to protein that have been observed in practice (Oldham, 1984; Waters, Dewhurst and Webster, 1987) than is the case with the calculation of protein needs based on ARC (1980 or 1984). There is now a wide variety of 'metabolizable protein' systems originating from working groups in different countries. These are described in EUR 10657 (1987) and some have been compared by Alderman (1987). While there is substantial agreement in concept amongst most published systems regarding the main principle of estimating separately the needs of rumen microbes and of the animal at tissue level, there is still disagreement in detail. But the philosophy of predicting (or trying to predict) amounts of an end-product of digestion by accounting for digestive events in the rumen is now well established and points the way to improvements to be expected in approaches to the estimation of all nutrient allowances.

Requirements, responses and allowances

Figure 10.1 is a simple diagrammatic representation of the relationship between requirement and response for a nutrient which is first limiting for animal performance. In order to achieve a certain level of performance (output), an input to the animal of a certain amount of the first limiting nutrient is needed. So for performance level A an input of nutrient a is needed. For performance level B an input of first limiting nutrient b is needed. The required amounts of nutrients (a and b) for performance levels A and B are therefore defined. It might be said that performance at level A for an input of a is an absolute response to that level of supply of a first limiting nutrient. Another important form of response is the incremental response, that is, the change in performance when input of a first limiting nutrient is changed; hence the incremental response to an increase in nutrient input from a to b would be $\Delta O/\Delta I$. In practice, this has particular relevance because the prediction of the consequences of change in nutrient supply is a major decision in feed management.

There is genetic variation in maintenance needs between animals and in limits to performance ('potential'). Taylor, Thiesson and Murray (1986) have recently indicated the variability between cattle in the efficiency of use of metabolizable energy for maintenance (hence maintenance needs) and phenotypic variability between ruminants in absolute levels of performance for prescribed allowances of food is also well recognized. The assumed linear relationship between input and output in *Figure 10.1*, which applies to an animal at a time, therefore needs to be elaborated in order to cope with between-animal variability. This was done elegantly by Fisher, Morris and

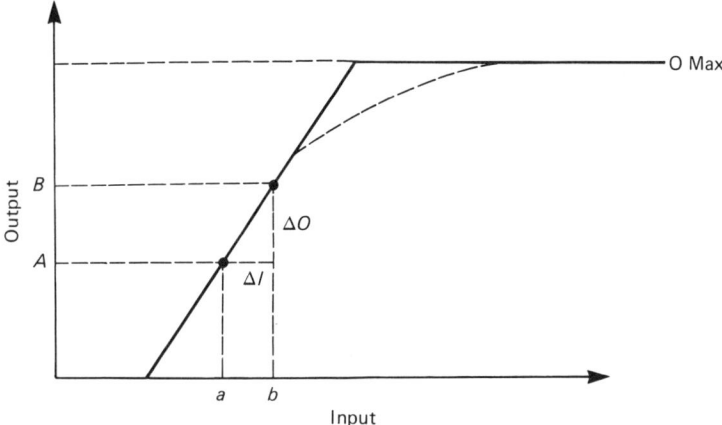

Figure 10.1 A simple model of the relationship between input of a limiting resource and output of the product for which that resource is used (*see* text for details)

Jennings (1973) who have offered a procedure for accommodating differences between individuals so that population response curves, of diminishing returns form, can be generated for flocks or herds. These procedures, originally developed for poultry, have equal relevance to ruminants and as long ago as 1959, Blaxter (Blaxter, 1959) suggested that 'safety margins' for assessment of allowances in relation to requirements should be based on a statistical appraisal of between-animal variability.

The Inter-Departmental Working Party on Energy Requirements has adopted a statistical appraisal of between-animal variability for giving guidelines on what it refers to as 'safety margins'. Given a normal distribution of performance in a group of animals offered a fixed allowance of feed (*Figure 10.2*) the assessment of variability

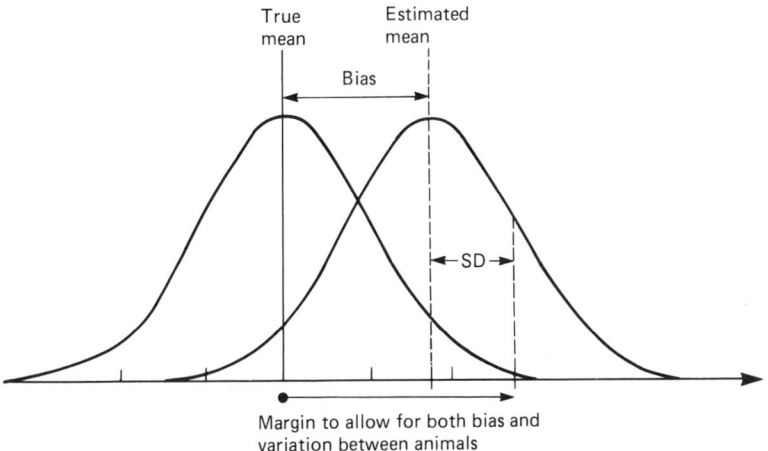

Figure 10.2 Statistical estimation of an allowance designed to meet the needs of a defined proportion of animals. The direction of bias indicates that estimated mean performance is greater than the true mean, i.e. requirements are underestimated. If performance of animals is normally distributed around the mean, then to meet the needs of all animals within 1 SD the marginal allowance is bias + 1 SD

152 *Nutrient allowances for ruminants*

can be used to identify the level of feeding at which the needs of a certain proportion of the group will be met. Hence, to meet the needs of 84% of animals in a group, the allowance would be the mean requirement plus a proportion equivalent to 1 SD from the mean.

In estimating requirements, the application of between animal variation to estimate the proportion of a group whose needs will be met at a given allowance is only valid if the procedures for estimating (the requirements 'system') has no inherent bias. If there is predictable bias (i.e. the requirement system consistently under- or over-predicts requirements) then this must be allowed for in estimating allowances (*see Figure 10.2*). The Energy Working Party has identified such a bias for the prediction of ME requirements of beef cattle (*Figure 10.3*) such that, at a given ME allowance, ARC (1980) overpredicts performance by about 15%. Hence, the allowance which is required for the needs of three-quarters of a group of animals to be met is, from *Figure 10.2*, ARC (1980) estimated ME requirements plus 15% (for bias) plus a further, approximately 15% to allow for between-animal variability, i.e. 30% more than calculated by the current requirement system.

This approach supposes that estimated requirements themselves are precise; that is, there is no substantial uncertainty or variability associated with the estimated requirement. This however is not the case and can be illustrated by a simple example.

ARC (1980) related the partial efficiency of use of ME for maintenance (k_m) and for fattening (k_f) to the metabolizability of a diet (q: ME as a proportion of gross energy). For values summarized across all of the diets used in the evaluation the relationships are:

$k_m = 0.35 q + 0.503$ (SD 0.064), and
$k_f = 0.78 q + 0.006$ (SD 0.097)

If we estimate, for example, the ME needs of a 300 kg steer (of medium mature size) gaining 1 kg/day while eating a diet with $q = 0.6$, the variability associated with estimates of k_m and k_f can be used to calculate the upper and lower limits to ME

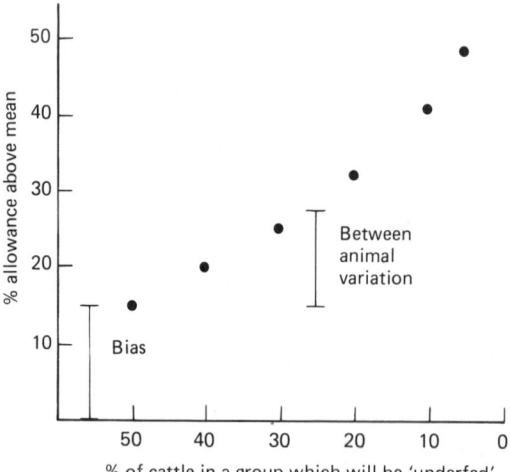

Figure 10.3 Estimated allowances above mean requirements (ARC, 1980) required to meet the ME needs of a defined proportion of beef cattle in a group. Additional allowance due to bias in the estimate of requirements is shown separately from that due to estimated between-animal variation. (Source: Inter-Departmental Working Party, 1988, Energy Group Report)

allowance which can be expected to satisfy that animal's requirements. Taking 1.5 SD either side of the mean (i.e. an ME allowance which would satisfy the needs of this sort of animal in seven instances out of eight) and using the double linear model in ARC (1980), then an allowance of 54–85 MJ ME/day could be relied on, in seven instances out of eight, to meet this animal's needs. If the exponential model (Blaxter and Boyne, 1978) is used the allowance would be 73–105 MJ ME/day. While there may be statistical satisfaction in describing an allowance of between 54 and 105 MJ ME/day to meet this animal's needs (seven times out of eight), for application the information has little value. It is, though, obviously relevant to investigate the origins of such imprecision and to identify approaches which will improve it. As discussed below, two major uncertainties are poor description of the nature of growth in growing ruminants (generalized estimates of the energy content of weight gain which are described only by animal weight and rate of increase in weight are not sufficient) and the manner in which different types of energy yielding nutrient can influence gain.

In lactation, the Energy Working Party have concluded that over complete lactations the ARC (1980) ME system accurately reflects total ME needs. This is consistent with other evaluations (e.g. Broster and Thomas, 1981) in which milk yield corrected for bodyweight change has been found to be predictable from measured ME consumption. However the Energy Working Party, using data from NIRD and the Langhill dairy herd (*Figure 10.4*) found that ARC (1980) consistently underpredicted milk yield and liveweight loss in early lactation and the converse in late lactation; that is, the ARC (1980) system for estimating requirement is not capable and, indeed, is not designed for accounting for variation in the partition of ME between milk and body tissue at different stages of the lactation cycle. But, in designing feed allowances this needs to be known and so developments from the current requirement system to enable partition to be accommodated are important.

Nutrients and the nature of growth

In estimating the precision of ARC (1980) for prediction of the ME needs for growth in cattle, the Energy Working Party (IDWP, 1988) refer to the data of Lonsdale for growth of cattle offered diets based on grass silage and in which energy retention was measured by carcass analysis. On the basis of measured ME intakes and change in liveweight, ARC (1980) grossly over-predicted liveweight change (at 740 g/day) compared with that observed (410 g/day). In these estimates, generalized values for the energy content of liveweight change were used; but if Lonsdale's own measured values for the energy content of liveweight change in these animals was adopted for the calculation of ME needs, the predicted liveweight change (using ARC, 1980) was much closer, at 500 g/day, to that observed. So in this instance, correct assessment of the nature of growth was important to the prediction of ME needs.

The nature of growth is readily manipulated in ruminants, and in other species, by nutritional means. For example, Gill *et al.* (1987) (*Table 10.1*) increased empty bodyweight gain in young cattle by giving supplements of fishmeal to a grass silage diet. Some animals had been implanted with oestradiol 17β but in both implanted and non-implanted cattle, fishmeal supplements increased rates of protein accretion with no effect on rate of fat accretion. So, in energy terms, the energy content of liveweight gain actually fell although liveweight gain itself increased. Applying standard predictions of the energy content of liveweight gain from ARC (1980) would not have predicted this.

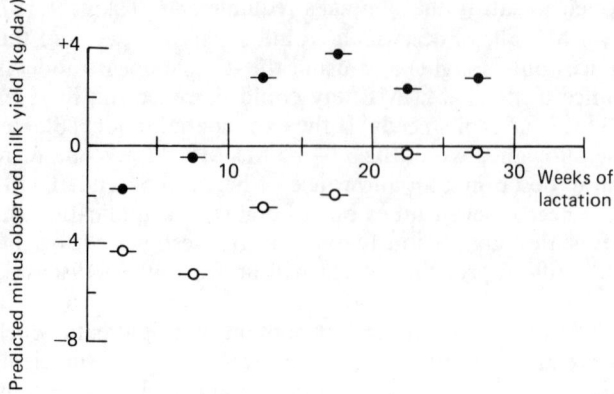
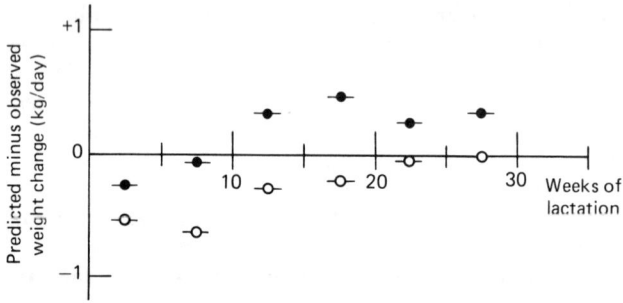

Figure 10.4 Change in calculated bias (predicted minus observed performance) in estimated requirements for ME of dairy cows over a complete lactation cycle. (● Langhill data; ○ NIRD data). Bias was estimated separately for milk yield and observed weight change (see text for details; Inter-Departmental Working Party 1988, Energy Working Party Report)

Table 10.1 BODY PROTEIN AND FAT GAINS AND THE ENERGY CONTENT OF GAIN IN YOUNG, GROWING FRIESIAN CASTRATES (120–210 kg) OFFERED GRASS SILAGE DIETS WITH OR WITHOUT FISHMEAL (0.15 OF SILAGE DM) (ANIMALS WERE IMPLANTED OR NOT WITH OESTRADIOL 17β). THE PREDICTIONS OF ENERGY CONTENT OF GAIN ARE BASED ON ARC (1980) FOR ANIMALS OF THIS WEIGHT GROWING AT THE RATES OBSERVED IN THE TRIAL

	Not implanted		Implanted	
	Silage	Silage + fishmeal	Silage	Silage + fishmeal
DE intake (MJ/day)	50.3	52.0	51.6	53.4
Gain (g/day):				
Protein	95	145	94	168
Fat	92	98	102	100
Energy content of empty-body gain (MJ/kg):				
Observed	10.2	9.1	10.8	8.8
Predicted	11.4	12.6	11.4	13.2

Source: Gill *et al.* (1987)

Protein supplements with poor quality forage diets have been found to influence growth substantially in cattle at apparently similar intakes of digestible organic matter (e.g. Smith, Broster and Hill, 1980; Oldham and Smith, 1982). Ortigues (1987) has investigated this phenomenon in heifers offered straw-based diets with supplements of barley and/or fishmeal by combining studies of growth with measurements of nutrients made available as end products of digestion in cannulated cattle offered the same diets. Fishmeal supplements increased weight change both in the presence and absence of barley supplements—though the effect of additional fishmeal was not so great when additional barley was given. Associated with both fishmeal and barley supplements, however, there were substantial increases in total amino acids apparently absorbed from the small intestine (*Figure 10.5*). A major influence of increased allowances both of 'protein' (fishmeal) and 'energy' (barley) was therefore an increase in the effective amino acid supply to the animals which were capable of responding in terms of essential protein accretion. These effects on efficiency of use of ME for growth were achieved with no detectable change in rumen VFA proportions.

In this trial we can see both an effect of dietary energy on N utilization (barley supplements increased absorbed amino acid supply) and an association between amino acid supply and the use of energy-yielding nutrients in the tissues (*Figure 10.5*).

The efficiency with which ME is used for growth has been recognized, from the inception of the ME system, as a function of the nature of the energy-yielding nutrients absorbed from the gastrointestinal tract (see ARC, 1980; MacRae and Lobley, 1982). The concept, that low efficiencies of energy utilization for growth may be related to the lack of glucogenic substrates to facilitate lipogenesis from acetate units, dates back to Martson (1948) and McClymont (1952). ARC (1980) recognizes that k_f varies not only with the metabolizable energy content of the diet (q) but also

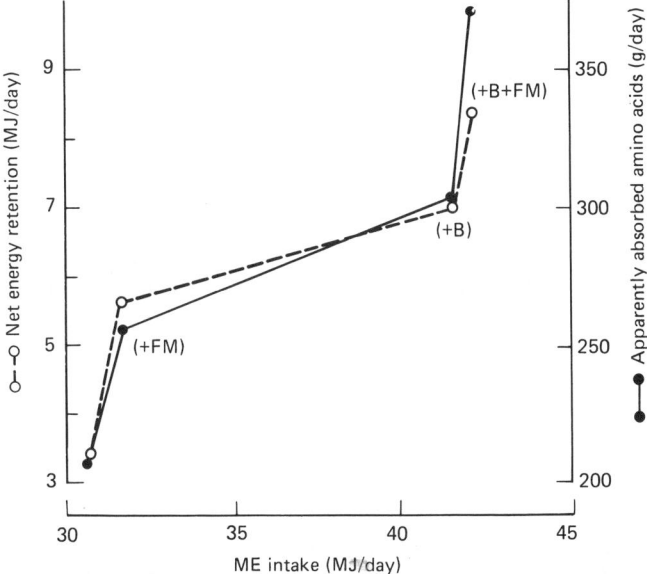

Figure 10.5 Associations between ME intake, estimated net energy retention and supply of apparently absorbed amino acids in young, growing heifers (160-230 kg) offered straw diets with barley (B) and/or fishmeal (FM) additions (Ortigues, 1987)

the overall nature of the diet, in that, five different relationships between k_f and q are given respectively for pelleted feeds, first growth forages, forage aftermaths, mixed diets and a generalized equation for all diets. For assessing allowances, however, both ARC (1980) and IDWP (Energy; 1988) adopt a generalized relationship for all diets.

MacRae and Lobley (1982) have suggested that amino acids as well as rumen propionate act as potential glucogenic substrates whose availability can influence the utilization of acetate units for fat synthesis. The different relationships between k_f and q might therefore reflect differences in amino acid supply which can be sustained by different classes of feed. That effective amino acid supply from the gut can influence the efficiency with which ME from autumn-harvested grass is used might be an illustration of this (MacRae et al., 1985; but see also Graham, 1986).

Taken together, these various observations demonstrate quite clearly that in order to describe feed allowances which will sustain predictable rates of carcass gain, a more integrated approach to the description of nutrient allowances will be needed than is currently used. Protein and energy allowances require to be considered together, not separately. Not only does energy intake influence protein utilization in the rumen (which is explicitly described in ARC (1980) and (1984) for estimations of rumen degradable protein requirements) but also protein supply to the body will influence use of energy yielding nutrients for growth.

As protein and lipid accretion can be varied separately from each other, descriptions of growth and, if appropriate, energy values for bodyweight change, must take this into account. The energy costs associated with units of body tissue gain are not the same for protein and fat so it is obvious that the overall energy cost of tissue gain will depend on the nature of that gain. Various estimates of the energy costs of lipid and protein accretion are available (Keilanowski, 1976; Pullar and Webster, 1977; Reeds, Wahle and Haggarty, 1982; Emmans, 1988) and the work associated with digestion, separate from maintenance and accretion can also be identified (Webster, 1980; Emmans, 1988). Assessment of nutrient allowances in relation to these animal functions would be a desirable development from our current estimates of ME and protein needs. In light of the increased need for production of carcasses with controlled lean and fat content, it would seem imperative that, both from the market viewpoint and that of improved scientific precision, such a development is made.

Nutrient use in lactation

While it is recognized (Broster and Thomas, 1981; IDWP, 1988) that for known levels of performance, estimates of ME requirements as given in ARC (1980) are accurate for lactating cattle, it is also well-recognized (Beever and Oldham, 1986; MacRae, Buttery and Beever, 1988) that our current systems for assessing nutrient requirements are not designed to allow prediction of responses to those nutrients in terms of milk constituents. Yet in practice, feed allowances should be designed to help control the production of milk products of commercial importance, i.e. milk fat, protein and lactose. Olham and Emmans (1988) have listed the important performance targets for lactating cattle as yield of milk and milk constituents, tissue protein and fat and rates of change of tissue protein and fat. Prediction of the partition of available nutrients between these functions is crucial to the estimation of allowances (*Figure 10.6*).

In growing ruminants the partition problem to be solved is that between use of nutrients for protein accretion, lipid accretion or catabolism with associated heat production. In lactation, critical partitions are, for energy yielding nutrients, between milk lactose, milk fat and body fat and, for protein yielding nutrients, between milk lactose, milk protein and body protein.

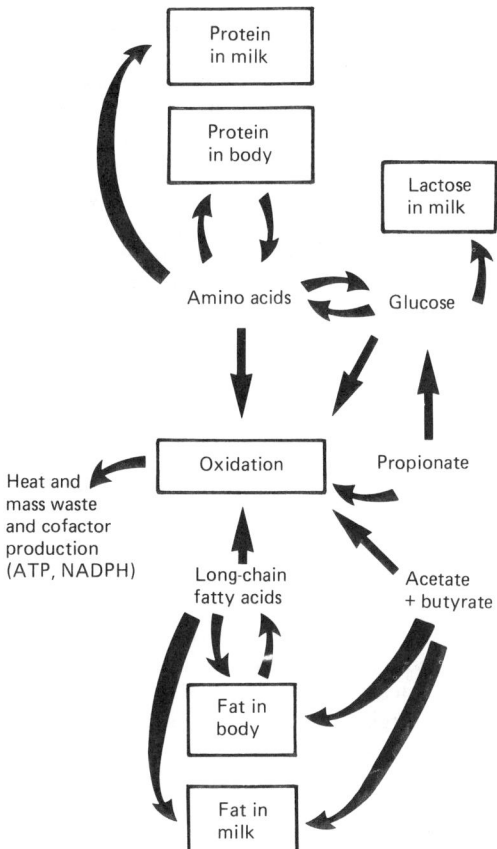

Figure 10.6 A representation of the relationships between major nutrients absorbed as end-products of digestion (amino acids, glucose, propionate, acetate plus butyrate and long-chain fatty acids) and their use for production of milk solids (fat, protein, lactose), protein and fat in the body or for catabolism resulting in production of heat and/or generation of necessary co-factors

Controlled increases in the allowance of particular nutrients are possible by abomasal infusion. When Oldham, Bines and MacRae (1984) infused iso-energetic amounts of casein or glucose into the abomasum of lactating cows in early lactation, the effects on milk constituent yield were such that, while casein increased yields of protein, lactose and fat, such that the apparent efficiency of use of ME from casein was > 1, glucose infusion depressed milk fat yield and milk energy although ME supply had been increased. Lees, Garnsworthy and Oldham (1982) changed dietary protein concentrations for cows in early lactation; milk lactose, protein and fat yields were all increased with a diet which was conducive to the maintenance of high milk fat content, but for milk fat depressing diet, additional protein had no effect on milk fat yield (though lactose and protein yields were increased). From these studies it is clear that the response to an allowance of a particular nutrient depends both on the nature of that nutrient and of the profile of other nutrients available. As with growth, so in lactation, performance responses to nutrient allowances can only be assessed in relation to the complete profile of nutrients made available as end products of digestion and the influences of each nutrient on the use of others requires that an integrated approach to the assessment of nutrient allowances needs to be taken.

Measurement of nutrient supply in ruminants depends on the use of intestinally cannulated animals with related technical uncertainties (Sutton and Oldham, 1977; Oldham *et al.*, 1988; Ortigues, 1987), or measurements of net portal venous absorption (e.g. Huntington, 1984; Huntington and Reynolds, 1986) or control of nutrient supply by complete intragastric infusion (Orskov *et al.*, 1979). Rarely have attempts been made to measure both the amounts of the end products of digestion which are made available for absorption (or are absorbed) and the synthesis of animal products from those nutrients. Perhaps the most complete data set relating nutrient supply to nutrient use is that for dairy cattle, summarized by Sutton (1985) based on earlier reports (Sutton *et al.*, 1980; Brumby *et al.*, 1979; Oldham, Sutton and MacAllan, 1979) (*Table 10.2*). This combines data from measurements of sites of digestion in cannulated cows with performance of cows offered the same diet in production trials. The diets were deliberately chosen to have a very high proportion of concentrates in order to study relationships between nutrient profile and milk fat production. It is clear from *Table 10.2* that the single major contributors to digestible energy supply are acetate plus butyrate and that wide variations in the contribution of these combined nutrients is possible. Similarly there can be wide variation in propionate supply and, proportionately, in the contribution to DE from glucose and long-chain fatty acids.

It is well-established that there is a strong association between the relative proportions of acetic + butyric acids and propionic acid generated in the rumen and milk fat content (Sutton, 1984; 1985). The influence of rumen VFA on partition of fat between milk and body tissue appears to be related to the influence of rumen

Table 10.2 DIGESTIBLE ENERGY (DE) INTAKE, NUTRIENT SUPPLY, MILK AND MILK CONSTITUENT YIELDS IN COWS OFFERED RATIONS CONTAINING HAY AND CONCENTRATES IN THE RATIO (DRY MATTER BASIS) 40:60 AND 10:90 AND IN WHICH THE CEREAL CONTENT OF THE CONCENTRATE IS EITHER ROLLED BARLEY OR GROUND MAIZE. DE INTAKE, NUTRIENT SUPPLY AND MILK CONSTITUENT YIELDS ARE ALL GIVEN AS MJ/DAY WITH NUTRIENT SUPPLY IN PARENTHESES, GIVEN AS MJ/100 MJ DE. NUTRIENT SUPPLY MEASURES WERE MADE IN COWS FITTED WITH INTESTINAL CANNULAE; MILK AND MILK CONSTITUENT YIELDS IN INTACT COWS OFFERED THE SAME RATION AT APPROXIMATELY THE SAME PLANE OF FEEDING

Hay:concentrate	40:60	10:90	40.60	10:90
Cereal	Barley		Ground maize	
DE intake (MJ/day)	157	164	158	158
Nutrient supply (MJ/day and in parentheses MJ/100 MJ DE)				
Acetate + butyrate (C2)	66 (42)	54 (33)	63 (40)	55 (35)
Long chain fatty acids (LCFA)	9 (5)	10 (6)	9 (6)	14 (9)
Amino acids	28 (18)	31 (19)	27 (17)	27 (17)
Propionate (C3)	24 (15)	44 (27)	24 (15)	28 (18)
Glucose (C6)	3 (2)	7 (4)	6 (4)	14 (9)
Milk yield (kg/day)	16.1	20.6	18.9	15.6
Milk constituent yield (MJ/day):				
fat	29	17	30	18
protein	12	15	13	13
lactose	12	15	14	11

Full details and assumptions are given by Sutton (1985)

propionate on insulin status (Hart, 1983). Oldham and Emmans (1988) pointed out a simple association between precursor ratio and product ratio for the data in *Table 10.2* which could account for the partition of fat precursors between milk and tissue in this particular data set. The association may be spurious and merits testing. It might, however, give some hope for relatively simple representations of partition which would facilitate developments towards improved descriptions of nutrient allowances and responses which follow therefrom.

Another attempt to relate nutrient supply to nutrient use was made under the auspices of the IDWP Protein Working Party. In order to test ARC (1980 and 1984) proposals for estimating protein needs, a coordinated UK trial was conducted in which cows were offered diets of calculated RDP and UDP content. Measurements of amino acid supply beyond the duodenum were made in cannulated animals and these measures were related to the performance of intact animals offered the same diets in a feeding trial. While increments of fishmeal or soyabean protein in food resulted in enhanced duodenal amino acid supply, the resulting influences on performance were less than would have been expected from calculations based on ARC (1980) (see Oldham, 1987; Cottrill, 1988 for more details). Increments of duodenal amino acid have generally been found to be used with a lower than expected 'efficiency' for milk production either when the increments result from dietary manipulation of protein or by abomasal infusion (Oldham, 1987). Whitelaw *et al.* (1986) infused cows in early lactation with increments of casein and found that each increment was approximately equally partitioned between additional milk protein and repletion of body nitrogen (*Figure 10.7*). It would therefore appear that partition of protein as well as of fat requires to be described in order to assess relationships between nutrient allowance and performance.

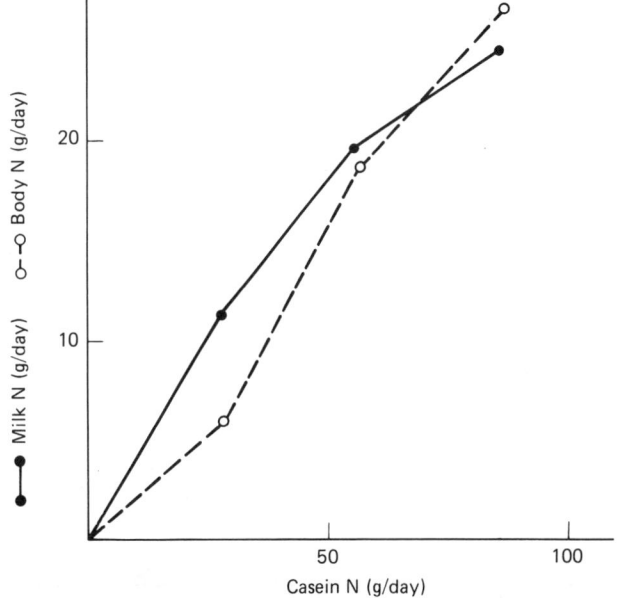

Figure 10.7 Response of cows in early lactation to abomasal casein infusion. Values for milk N and body N are incremental responses to additions of casein N (Whitelaw *et al.*, 1986)

Nutrient allowances—possible frameworks for response prediction

It should be clear from the above that there are now strong reasons for moving rapidly towards the development of frameworks which will allow prediction of the responses of ruminants to allowances of food. Where all or part of the total food allowance is available *ad libitum* a major response, of course, is voluntary consumption of that food. The rationale for developing from requirements to response is not specific to ruminants and the establishment of TCORN with a purview of all farm livestock reflects this. Fisher (1986), Emmans and Fisher (1986), Oldham and Emmans (1988) and the report of the TCORN Working Party on Prediction of Response of cows to variation in nutrient supply (AFRC, TCORN, 1988) deal with many of the important attributes which must be considered in developing frameworks for the prediction of responses and hence definition of food allowances which will allow control of animal responses to available nutrients.

As with systems for estimating animal requirements there are two major aspects. One is the nature of the animal and the resources which it needs (requires) in order to fulfil its performance targets; the other is a description of feedstuffs in terms which allow prediction of the resources which the animal needs. In order to describe feedstuffs in terms of their potential nutrient yield, where nutrients which are likely to be described as end products of digestion, i.e. volatile fatty acids, absorbable amino acids yield, long-chain fatty acids and glucose, new approaches to food characterization will be needed and this is dealt with in Chapter 11.

ANIMAL CHARACTERISTICS

In nutritional systems for ruminants, the characterization of animal genotypes has generally received much less attention than has description of food, yet it is equally important for assessing nutrient allowances to achieve prescribed levels of performance or for estimating the extent and manner in which an animal in a given state will respond when allowances are changed. ARC (1980) describes cattle as early, medium or late maturing, and as male, female or castrate. Sheep are described only as male or female! If, as has been argued here, future descriptions of nutrient allowance should be geared to the generation of body fat, body protein, milk fat, milk protein and milk lactose, then descriptors of animal growth and lactational performance, in these units, obtained under non-limiting conditions would be the required future descriptions of genotype. Taylor (1985) has suggested scaling rules which may be appropriate for description of rates of body constituent maturation in the growth cycle and Wood (1976) and Emmans, Neilson and Gibson (1983) have been given descriptions of lactation potential which can be defined by genetic parameters.

Figure 10.8 demonstrates how attributes of the animal, its food and its environment can be brought together.

The animal is described in terms of its current state and its 'potential'. In order for the animal to achieve its potential, resources/nutrients are required in amounts which can be calculated from knowledge of nutritional constants. Very often this will be based on stoichiometric relationships, e.g. amounts of amino acid and energy needed to synthesize a unit weight of protein. The resources/nutrients needed come from food. As an article of faith (Emmans and Fisher, 1986) the animal is assumed to try to eat sufficient food in order to meet its needs. Hence, from the estimate of resource/nutrient needed and the yield per unit of food, a 'target' food intake can be calculated.

A second element to the description of the animal is its capacity to cope with constraints which relate to attributes of food or the environment. Thus, there may be a limit to the physical capacity of an animal to ingest low nutrient density feeds; or a

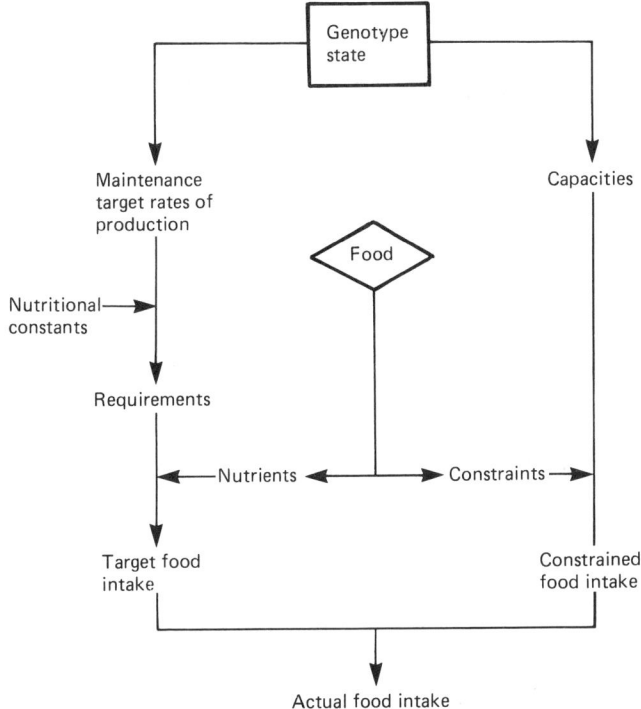

Figure 10.8 A generalized framework to enable prediction of responses. Genotype and animal state determine target rates of production and capacity of the animal to cope with nutritional and environmental constraints. Food attributes allow prediction of nutrient supply and assessment of food-related constraints. Actual food intake is the lesser of target food intake and constrained food intake. The physical environment will also have a constraining influence (especially in relation to capacity of the animal to dispose of heat) but this is not shown on the diagram

limit to the capacity of the animal to lose heat to its environment. These capacities are important and require to be measured. Where the food consumed has attributes for which the animal only has a limited capacity, then those food attributes might constrain food consumption to be less than the animal's target (if it were to achieve its potential). Thus, food consumed will be the lesser of 'target' and 'constrained' intake. In this way a nutrient allowance framework can be used to estimate voluntary food consumption as a response.

When food consumed is less than the animal needs to achieve its potential, then a set of rules is needed to describe how the animal copes with a nutrient shortage. That is, how can we predict the manner in which an animal (e.g. dairy cow) partitions a scarce resource/nutrient between different functions?

NUTRIENT PARTITION

The issue of nutrient partition is at the core of any system designed to estimate the manner in which ruminants respond to nutrient allowances. Of the various approaches which have been taken to predict partition in lactating cattle none, thus far, has met with great success. Blaxter (1959) introduced the concept that the partition of increments of ME towards milk was a predictable function of current yield for cows fed to 'Woodman standards'. This principle is still accepted in ARC (1980) with no apparent qualms about the incongruity of relating responses to ME to 'standards' which have been superseded by ME. In general, however, this relationship is not acceptable as it takes no account of the influence of body condition (Garnsworthy, 1988) nor of the nature of supplied nutrients (pp. 154–157) on observed responses.

Wood (1979) provided a framework in which ME needs for lactation would be partitioned between milk and tissue in proportion to 'potential demand'. In a somewhat analogous approach to the prediction of partition, Bruce, Broadbent and Topps (1984) used a principle of equal deficits, whereby each potential (for lactation or growth) is reduced by the same amount of energy during a shortage. In a more elaborate description of nutrient utilization, using principles based on those of enzyme kinetics, Baldwin, France and Gill (1987) chose to use a representation of glucose status to govern partition of available fat precursors between milk and tissue fat formation. This was partly successful. Yet another option might be to partition nutrients according to a hierarchy of nutrient use—as used by Emmans (1981) to predict growth in poultry. Oldham and Emmans (1988) gave an illustration of how this might apply to the partition of the major nutrients made available for use in dairy cows (*Figure 10.9*).

Imaginative developments from some of these ideas can be expected to allow useful progress towards applicable systems for predicting performance responses to allowances of food in the relatively near future.

Concluding remarks

In the UK the ARC (1980 and 1984) publications continue to be the basis for estimating nutrient and energy allowances for ruminants. The next stages of development will be to move from calculation of requirements towards use of that knowledge about requirements to enable the prediction of responses to nutrients. This will, inevitably, bring voluntary food consumption into true focus as a response to be predicted, rather than as an input to be assumed. The description of animal performance to be predicted should be that which is both biologically and economically relevant. In growth this means description of protein (lean tissue) and fat—for which better methods for *in vivo* assessment are essential for improvement in testing the precision of predictions. In lactation milk fat, protein and lactose production as well as body state change are the targets. In giving more attention to the interactions between nutrient use, genotype and current animal state, closer links between ruminant nutritionists and applied geneticists are to be encouraged. New methods for the estimation of important attributes of food will be called for.

It must not be forgotten that a main test of any system for predicting necessary allowances of nutrients, or the responses to nutrient supply, is that the system is consistent with the laws of thermodynamics, i.e. that energy balance can be predicted. Existing and future data sets which allow precise measurement of energy retention in tissue and secreted products, and of heat production will therefore remain central to future developments as they have been in the past.

Nutrition is a major and long-established biological technology. It is in a stage of transition, from calculation of requirements to prediction of responses, which is as intellectually stimulating, demanding and exciting as many of the other so-called, 'biotechnologies' which have recently become more fashionable. It seems reasonable to expect that the next few years will see substantial and practically effective

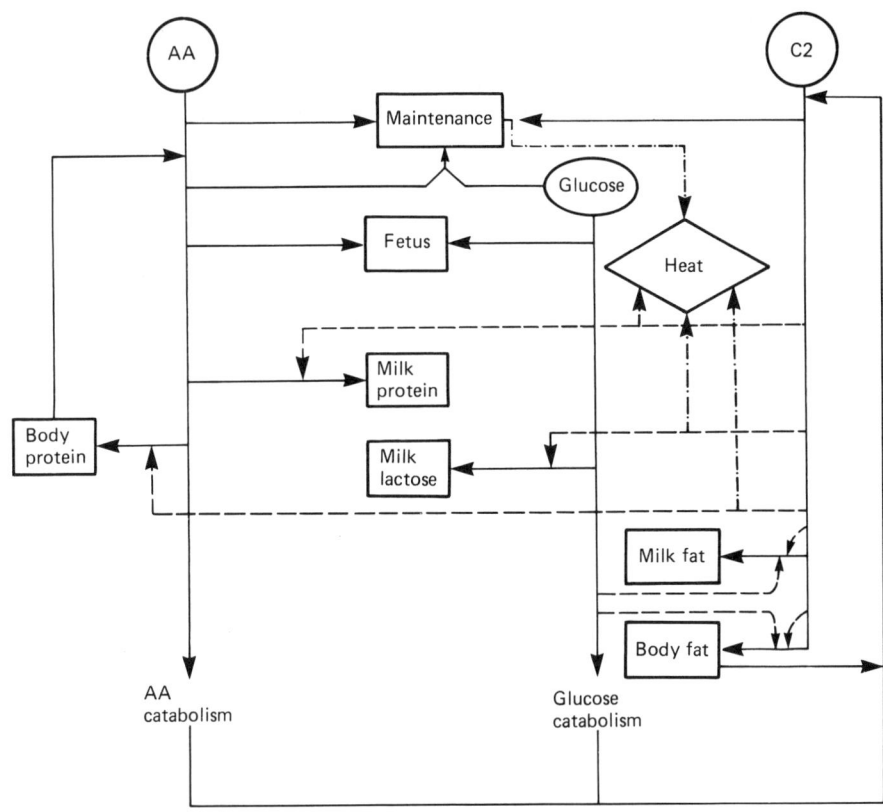

Figure 10.9 A framework to describe the manner in which major classes of nutrient are used for certain animal functions according to an order of supposed priorities. Resources supplied are in circles. They are AA (total amino acids), glucose (a combination of absorbed glucose (C6) and absorbed propionate (C3)) and C2 (absorbed acetic and butyric acids and long chain fatty acids (LCFA)). Functions for which these resources are used are indicated in rectangular boxes. Resources are used in priority order, thus, for example, the first call on AA is to supply amino acids to meet N maintenance, then to help meet obligatory glucose needs (if glucose supply is insufficient), followed by use for development of fetus, generation of milk protein and replenishment of labile body protein 'reserves'. Any amino acids which remain after maintenance needs and target rates of use of AA for fetus, milk protein and body protein have been met, will be catabolized and used as C2. Solid lines indicate main pathways of use for meeting defined animal functions. The dotted lines indicate use of resources (glucose and C2) in support of synthetic processes. The hatched lines indicate pathways of use which result in heat production

modifications to the manner in which nutrient allowances for ruminants will be assessed.

Acknowledgements

I am grateful to Dr M. Lewis and Mr G.C. Emmans for useful discussions about concepts used in this chapter and to Dr R.A. Edwards, Mr G. Alderman and Professor P.C. Thomas, Chairmen respectively of the IDWP Energy and Protein Working Parties and the TCORN Dairy Cow Nutrient Response Working Party, for permission to refer to the work of these groups before publication.

References

AFRC TECHNICAL COMMITTEE ON RESPONSES TO NUTRIENTS (TCORN) (1988). Report of the Working Party on 'Responses in the yields of milk constituents to the intake of nutrients by dairy cows' (in press)

AGRICULTURAL RESEARCH COUNCIL (1965). *The Nutrient Requirements of Farm Livestock, N. 2 Ruminants*, ARC, London

AGRICULTURAL RESEARCH COUNCIL (1980). *The Nutrient Requirements of Farm Livestock*. Commonwealth Agricultural Bureaux, Farnham Royal

AGRICULTURAL RESEARCH COUNCIL (1984). *The Nutrient Requirements of Ruminant Livestock, Supplement No. 1.* Commonwealth Agricultural Bureaux, Farnham Royal

ALDERMAN, G. (1987) In *Feed Evaluation and Protein Requirement Systems for Ruminants*, pp. 47–53. Ed. Jarrige, R. and Alderman, G. ECSC-EEC-EAEC, Brussels

BALDWIN, R.L., FRANCE, J. and GILL, M. (1987). *Journal of Dairy Research*, **54**, 77

BEEVER, D.E. and OLDHAM, J.D. (1986). In *Principles and Practice of Feeding Dairy Cows*, pp. 45–72. Eds. Broster, W.H., Phipps, R.H. and Johnson, C.L. National Institute for Research in Dairying, Technical Bulletin No. 8

BLAXTER, K.L. (1959). In *Scientific Principles of Feeding Farm Livestock*. pp.21–36. Farmer and Stockbreeder Publications, London

BLAXTER, K.L. (1986). *Proceedings of the Nutrition Society*, **45**, 177–183

BLAXTER, K.L. and BOYNE, A.W. (1978). *Journal of Agricultural Science, Cambridge*, **90**, 47

BROSTER, W.H. and THOMAS, C. (1981). In *Recent Advances in Animal Nutrition—1981*, pp. 49–67. Ed. Haresign, W. Butterworths, London

BRUCE, J.M., BROADBENT, P.J. and TOPPS, J.H. (1984). *Animal Production*, **38**, 351–362

BRUMBY, P.E., STORRY, J.E., SUTTON, J.D. and OLDHAM, J.D. (1979). *Annales de Recherches Veterinaires*, **10**, 310–319

COTTRILL, B.R. (1988). *The Feed Compounder*, **8** No. 1, 112–116

EMMANS, G.C. (1981). *Occasional Publication of the British Society of Animal Production, No. 5*, 103–110

EMMANS, G.C. (1988). *Journal of Agricultural Science, Cambridge* (submitted)

EMMANS, G.C. and FISHER, C. (1986). Problems in nutritional theory. In *Nutrient Requirements of Poultry and Nutritional Research*, pp. 9–39. Ed. Fisher, C., and Borman, K.N. Butterworths, London

EMMANS, G.C., NEILSON, D.R. and GIBSON, A. (1983). *Animal Production*, **36**, 546

EVANS, R.E. (1960). *Rations for Livestock*, MAFF Bulletin 174. HMSO, London

EUR 10657 (1987). *Feed Evaluation and Protein Requirement Systems for Ruminants.* Ed. Jarrige, R. and Alderman, G. ECSC-EEC-EAEC, Brussels

FISHER, C. (1986). In *Proceedings of the XIIIth International Congress of Nutrition*, pp. 437–442. Ed. Taylor, T.G. and Jenkins, N.K. John Libbey, London

FISHER, C., MORRIS, T.R. and JENNINGS, R.C. (1973). *British Poultry Science*, **14**, 469–484

GARNSWORTHY, P.C. (1988). In *Nutrition and Lactation in the Dairy Cow*. Ed. Garnsworthy, P.C. pp. 157–170. Butterworths, London
GILL, M., BEEVER, D.E., BUTTERY, P.J., ENGLAND, P., GIBB., M.J. and BAKER, R.D. (1987). *Journal of Agricultural Science, Cambridge*, **108**, 9–16
GRAHAM. N.MCC. (1986). *British Journal of Nutrition*, **56**, 315
HARKINS, J., EDWARDS, R.A. and MCDONALD, P. (1974). *Animal Production*, **19**, 141–148
HART, I.C. (1983). *Proceedings of the Nutrition Society*, **42**, 181–194
HUNTINGTON, G.B. (1984). *Journal of Dairy Science*, **67**, 1919–1927
HUNTINGTON, G.B. and REYNOLDS, P.J. (1986). *Journal of Dairy Science*, **62**, 2428–2436
INTER-DEPARTMENTAL WORKING PARTY (IDWP) (1988). *Report on the Nutrition Requirements of Ruminant Animals—Energy*. HMSO, London
KIELANOWSKI, J. (1976). In *Protein Metabolism and Nutrition*, p. 207. Eds. Cole, D.J.A., Boorman, K.N., Buttery, P.J., Lewis, A., Neale, R.J. and Swan, H. Butterworths, London
LEES, J.A., GARNSWORTHY, P.C. and OLDHAM, J.D. (1982). *Occasional Publication of the British Society of Animal Production*, **6**, 157–159
MARSTON, H.R. (1948). *Australian Journal of Scientific Research, Series B*, **1**, 93–129
MCCLYMONT, G.L. (1952). *Australian Journal of Scientific Research, Series B*, **5**, 374–383.
MACRAE, J.C. and LOBLEY, G.E. (1982). *Livestock Production Science*, **9**, 447–456
MACRAE, J.C., BUTTERY, P.J. and BEEVER, D.E. (1988). In *Nutrition and Lactation in the Dairy Cow*, Ed. Garnsworthy, P.C. pp. 55–75. Butterworths, London
MACRAE, J.C., SMITH, J.S., DEWEY, P.J.S., BREWER, A.C., BROWN, D.S. and WALKER, A. (1985). *British Journal of Nutrition*, **54**, 197–209
MAFF, DAFS, DANI, UKASTA, BVA (1983). *Mineral, Trace Element and Vitamin Allowances for Ruminant Livestock*. MAFF, ADAS
MINISTRY OF AGRICULTURE, FISHERIES AND FOOD (1975). *Technical Bulletin No. 33*. HMSO, London
OLDHAM, J.D. (1984). *Journal of Dairy Science*, **67**, 1090–1114
OLDHAM, J.D. (1987). In *Feed Evaluation and Protein Requirement Systems for Ruminants*, pp. 171–186. Eds. Jarrige, R. and Alderman, G. ECSC-EEC-EAEC, Brussels
OLDHAM, J.D. and SMITH, T. (1982). In *Protein Contribution of Feedstuffs for Ruminants*, pp. 103–130. Eds. Miller, E.L., Pike, I.H. and Van Es,, A.J.H. Butterworths, London
OLDHAM, J.D. and EMMANS, G.C. (1988). In *Nutrition and Lactation in the Dairy Cow*, pp. 76–96. Ed. Garnsworthy, P.C. Butterworths, London
OLDHAM, J.D., BINES, J.A. and MACRAE, J.C. (1984). *Proceedings of the Nutrition Society*, **43**, 65A
OLDHAM, J.D., SUTTON, J.D. and MCALLAN, A.B. (1979). *Annales de Recherches Veterinaires*, **10**, 290
OLDHAM, J.D., NAPPER, D.J., JACOBS, J.C. and PHIPPS, R.H. (1988). In *Proceedings 5th EAAP Symposium on Protein Nutrition and Metabolism* (in press)
ORSKOV, E.R., GRUBB, D.A., WENHAM, G. and CORRIGALL, W. (1979). *British Journal of Nutrition*, **41**, 553–558
ORTIGUES, I. (1987). Nutrient supply, growth and calorimetric efficiency in heifers offered straw rich diets. PhD Thesis, University of Reading
PULLAR, J.D. and WEBSTER, A.J.F. (1977). *British Journal of Nutrition*, **37**, 355–363
REEDS, P.J., WAHLE, K.W.J. and HAGGARTY, P. (1982). *Proceedings of the Nutrition Society*, **41**, 155–159

SMITH, T., BROSTER, V.J. and HILL, R.E. (1980). *Journal of Agricultural Science, Cambridge*, **95**, 687–695

SCIENTIFIC PRINCIPLES OF FEEDING FARM LIVESTOCK (1959). Proceedings of a Conference held at Brighton. London, Farmer and Stockbreeder Publications Ltd

SUTTON, J.D. (1984). *Occasional Publication of the British Society of Animal Production*, **9**, 43–52

SUTTON, J.D. (1985). *Journal of Dairy Science*, **68**, 3376–3393.

SUTTON, J.D. and OLDHAM, J.D. (1977). *Proceedings of the Nutrition Society*, **36**, 203–209

SUTTON, J.D., OLDHAM, J.D. and HART, I.C. (1980). In *Energy Metabolism*, pp. 303–306. Ed. Mount, L.E. Butterworths, London

TAYLOR, ST. C.S. (1985). *Journal of Animal Science*, **61**, 118–143

TAYLOR, ST. C.S., THIESSON, R.B. and MURRAY, J. (1986). *Animal Production*, 37–61

TYLER, C. (1959). In *Scientific Principles of Feeding Farm Livestock*, pp. 8–18. Farmer and Stockbreeder Publications Ltd

WATERS, C.J., DEWHURST, R.J. and WEBSTER, A.J.F. (1987). *Animal Production*, **44**, 475

WEBSTER, A.J.F. (1987). In *Feed Evaluation and Protein Requirement Systems for Ruminants*, pp. 47–53. Eds. Jarrige, R. and Alderman, G. ECSE-EEC-EAEC, Brussels

WEBSTER, A.J.F. (1980) In *Digestive Physiology and Metabolism in Ruminants*, pp. 469–484. Eds. Ruckenbosch, Y. and Thivend, P. MTP Press, Lancaster

WOOD, P.D.P. (1976). *Animal Production*, **22**, 35–40

WOOD, P.D.P. (1979). *Animal Production*, **28**, 55–63

WHITELAW, F.G., MILNE, J.S., ORSKOV, E.R. and SMITH, J.S. (1986). *British Journal of Nutrition*, **55**, 537

11

ALTERNATIVE APPROACHES TO THE CHARACTERIZATION OF FEEDSTUFFS FOR RUMINANTS

A. J. F. WEBSTER, R. J. DEWHURST and C. J. WATERS
Department of Animal Husbandry, University of Bristol, Langford, Bristol, UK

Introduction

The remit of this chapter is to consider alternative, practical approaches to the characterization of feedstuffs for ruminants other than those based simply upon the apparent digestibility of proximate constituents (the Weende system; Henneberg and Stohmann, 1860) or, even more simply, measurement of a single dietary constituent, such as the modified acid-detergent fibre method (MAD fibre; Barber, Adamson and Altman, 1984) for prediction of the metabolizable energy (ME) concentration of forages. The objective is to predict the supply of available nutrients from robust measurements of feed chemistry yet in such a way as to recognize the essential principles of ruminant and post-ruminant digestion. This pursuit can be likened to that of the circus performer who rides standing astride two horses; a tricky business at the best of times and further complicated by the fact that one horse, feeding practice, is a plodder, while the other, nutritional science, is a spirited thoroughbred able to take off at high speed but with no great sense of direction. Any system of feed evaluation that can bring these two into line must, ideally, meet the following criteria:

(1) It should be based on measurements of feed chemistry, physical form or *in vitro* digestion that can be adopted as routine by the feed compounder.
(2) It should be deterministic, rather than empirical and sufficiently descriptive of ruminant physiology to be able to incorporate essentials of present and future knowledge.
(3) It should predict the yield of the major truly-absorbed substrates for energy and protein metabolism.
(4) It must, when tested in production trials, be demonstrably better than existing empirical systems.

This chapter will review, very briefly, the historical development of feed evaluation systems for ruminants then propose and discuss an alternative approach which may be simple enough yet deterministic enough (preserving our simile) to allow both circus horses to progress together.

Development of systems for food evaluation

PROXIMATE ANALYSIS

The history of feed evaluation systems has been reviewed by Blaxter (1962, 1986) and Van Soest (1982). The earliest attempts were strictly empirical and categorized feeds in terms of *hay equivalent*. The scientific foundation of feed evaluation was laid down in the mid-nineteenth century by Liebig and his school who recognized that feeds could be categorized into digestible nutrients and that by far the major part of these digestible nutrients serves as a source of energy (ME) to do work, the energy being dissipated as heat. This approach established for the first time the two essentials of any feed evaluation system.

(1) a feed description which approximates to the yield of absorbed nutrients,
(2) precise techniques for measurement of the capacity of absorbed nutrients to support the metabolic requirement of animals for maintenance and synthetic functions like growth and lactation.

The proximate system for analysis of feedstuffs has changed little in the last 130 years (Wolff, 1856, cited by Van Soest, 1982). The energy and protein-yielding components of the dry matter (DM) are described by ether extract (EE), crude protein (CP), crude fibre (CF) and nitrogen-free extract (NFE). The division of the carbohydrate fraction into NFE and CF was designed to reflect, respectively, digestible and indigestible carbohydrate. While this division may be a reasonable, empirical approximation to the truth for cereals and other concentrate feeds given to simple-stomached species, it has long been accepted that it does not relate to the capacity of ruminants to digest carbohydrates. More appropriate techniques exist for analysing cell wall carbohydrates according to solubility in neutral or acid detergent (Goering and Van Soest, 1970). The residue, neutral detergent fibre (NDF), corresponds approximately to cellulose + hemicellulose + lignin and acid detergent fibre (ADF) to cellulose plus lignin only. MAD fibre (Clancy and Wilson, 1966) is used by MAFF as an (imperfect) predictor of ME. Yet, amazingly, measurement of CF remains a statutory requirement.

Workers at the Oscar Kellner Institute, over many years, measured the apparent digestibilities of the proximate constituents of feedstuffs for ruminants and used the data to generate preferred values for ME and digestible crude protein (DCP) for different classes of feedstuffs (Nehring, 1969). The Ministry of Agriculture, Fisheries and Food (MAFF, 1975) assumed constant digestibilities of proximate constituents to generate values of ME and DCP in current tables of feed composition.

ENERGY

Most published tables describing the capacity of feeds to provide energy have been based on proximate analysis, whether energy is defined by total digestible nutrients (TDN), which corresponds approximately to ME, or in terms of net energy (NE) which included starch equivalent. The ME system as formulated by Blaxter (1962) and adopted by the Agricultural Research Council (1965) differs from these in only one respect, namely that calorimetry was used to measure all aspects of the energy exchange, the heats of combustion of feeds and excreta that determine ME, and metabolic heat production to describe NE requirement and supply. This apart, the ME system is conceptually identical to NE systems when the efficiency of utilization of ME (k_f, k_l, etc) is properly taken into account.

Although ME values were still predicted from proximate analysis when the ME system was first adopted (MAFF, 1975) more recent studies at the Feed Evaluation Units of AFRC (Wainman, Dewey and Boyne, 1975, 1978, 1981; Wainman, Dewey and Brewer, 1984) and MAFF (Barker, Adamson and Altman, 1984) have measured ME values directly.

The use of energy units (MJ) to describe the supply and utilization of energy-yielding nutrients has the one overwhelming merit that all elements of the energy balance equation can be measured with precision. It is also valid in a strictly biological sense since most energy-yielding substrates absorbed into the body are used to drive ATP synthesis, do work and generate heat. Nevertheless there are inherent limitations to the ME system as it stands (see AFRC, 1987).

Formulation of feeds on the basis of ME has a number of implicit features:

(1) It assumes additivity, which may not always be correct, e.g. when starch and cellulose are fed together.
(2) It does not distinguish between fermentable and unfermentable substrates (e.g. cellulose and volatile fatty acids in silage). This does not necessarily affect ME values but can affect microbial protein synthesis.
(3) It does not distinguish between patterns of fermentation, leading, e.g. to differing proportions of acetate and propionate and so differences in efficiency of utilization of ME (perhaps) or in the composition of milk.
(4) It does not permit a logical interpretation of the effect of plane of nutrition on the extent of fermentation and thus on ME yield.

However, the amount of published information to describe the energy requirements of ruminants in terms of ME is so enormous (ARC, 1980) that, despite these limitations, the ME system is quite precise enough for the practical purpose of feeding farm animals *so long as the ME value of the feed is known with equal precision*. Here lies the rub. Since it is impossible to conduct feeding trials to determine the ME value of every forage and compound feed, this must be predicted from feed chemistry or an *in vitro* measurement of digestibility. It is undeniable that, at present, we are better able to predict the requirement of a ruminant for ME than the capacity of a diet to provide it. It may be that workers have found cattle and sheep to be so much more interesting than grasses and cereals.

The Feedingstuffs Evaluation Unit at the Rowett Institute, Aberdeen, was set up to provide a comprehensive body of data to improve the prediction of ME from feed chemistry or simple *in vitro* digestibility measurements. When three successive sets of silages and dried grasses were tested (Wainman, Dewey and Boyne, 1975, 1978; Wainman, Dewey and Brewer, 1984) the best single predictor differed for each set between MADF, ADL (sulphuric acid–detergent lignin) and CP (*Table 11.1*). Since the best predictors were not consistent even within one class of feed and could, of course, only be determined retrospectively they offer no guidance as to which, if any, could be used as a general predictor even for conserved grasses. Moreover, the observed values for ME of these silages from Scotland and the north of England were, on average, 2.9 MJ/kg DM greater than would be predicted from using the MADF equation of Barber, Adamson and Altman (1984) which was derived from measure-

Table 11.1 A COMPARISON OF VARIOUS METHODS FOR PREDICTING TRUE ME IN FORAGES TESTED AT THE ROWETT FEED EVALUATION UNIT. THE APPROACHES INCLUDE THE BEST SINGLE PREDICTORS WITHIN THE ROWETT DATA SETS, PREDICTION USING 'MENTOR' AND PREDICTIONS FROM MADF EQUATIONS

Forage type	Rowett correlations		MENTOR	Mean difference (MJ/kg DM) between observed and predicted values		
	Best predictor	r	r	MADF Rowett	Drayton[a]	MENTOR
Dried grass (DG)						
DG1	ADL	−0.93	0.95	+0.59	+1.51	+0.22
DG2	CP	0.79	0.79	+0.88	+1.97	+0.17
DG3	MADF	−0.94	0.93	—	+0.75	+0.16
Silage (S)						
S1	MADF	−0.78	0.72	—	+0.82	−0.34
S2	ADL	−0.76	0.80	+0.17	+2.97	−0.31
S3	CP	0.67	0.70	+0.24	+3.03	−0.16

[a]Barber, Adamson and Altman (1984)
This table has been condensed from Tables 3 and 4 in Dewhurst *et al.* (1986)

ments made at the Agricultural Development and Advisory Service (ADAS) Feed Evaluation Unit, Drayton on silages obtained largely from southern England.

Alderman (1985) has described the approach taken by a joint working party of the United Kingdom Agricultural Supply Trade Association (UKASTA), ADAS and the Council of Scottish Agricultural Colleges (COSAC) to the improved prediction of compound feeds. The data used to derive prediction equations for ruminants are those of Wainman, Dewey and Boyne (1981). The approach was intentionally empirical, using statistical analysis to generate the most precise multiple regression equations which were:

$$\text{ME(MJ/kg DM)} = 11.78 + 0.0654\text{CP} + 0.0665\text{EE} - 0.0414\text{EE} \times \text{CF} - 0.118\text{TA} \quad (\text{U1})$$

$S'' = 0.36$

$$\text{or ME} = 11.56 - 2.375\text{EE} + 0.030\text{EE}^2 + 0.030\text{EE} \times \text{NCD} - 0.034\text{TA} \quad (\text{U2})$$

$S'' = 0.32$

$$\text{or ME} = 13.8 - 0.488\text{EE} + 0.0394\text{EE} \times \text{CP} - 0.0085\text{MADF} \times \text{CP} - 0.138\text{TA} \quad (\text{U3})$$

$S'' = 0.35$

All values other than ME are expressed in percentage units. Abbreviations not defined already are TA = total ash; NCD = organic matter digested by neutral detergent and cellulase; S'' is the prediction error (for further explanation of NCD and S'' see Alderman, 1985).

Equation U1 is intended for purposes of legislation, being based on proximate constituents only. The 'best' equation is U2 which incorporates a simple measurement of fibre digestibility (NCD) that has a wider applicability than the two-stage *in vitro* digestibility method of Tilley and Terry (1963) since it does not require a supply of fresh rumen contents. The usefulness of Equation U1 is limited, of course, by the unsuitability of CF as a predictor of indigestible carbohydrate and seriously distorts ME values for some compound feeds, particularly those containing substantial amounts of digestible fibre and fat. Equation U2 is better in this respect. However, the main factor limiting the general applicability of these equations was the relatively narrow range of chemical composition in the 24 diets tested, in particular the narrow range in fibre (164–325 g/kg NDF for 22/24 feeds). An improved empirical prediction equation obtained from a wider range of diet types has been presented in Chapter 9 by Thomas and Livingstone.

PROTEIN

The limitations to DCP as a description of the protein value of feeds for ruminants have been discussed widely (ARC, 1980, 1984; INRA, 1978; Owens, 1982) and do not need to be repeated here. All modern alternatives to the DCP system distinguish between nitrogenous compounds which are degraded in the rumen and incorporated into microbial protein at a rate determined by the capacity of the rumen micro-organisms to ferment energy-yielding substrates, and those which escape degradation but are available for acid digestion. The ARC system originally proposed in 1980 and modified in 1984, is based on an amalgamation of observations and theories concerning the physiology of digestion in the ruminant (proportion of digestible organic matter (DOM) apparently digested in the rumen, microbial N yield/kg OM apparently digested in the rumen, etc). Although these proposals, considered step by step, appear sound, they had not been tested in production experiments at the time of publication and were viewed with suspicion by practical feeders since they gave estimates of protein requirement (at optimal rumen degradability) that were conspicuously lower than those generated by working groups from other nations (*Table 11.2*) or accepted for use in commercial practice.

In 1981 an Inter-Departmental Working Party (IDWP) was formed in the UK and charged with the task of considering the ARC (1980, 1984) proposals and making recommendations on its suitability (with appropriate modifications) for adoption as the official UK system for assessing the protein needs of ruminant livestock. The committee has not yet reported although its provisional proposals have been

Table 11.2 COMPARISON OF THE PROTEIN REQUIREMENTS OF A 600 kg COW YIELDING 30 kg MILK AND LOSING 0.5 kg LIVEWEIGHT/DAY CALCULATED USING DIFFERENT PROTEIN SYSTEMS

	UK	France	System Germany	Denmark	USA
DM intake (kg/day)	17.5	17.5	17.5	17.5	17.5
ME intake (MJ/day)	200	200	200	200	200
RDP requirement (g/day)	1680	1688	2118	1863	1757
UDP requirement (g/day)	572	1142	671	888	1186
CP requirement (g/day)	2252	2830	2789	2751	2943
Optimal degradability (%)	75	60	76	68	60
CP concentration in diet (g/kg)	129	162	159	157	168

From Alderman, 1987
RDP and UDP are rumen degradable and undegradable protein respectively

172 *Alternative approaches to the characterization of feedstuffs for ruminants*

presented in outline (Webster, 1987a,b) and Oldham (1987) has discussed the first production trials designed to test the system. The outline is illustrated in *Figure 11.1* (from Webster, 1987b) but it will not be discussed further at this stage since it is incorporated within the more general model of energy and nitrogen supply to ruminants ('MENTOR') presented in detail in the next section.

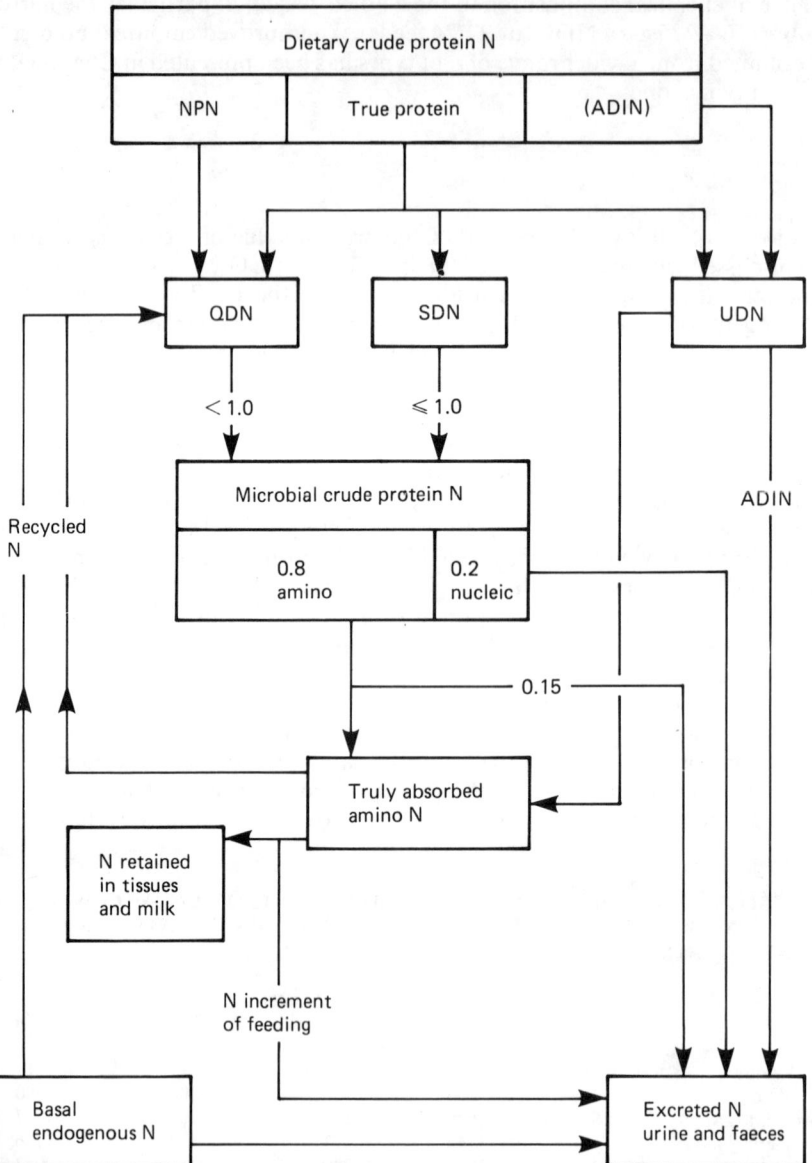

Figure 11.1 Factors determining the supply of truly absorbed amino N. ADIN = acid-detergent insoluble N, NPN = non-protein N, SDN = slowly degraded N, QDN = quickly degraded N, UDN = undegraded N (from Webster, 1987b)

MODELLING RUMINANT DIGESTION AND METABOLISM

If our understanding of the physiology of digestion and metabolism in the ruminant is ever to be incorporated into a practical system for feed evaluation then the empirical, statistical approach to feed characterization must be abandoned, however elaborate *and precise* the regression equation, in favour of an approach that reflects more closely the dynamic, two stage nature of digestion in the ruminant. Elaborate models such as those of Baldwin, Koong and Ulyatt (1977), Beever, Black and Faichney (1981), Black *et al.* (1981) and Gill *et al.* (1984) have been developed which elegantly describe the digestion and metabolism of a wide range of defined substrates. None of these models is ideal but all have the merit that they are able to incorporate new knowledge as it occurs. A simple, empirical Feed Evaluation system based, say, on MADF is inherently unimprovable however many thousand more determinations are introduced into the data set. The disadvantage of the elaborate models is that they require more information than the feed compounder can provide, information which, in some cases, can only be obtained from surgically prepared animals. In consequence, they are of more help to the research worker than to the feeder and so increase still further the painful gap between the two circus horses. In the next section attempts will be made to draw them closer together by proposing a model 'MENTOR' which predicts the yield of truly metabolizable energy (ME_t) and metabolizable protein (MP) (truly absorbed amino nitrogen (TAAN) × 6.25) from ruminant diets. The model seeks to meet the criteria for an ideal feed evaluation system outlined in the introduction—although it does not achieve them all. It is designed to be no more complicated than absolutely necessary and based on measurements that can readily be incorporated into practical ration formulation.

'MENTOR': an approach to feed evaluation

MODEL DESCRIPTION

It is assumed that, in making substrates available for energy and protein metabolism, the processes of ruminant digestion distinguish four fractions of the feeds (*Figure 11.2*). These are:

(1) Material which is quickly and completely fermented or degraded in the rumen. Material (carbohydrate and protein) which is fermented to energy-yielding substrate as volatile fatty acids (VFA) is termed *quickly fermentable energy* (QFE). Organic nitrogenous material that is rapidly degraded to ammonia is *quickly degradable nitrogen* (QDN).
(2) Material which is slowly (S) and thus incompletely fermented or degraded in the rumen is termed SFE or SDN.
(3) Material which is unfermented or undegraded in the rumen but subsequently digested is termed unfermentable, digestible energy (UDE) or undegradable, digestible nitrogen (UDN).
(4) The final fraction is that which is neither fermented nor digested. Since this does not contribute to the supply of ME_t or TAAN (MP) it requires no code.

Fractions 2–4 are all influenced by physiological factors such as rumen retention time which itself reflects input variables such as plane of nutrition and the physical

174 Alternative approaches to the characterization of feedstuffs for ruminants

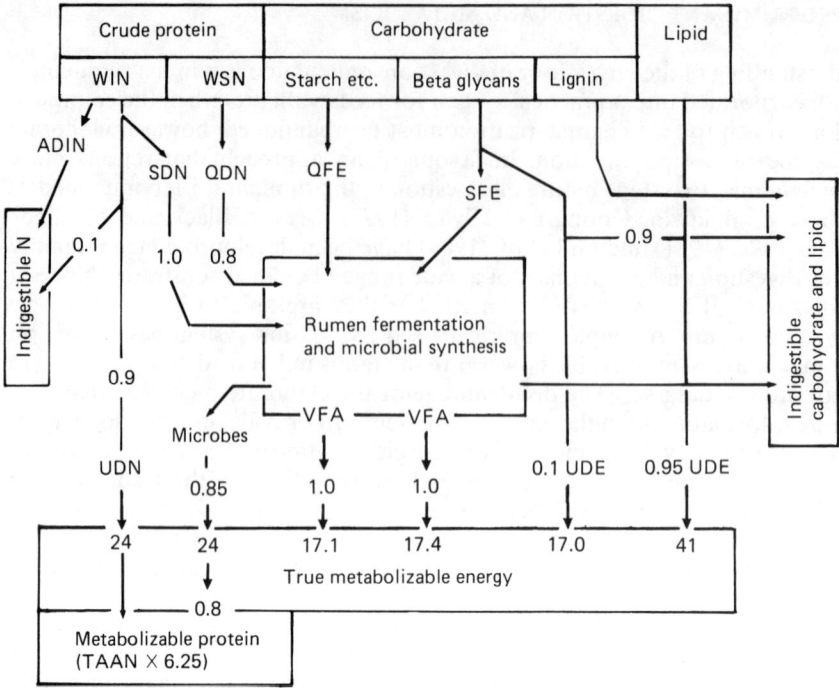

Figure 11.2 MENTOR, a model of metabolizable energy and metabolizable protein supply to ruminants. QFE, SFE and UDE are quickly fermented, slowly fermented and unfermented digestible energy respectively; QDN, SDN and UDN are quickly degraded, slowly degraded and undegraded digestible N; WIN and WSN are water-insoluble and water-soluble N

form of the diet. The use of the term UDN to describe undegradable, digestible N is inevitably confusing for those who have become accustomed to using UDN to describe total undegradable dietary nitrogen whatever its subsequent digestibility (ARC, 1980, 1984). It is therefore proposed that total undegradable N in future be termed UN and that the former use of UDN be abandoned, partly because it contains a superfluous letter and partly because it does not take into account variations in the digestibility of undegraded N (Webster, Simmons and Kitcherside, 1982).

The three contributors to ME_t (QFE, SFE and UDE) and TAAN (QDN, SDN and UDN) constitute a sufficient description of the capacity of feeds to supply energy and protein yielding substrates. These definitions are also unlikely to require significant alteration in the light of new research though it may sometimes be useful to partition ME_t according to the major classes of absorbed substrates. The model can accommodate this. Its success or failure will depend on how precisely it is possible to

(1) describe truly absorbed energy and nitrogen fractions in relation to feed chemistry and the physiological state of the animal,
(2) describe the interactions between fermentation, degradation and microbial synthesis in the rumen that determine the utilization of QDN, SDN, QFE and SFE and the supply of nutrients as VFA, UFE, microbial protein N and UDN.

PREDICTION OF TRULY METABOLIZABLE ENERGY

The approach used by the model in prediction of truly metabolizable energy for grass forages has been described in detail by Dewhurst et al. (1986) and will be recapitulated here only briefly.

True metabolizable energy (ME_t) equals gross energy less energy losses as methane, fermentation heat and faecal losses of dietary origin. Urinary energy losses are not included because they are an end-product of the incomplete catabolism of absorbed nutrients and depend in large part on the physiological state of the animal rather than attributes of feed chemistry. The description of feeds, at present, is based in part on Weende proximate analysis but incorporates the detergent analyses of Goering and Van Soest (1970). The following analyses are required as inputs to the model when used to evaluate forage; TA, CP, EE, NDF, acid-detergent lignin (ADL) and, for silages, total VFA and lactic acid. The carbohydrate fraction is then subdivided into

(1) β-glycans (NDF-ADL); sources of SFE.
(2) simple sugars, pectin, starch, and other α-glycans, sources of QFE.

Current assumptions concerning the fermentation and digestion of energy yielding substrates and the enthalpy of absorbed nutrients are outlined in *Table 11.3*. Sugars, α-glycans and all sources of QFE are assumed to be completely fermented at all rumen solid-phase outflow rates. The effective fermentability of β-glycans (P_g) is predicted from Equation (11.1) (Waldo, Smith and Cox, 1972).

$$P_g = b_g c_g / (c_g + k) \tag{11.1}$$

where b_g is the fermentability of β-glycans at infinite time, c_g is fermentation rate (/h) and k is solid-phase outflow rate from the rumen (/h). Values for b_g and c_g for grasses were taken from Smith, Goering and Gordon (1972). A further 0.09 of previously unfermented β-glycans are assumed to ferment to VFA in the hind gut (Zinn and Owens, 1982).

Protein degradation (P_n) may be described, where possible, using Equation (11.2) (Ørskov and McDonald, 1979)

$$P_n = a_n + b_n c_n / (c_n + k) \tag{11.2}$$

where a_n is the quickly degraded fraction (QDN), b_n is SDN at infinite time, c_n is degradation rate for SDN (/h), and k is the solid-phase outflow from the rumen (/h). Where this is not possible, our best available description of P_n for forages is from Webster, Simmons and Kitcherside (1982) given in Equation (11.3).

$$P_n = (0.9 - 2.4k)(CP - 0.059\ NDF)/CP \tag{11.3}$$

Lactic acid is assumed to have an effective fermentability of 1.0 but yield no ATP to the microbes. Lipids and dietary VFA are assumed to be unfermented and thus a source of UDE; lignin is assumed to be unfermentable and indigestible.

The molar proportions of VFA in *Table 11.3* are derived from Murphy, Baldwin and Koong (1982). The ATP yield from the fermentation of carbohydrate and proteins (27.5 and 12.5 mol/kg respectively, Tamminga 1982) is used for microbial maintenance and growth. The proportion of fermented organic matter used for microbial maintenance and the supply of microbial dry matter and N to the abomasum are both dominated by outflow rate from the rumen. Microbial yield has been related to fermentation rate *in vitro* by Stouthamer and Bettenhausen (1973).

Table 11.3 PRINCIPAL ASSUMPTIONS CONCERNING THE FERMENTATION, DIGESTION AND ABSORPTION OF SUBSTRATES CONTRIBUTING TO TRUE METABOLIZABLE ENERGY

Dietary substrate	Fermentability	Products of fermentation				True absorbability	Enthalpy or absorbed nutrients (MJ/kg)
		(Mol VFA/mol substrate)			(Mol ATP/kg substrate)		
		Acetate	Propionate	Butyrate			
Sugars }	1.0	1.28	0.37	0.17	27.5	1.0	17.1
α-glycans							
β-glycans	$b_g c_g/(c_g + k) + 0.09R$	1.23	0.30	0.23	27.5	1.0	17.4
Lactic acid	1.0	nil	0.60	0.30	nil	1.0	22.5
VFA	0 ─────→					1.0	16.3
QDN	a_n ─────→	1.0 microbial DM				1.0	17.1
SDN	$b_n c_n/(c_n + k)$		0.57 {0.49 CP	0.04	13.5	0.8	21.0
			0.10 EE + 0.35 de novo				
Ether extract	0					}0.95	42.1
UN	0					0.9(UN − ADIN)	21.0

Adapted from Dewhurst et al., 1986

Adapting their approach to predict yield of microbial DM (or N; g M_iDM/h) *in vivo* generates Equation (11.4).

$$\frac{g\ M_iDM}{(QFE + SFE)} = \frac{1}{(1/M_iDM,max + M_i\ \text{maintenance/outflow rate})} \tag{11.4}$$

where QFE and SFE are the quickly and slowly fermentable sources of energy expressed as MJ energy fermented/h. M_iDM,max is the asymptotic maximum microbial DM yield expressed in g/h and M_i maintenance is the amount of fermented energy required in MJ/g M_i DM.h.

Assumptions concerning the true absorbability of the nutrients contributing to ME_t are also listed in *Table 11.3*.

Since the yield of ME_t is so dependent on rumen solid phase outflow rate (k), the usefulness of the model is enhanced if k can be predicted from DM intake and/or other food attributes since it is then possible to incorporate effects of plane of nutrition (or physical form of the diet) on ME_t. Since the model predicts the rate at which material leaves the rumen, it is only necessary to introduce constants for rumen volume and dry matter concentration to predict interactions between dry matter intake and ME_t on the simple and eminently defensible premise that what goes in equals what comes out.

TESTING THE PREDICTION OF ME_t

The use of the model to predict ME_t for grass-based forages was tested using data from a total of 121 samples of forage comprising silage (46), dried grass (46), hay (21) and straw (8) that were comprehensively evaluated at the Rowett Research Institute (Wainman, Dewey and Boyne, 1975, 1978; Wainman, Dewey and Brewer, 1984). In comparing predictions of ME_t with the observed values for apparent ME, correction was made for urinary energy losses and it was assumed that, at maintenance, endogenous losses of energy into the faeces are approximately equal to fermentation heat loss and so cancel one another out (Dewhurst *et al.*, 1986). Rumen solid phase outflow rate was assumed to be 0.03/h. The agreement between observed and predicted ME_t is illustrated in *Figure 11.3* (Dewhurst *et al.*, 1986). Re-inspection of *Table 11.1* reveals that 'MENTOR' nearly always predicted ME_t as well as, or better than, the best single predictor derived retrospectively from the Rowett data sets. There was some systematic error in the model predictions, with silages slightly overvalued and dried grasses and hays slightly undervalued, but the discrepancy was far less than that created by applying the MADF equation of Barber, Adamson and Altman (1984) to the Rowett forages (*Table 11.1*).

Recently 'MENTOR' has been used to examine the Drayton forages (R.J. Dewhurst and D.I. Givens, unpublished data). The correlations were not as good as for Rowett forages but the systematic discrepancy between northern and southern forages largely disappeared, presumably because MENTOR incorporates differences attributable to climate in ADL/NDF ratios.

Figure 11.4 illustrates the effect of k in ME_t for four forages across the range examined. The reduction in ME_t with increasing k increases as ME_t at maintenance declines due mainly to the effect of increasing k on SFE.

MENTOR, in its current form, has also been examined as a predictor of ME_t in

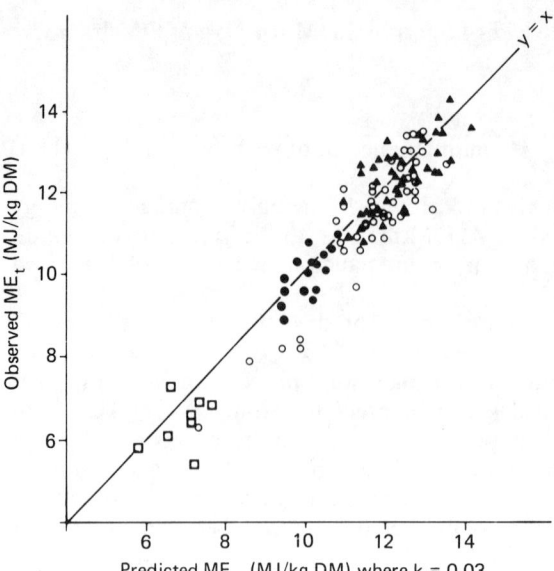

Figure 11.3 Regression of observed true metabolizable energy (ME$_t$) against predicted ME$_t$ for 121 forages tested at the Rowett Research Institute. ○ = dried grass, ● = hay, ▲ = silage, □ = straw (from Dewhurst *et al.*, 1986)

compound feeds, including the 24 tested at the Rowett Institute (Wainman, Dewey and Boyne, 1981) and a further 18 from our laboratory. This generated Equation (11.5)

$$\text{ME}_t, \text{ predicted} = 3.03 + 0.846 \text{ ME}_t, \text{ observed} \tag{11.5}$$

The correlation coefficient $(r) = 0.89$, with an $SD = 0.76$ which is as good as for forages. Empirically therefore the model is reasonably satisfactory. However, the fact that the intercept and regression coefficient depart significantly from 0 and 1.0 respectively indicates that current assumptions as to the dietary contributors to QFE, SFE and UDE based almost entirely on forage data cannot simply be extrapolated to the description of compound feeds fed alone or in combination with forage. It is not yet possible to modify assumptions as to the efficiency of fermentation and digestion of compound feeds, partly because the Rowett data set does not contain a sufficiently wide range of sources of QFE, SFE and UDE. However, this systematic overprediction of ME$_t$ by the model does not appear to be related to the proportion of starch or fat in the diet, nor is the prediction of ME$_t$ in compound feeds very sensitive to variations in k within the range 0.02–0.07/h. At this stage therefore it appears that either the fermentability of the fibre fraction in compound feeds is being overpredicted or there is a failure to account for associative, inhibitory effects, e.g. that of starch on fibre digestion.

PREDICTION OF TAAN OR METABOLIZABLE PROTEIN

The approach within the model to the prediction of the supply (and metabolism) of TAAN (*Figure 11.1*) has been outlined previously (Webster, 1987a). In essence, it adopts the ARC (1980) distinction between degradable and undegradable dietary N

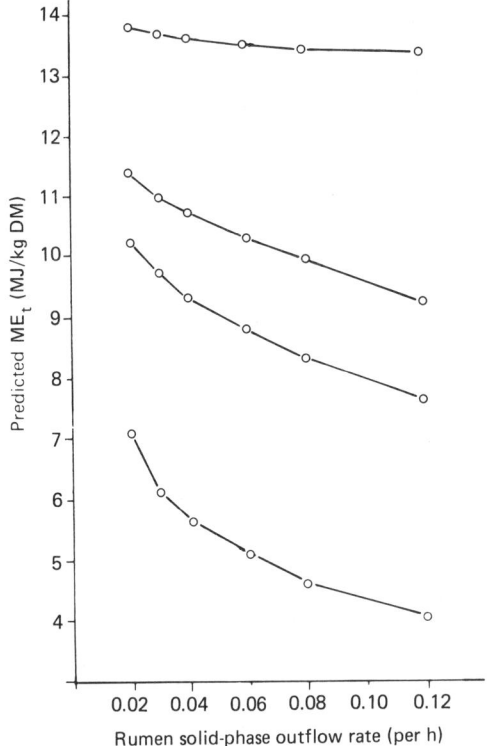

Figure 11.4 Effect of increasing rumen solid-phase outflow rate (k h^{-1}) on predicted true metabolizable energy (ME$_t$) for four silages varying especially in SFE (from Dewhurst et al., 1986)

and the ARC (1984) proposal to use true rather than apparent digestibility. The only differences relate to the efficiency of utilization of QDN, the efficiency of microbial protein synthesis and the digestibility of UN, as indicated below:

(1) QDN (g/day) is considered to be equal to water-soluble nitrogen. In practice this is measured by submitting material in a porous synthetic fibre (PSF) bag (42 μm pore diameter) to a 35 min cold rinse in a commercial washing machine.

(2) SDN (g/day) has to date been derived from the results of PSF bag incubations in the rumen using the modified equation of McDonald (1981) to incorporate any lag phase prior to the onset of the exponential phase of the degradation process. Where there is a significant lag and a_n in Equation (11.2) differs substantially from water-soluble N, the latter is still used to define QDN. Where PSF bag degradation data are unavailable, SDN can be obtained from $(P_n - QDN)$ using Equation (11.3).

(3) Undegraded N is assumed to have two components; that which is linked to structural carbohydrate and the remainder which has a digestibility close to but less than 1.0. The former, namely the nitrogen in acid-detergent fibre, or acid-detergent insoluble nitrogen (ADIN, *Figures 11.1* and *11.2*) is assumed to be completely indigestible. This claim by Van Soest (1982) and Wilson and Strachan (1980) has largely been borne out by digestibility trials at Bristol in which it was found that ADIN in food was very similar to that in faeces except for a few feeds containing significant amounts of raw materials containing

antinutrient substances (e.g. Shea nut). In these cases ADIN in faeces greatly exceeded that in the food. Webster *et al.* (1984) briefly outlined the results of measurements made of the pepsin solubility of the UN residues of 18 raw materials incubated in PSF bags for 18 h. The true digestibility of the UN component unlinked to structural carbohydrate was obtained by regressing $y = $ (pepsin soluble UN − ADIN) against $x = $ (UN − ADIN) which gave Equation 11.6

$$y = 0.963x - 1.42, \quad r = 0.997 \tag{11.6}$$

Constraining this equation to pass through the origin gives a regression coefficient of 0.88. It is therefore proposed that a realistic approximation is given by Equation (11.7)

$$\text{UDN} = 0.9(\text{UN} - \text{ADIN}) \tag{11.7}$$

Table 11.4 compares some estimates of UDN (i.e. true digestibility of UN) with measurements of apparent digestibility of total N (which corresponds to DCP) for a selection of diets based on high protein raw materials supplemented by nutritionally-improved straw and starch so that the test protein usually contributed more than 90% of total protein in the diet. This selection from a much wider range of feeds illustrates differences in UN, ADIN and estimated UDN. Considerable variation was observed within each feed type and these values should not be considered as definitive. They do show that the estimates of the true digestibility of UN are close to the fixed value of 0.85 assumed by ARC (1984) when ADIN content is low relative to total UN (e.g. fishmeal, protected soya, maize gluten). However, when ADIN is relatively high (e.g. extracted rice bran) or UN is very low (e.g. soya) the true digestibility of UN falls substantially below 0.85. Moreover this effect of increasing ADIN on the estimated true digestibility of UN is entirely consistent with the decline in the apparent digestibility of total dietary N indicated in *Table 11.4*.

MICROBIAL PROTEIN SYNTHESIS

In calculating the yield of microbial protein N the model assumes in accordance with

Table 11.4 APPARENT DIGESTIBILITY OF TOTAL N AND ESTIMATED TRUE DIGESTIBILITY OF UN IN DIETS IN WHICH OVER 90% OF TOTAL PROTEIN WAS OBTAINED FROM THE SINGLE SOURCES LISTED BELOW

Test feed	Composition of test protein (mg N/g dietary N)		Digestibility	
	UN	ADIN	Total N apparent	UN true
Soya	112	24	0.81	0.71
Protected soya	497	25	0.82	0.85
Fishmeal	573	26	0.80	0.86
Maize gluten 20	388	34	0.71	0.82
Maize gluten 60	856	20	0.82	0.88
Stimuflav	415	117	0.51	0.65
Grain dregs	719	193	0.61	0.66
Extracted rice bran	651	232	0.46	0.58

ARC (1984) a true digestibility of 0.85 and that microbial amino N = 0.8 total microbial N (*Figure 11.2*). The model differs however in its estimates of the requirements of the rumen microbes for degradable N and their capacity for microbial protein symthesis. ARC (1980) assumed a fixed relationship between ME intake and microbial protein yield and that all degradable N could be utilized with an efficiency of 1.0 unless present in excess. Thus, the RDP requirement could be calculated using Equation (11.8):

$$\text{RDN requirement} = 1.25 \, \text{g/MJ ME} \tag{11.8}$$

This relationship, which was largely derived from trials with wether sheep fed at levels close to maintenance, is considerably below values of 1.5–1.6 proposed in the German (Rohr *et al.*, 1986) and USA (NRC, 1985) systems which were based largely on work with dairy cattle fed at close to three times maintenance.

All models of biological systems are distortions of reality. However, the ARC (1980) assumptions behind Equation (11.8) have a number of inherent problems:

(1) They do not distinguish between fermentable and unfermentable energy.
(2) They do not account for increased microbial yield with increasing k, due to increased plane of nutrition (etc).
(3) They do not recognize that, despite the capacity of the ruminant to recycle urea to the rumen, nitrogenous compounds degraded to ammonia more rapidly than they can be incorporated into microbial protein cannot be utilized with an efficiency of 1.0 unless, at other times of the day, there is a relative deficiency of degradable N. In other words, in a diet that is balanced for QDN, SDN, QFE and SFE, the efficiency of utilization of QDN must be less than 1.0.

At this stage it is not possible to give a definitive value for the efficiency of utilization of QDN, or even assume it to be a constant. Pending better evidence it is suggested that a value of 0.8 be used, which is that proposed by ARC (1980) for non-protein N, since its fate in the rumen is almost identical to that of water-soluble N.

The amount of degradable N available for microbial protein synthesis is therefore determined as 0.8 QDN + SDN. Microbial protein is constrained, of course, both by the supply of degradable N and fermentable energy. Equation (11.9), which expresses the amount of microbial nitrogen (M_iN) synthesized/MJ ME, is the one currently in use to describe this relationship and is a form of Equation (11.4):

$$\frac{g \, M_i N}{ME} = \frac{1}{(0.36 + [0.019/(OMI - 0.416 MEI)/0.195 W^{1.05}])} \tag{11.9}$$

where OMI = organic matter intake (kg/day), MEI = ME intake (MJ/day), W = bodyweight (kg) and 0.019 is the ME requirement of the microbes (MJ ME/gM_iN.h). A full explanation of the derivation of this equation is beyond the scope of this review. Very briefly the following points are worthy of note.

(1) $1/0.36 = 2.77$ = maximum M_iN yield/MJ ME
(2) 0.019 = maintenance coefficient ME(MJ ME/gM_iN.h)
(3) $(OMI - 0.416 \, MEI)/0.195 W^{1.05}$ is an estimate of the rate of organic matter leaving the rumen/h (microbes plus undigested material). The expression predicts a decline in OM leaving the rumen with increasing metabolizability of

the diet. It also predicts a small increase in rumen volume as a proportion of bodyweight with increasing bodyweight, being based on the exponent $W^{1.05}$ obtained by Van Soest (1982).

As indicated earlier, the incorporation of a fixed value for the mass of OM in the rumen enables k to be related to plane of nutrition for feeds of different fermentability. *Figure 11.5* uses Equation (11.9) to estimate the range of values for M_iN/total ME intake (so as to correspond to the ARC (1980) Equation 11.8) that may occur for sheep and cattle fed diets within the range 10–12 MJ ME/kg DM and at planes of nutrition from 0 to 4 times maintenance. The values for sheep are higher than for cattle at any plane of nutrition because rumen volume is related to $W^{1.05}$ but maintenance requirement for ME to $W^{0.75}$, thus outflow rate is greater for the smaller animal at any plane of nutrition scaled with respect to maintenance. *Figure 11.5* predicts that for a sheep eating a diet with an ME concentration of 10 MJ/kg DM at the maintenance level of nutrition, microbial protein N yield is 1.15 g/MJ ME; for a dairy cow eating a ration having 11 MJ ME/kg DM at three times maintenance, microbial protein yield is 1.45 g/MJ ME. The figure for sheep at maintenance corresponds very closely to the constant value proposed by ARC (1980), that for cattle at three times maintenance corresponds closely to the USA (NRC, 1985) and German (Rohr *et al.*, 1986) estimates based on cattle data.

TESTING THE METABOLIZABLE PROTEIN SYSTEM

It is inherently more difficult to measure the supply of TAAN (or MP) to an animal than it is to measure ME_t since a large and variable proportion of TAAN is deaminated, catabolized and excreted and this excretion is variably distributed between urine and faeces. ARC (1980, 1984) assume that the capacity of the diet to supply TAAN as microbial N and UN can be determined by measuring the flow of non-ammonia nitrogen (NAN) at the abomasum or duodenum. Quite apart from the fact that these two sites do not give the same answers, this assumption is only valid if

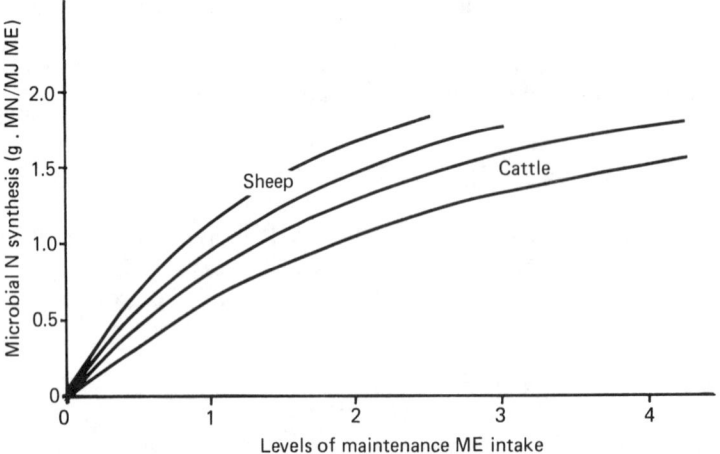

Figure 11.5 Predicted effect of plane of nutrition (ME relative to maintenance) on the energetic efficiency of microbial N synthesis in cattle and sheep (derived from Equation 11.9)

the contribution of endogenous protein to microbial synthesis in the rumen is small and constant.

Oldham (1987) has presented preliminary results of a large combined trial at the former AFRC Grassland Research Institute and National Institute for Research in Dairying, designed to test the ARC (1980) protein system for dairy cows. Control diets containing 120–140 g/kg crude protein were supplemented with highly degradable soyabean meal or undegradable fishmeal at two levels of incorporation. *Figure 11.6a* shows that there was no clear relationship between NAN flow at the duodenum and milk yield, largely because there was no correlation between observed and predicted NAN flow, i.e. the contribution of endogenous protein was large, variable and unpredictable. *Figure 11.6b* plots the milk yield of the cows on these trials simply as a function of the CP concentration (g/kg DM). Here there is a clear response to increasing CP concentrations up to 170 g/kg *irrespective of protein quality* with the possible exception of one group of cattle with high soya supplementation. This result

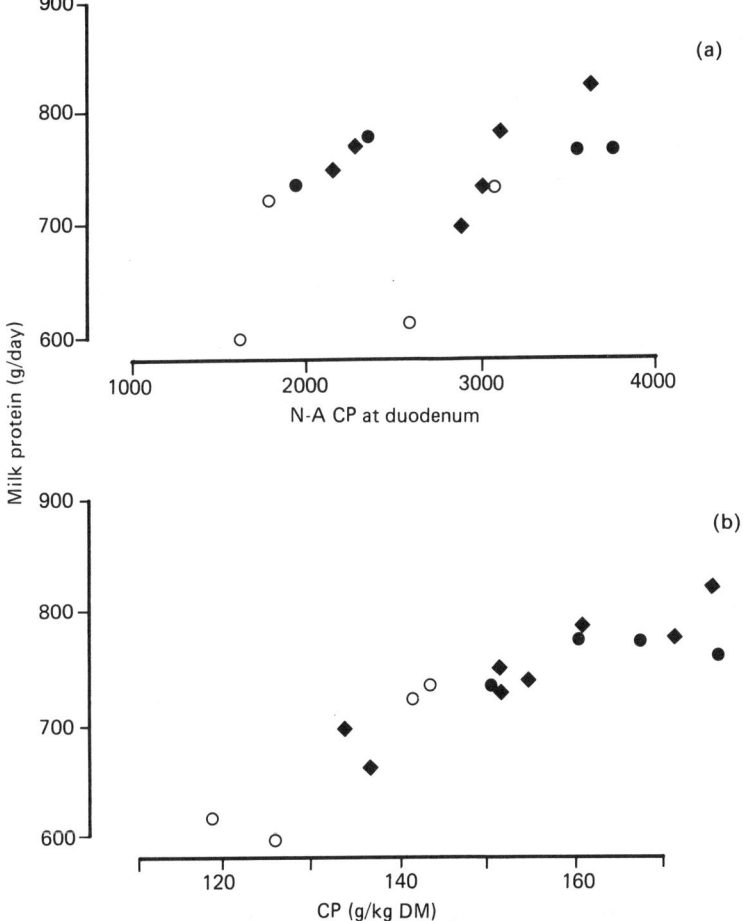

Figure 11.6 The relationship between yield of milk protein (*y*) and (a) non-ammonia crude protein (CP) flow at the duodenum, and (b) CP concentration in the diet (g/kg DM) ○ = control diets, ● = soyabean supplements, ◆ = fish meal (adapted from Oldham, 1987)

is completely inconsistent with the ARC (1980, 1984) models since they predict no response to increasing CP above 130 g/kg (*Table 11.2*). The fact that there was a response to increasing CP concentration and that it was largely independent of degradability strongly suggests that this was achieved by increased microbial protein synthesis since excess degradable protein can never serve as a source of TAAN, but excess UDN can be catabolized and recycled to the rumen as urea to support microbial protein synthesis. This implies that NAN flow at the duodenum cannot, on its own, be used as an indicator of TAAN input to the ruminant since it includes a large and uncertain element of output, namely endogenous N supply to the rumen microbes. This leads to a blinding glimpse of the obvious, namely that the best predictor of the protein supply is the amount the animal eats, properly defined with respect to other nutrients such as fermentable energy. This is what the proposed model attempts to do. However, since there is not yet a cost-effective way of measuring TAAN directly, there is no option but to test these assumptions in production trials, something that is required eventually in any system of modelling anyway.

The combined production trial described by Oldham (1987, *Figure 11.6a*) supports the contention that the main reason why the ARC (1984) appears to underestimate the protein requirements of dairy cattle (*Table 11.2*) is that it overestimates the capacity of feeds to supply degradable nitrogen and underestimates the capacity of micro-organisms to use it.

The discrepancy between prediction of MP yield from ARC (1984) and the proposed model is particularly great for silages since they contain relatively large amounts of QDN and a high ADIN/UDN ratio. This is illustrated in *Table 11.5* for four silages tested at the University of Bristol in which the prediction of TAAN is 0.68 to 0.81 that of ARC (1984). Waters, Dewhurst and Webster (1987) have briefly described a production trial with 33 dairy cows designed to compare ARC (1984) with MENTOR. This is summarized in *Table 11.6*. Concentrate rations were devised to meet TAAN requirement when fed in association with a silage (previously analysed). The ARC (1984) system called for a concentrate containing 100 g/kg CP at a degradability of 0.75, the MENTOR system called for 180 g/kg CP at a degradability of 0.65. Since these were so far apart it was decided to include an intermediate group fed equal amounts of each concentrate. The concentrations were isoenergetic

Table 11.5 CHEMICAL COMPOSITION (g N/kg DM) AND ESTIMATED YIELDS OF METABOLIZABLE PROTEIN FOR FOUR SILAGES AT A DRY-MATTER OUTFLOW RATE FROM THE RUMEN OF 0.08/h

	Silage			
	1	2	3	4
Total N (g/kg)	33.6	30.9	25.5	22.4
ADIN	3.0	3.7	3.4	4.3
QDN	18.1	21.9	13.8	14.1
SDN	8.3	3.0	4.8	1.4
UDN	3.9	2.1	3.2	2.3
Predicted metabolizable protein (g/kg DM)[a]				
ARC (1984)	152	138	116	101
MENTOR	124	102	89	68
MENTOR: ARC (1984)	0.81	0.74	0.77	0.68

[a] when constrained by 0.8QDN + SDN

Table 11.6 MEAN YIELDS OF MILK AND MILK PROTEIN IN COWS FED CONTROLLED AMOUNTS OF CONCENTRATE AT THREE PROTEIN CONCENTRATIONS PLUS *AD LIBITUM* GRASS SILAGE

	ARC	Intermediate	Mentor	SE
CP in concentrate (g/kg)	100	140	180	
Concentrate intake (kg/day)	9.5	9.8	9.8	
Milk yield (kg/day)				
period 1	24.8^a	26.4^b	26.9^b	0.37
period 2	24.1^a	26.1^{a,b}	27.9^b	0.77
Milk protein (kg/day)				
period 1	683^a	729^{a,b}	767^b	19.3
period 2	733^a	823^{a,b}	852^b	32.5

From Waters, Dewhurst and Webster (1987)
^{a,b}Values on the same line with different superscript letters are significantly different ($P < 0.05$) from each other

(13.5 MJ ME/kg DM). Overall concentrate consumption was similar for all groups (*Table 11.6*). There was a statistically significant increase in total milk yield and protein yield in response to both increments of CP, a result consistent with that described by Oldham in Chapter 10, consistent with the model's assumptions as to the capacity of feeds to support microbial protein synthesis, consistent with accepted feeding practice, but totally at variance with ARC (1984).

Discussion of the merits and shortcomings of 'MENTOR'

The results of attempts to test 'MENTOR' as a predictor of true metabolizable energy and metabolizable protein supply to ruminants give accurate values for ME_t of forages, usefully precise but illogical values for ME_t in compound feeds (Equation (11.5)) and sensible, but as yet insufficiently tested values for MP. At present therefore it is far from ideal. However the model is designed so that it can incorporate improvements to knowledge without the necessity to change its essential structure. In other words, the essential compartments and directions of flow illustrated in *Figure 11.2* need not change, only the numbers. Matters requiring the most urgent attention are:

(1) improved description of SFE,
(2) incorporation of associative effects between the fermentation of QFE and SFE,
(3) improved prediction of microbial protein synthesis,
(4) estimation of the efficiency of incorporation of QDN into microbial protein.

IMPROVED DESCRIPTION OF SFE

Prediction of SFE from (NDF-ADL) assumes that the rate and extent of fermentation of plant cell walls are determined entirely by the extent of lignification and is based on experimental evidence obtained for grasses (Smith, Goering and Gordon, 1972). The overestimation of ME_t in compound feeds can probably be attributed to an overestimation of SFE. The model may also underestimate ME_t for dicotelydonous green feeds (e.g. lucerne) which have a higher ratio of ADF to NDF than grasses of comparable digestibility (Van Soest, 1982, 1985).

The most obvious limitation of (NDF-ADL) as a predictor of SFE is that it does not distinguish between cellulose and hemicelluloses. *Table 11.7* lists some approximate measurements of cellulose (ADF-ADL), hemicellulose (NDF-ADF) and lignin (ADL) in a variety of raw materials. MAFF (1980) values for the digestibility of CF (where available) are, as expected, inversely related to the degree of lignification. However there are wide differences between feeds in the ratio of hemicellulose:lignin and in the ratio (hemicellulose + lignin): cellulose. The structure of the plant cell wall may be considered, simply, as a three-dimensional matrix of hemicellulose and lignin with strands of cellulose running through it. Provided that cellulose can be made available to the rumen microbes the rate and extent of its fermentation (thus its contribution to SFE) may prove to be reasonably invariant or predictable using the NCD technique (Alderman, 1985). However, different components of the hemicellulose matrix vary in their fermentability from very rapid to zero (Van Soest, 1985). When hemicellulose is analysed into its component monosaccharides (Chesson, 1986; English and Cummings, 1984) xylose usually contributes 60–80% to the total, followed at some distance, by arabinose and uronic acids with only traces of galactose, mannose and rhamnose (Ben-Ghedalia and Rubinstein, 1984). Measurements of the rates of fermentation of the main carbohydrate constituents of cell walls in a range of fibrous raw materials using PSF bags is currently being undertaken at the University of Bristol and this may permit the prediction of SFE from their monosaccharide composition. However this chemical approach may fail to achieve a proper description of the effect of the biological structure of the plant cell wall on the rate and extent of its degradation by rumen microbes (Akin, 1986), in which case, it may be necessary to have different prediction equations for SFE for different classes of feeds based either on feed chemistry or an *in vitro* estimate of digestibility of 'fibre' such as NCD.

Table 11.7 APPROXIMATE PROPORTIONS OF CELL WALL CARBOHYDRATES IN A SELECTION OF FEEDS

	Cell wall (g/kg DM)	Composition of cell wall			Digestibility of CF (%)[d]
		Cellulose	Hemicellulose	Lignin	
Ryegrass silage[a]	593	452	402	146	76
Lucerne silage[b]	584	458	267	275	42
White clover[b]	300	710	160	130	60
Oat straw[a]	828	420	400	180	54
Nutritionally improved straw[c]	740	555	320	125	—
Sugar beet pulp[b]	510	610	350	40	90
Brewers' grains[b]	460	390	480	130	48
Coffee grounds[b]	740	350	120	270	—
Kapok meal[c]	448	275	260	465	—
Rice bran[c]	340	185	640	175	—

Sources of data:
[a]Ben Ghedalia and Rubinstein (1984)
[b]Van Soest (1985)
[c]University of Bristol (unpublished data)
[d]MAFF (1980)

ASSOCIATIVE EFFECTS OF STARCH AND CELLULOSE

It is clearly recognized that increasing the concentration of starches or sugars in a ration for ruminants beyond a certain level inhibits the fermentation of cellulose (AFRC, 1987; Russell and Hino, 1985). The proposed model would accommodate this by assuming that QFE remained unchanged but introducing a lag phase into Equation (11.1) so as to reduce SFE for those diets where this inhibitory effect may be expected. Of course, it makes more sense to formulate diets to prevent this inhibition occurring except in special circumstances such as cereal beef systems.

NITROGEN DEGRADATION AND MICROBIAL PROTEIN SYNTHESIS

The efficiency of capture of degradable N by the rumen micro-organisms and the supply of microbial protein N relative to ME as VFA remain two of the greatest elements of uncertainty attached to this, or any other, system of feed characterization for ruminants since they have a major direct bearing on CP requirement and optimal degradability and indirect effects on fibre fermentation and dry matter intake. Once again, it is not possible at this stage to make definitive recommendations, merely to reaffirm that MENTOR can accommodate new and better particulars as they arise.

LEGISLATION

The AFRC technical committee on responses to nutrients (AFRC, 1987) has made a number of specific recommendations with respect to the characterization of feedstuffs for ruminants. The first is that the use of NFE and CF should be discontinued, the latter to be replaced by NDF according to the method of Wainman, Dewey and Boyne (1981) in which NDF is measured after pretreatment with α-amylase to remove starch. They do not recommend any specific technique for the determination of lignin but propose that 'efforts should be concentrated on the development of routine methods for the characterization of the phenolic contents of feeds'.

It is the responsibility of feed chemists to say which methods are best for purposes of legislation, but the suggestions of the AFRC technical committee that NDF and 'lignin' are far better descriptions of digestible and indigestible cell walls for ruminants than is the single measurement of CF is very valid. Currently there appears to be no pressing need to change any other element of the proximate analysis now used as the basis for the statutory description of ruminant feeds, particularly since most feeders and feed manufacturers will require a more complete feed categorization anyway.

Conclusions

An alternative approach to the categorization of feedstuffs for ruminants has been proposed which recognizes three distinct sources of true metabolizable energy and truly absorbed amino nitrogen or metabolizable protein (QFE and QDN, SFE and SDN, UDE and UDN). By the standards of the criteria for an ideal system outlined in the introduction it is possible to conclude that:

(1) 'MENTOR' is largely based on measurements of feed chemistry that are robust and available now to the feed compounder. Some inputs to the model depend on results obtained from incubating materials in PSF bags. Moreover the description of cell wall carbohydrates needs to be improved, especially for compound feeds.
(2) The model is deterministic and distinguishes properly between ruminant and post-ruminant digestion. The largest elements of uncertainty are attached to estimates of outflow rate of solids and microbes from the rumen, microbial protein yield and the extent to which these may be affected by factors other than feed chemistry.
(3) The model predicts truly absorbed metabolizable energy and metabolizable protein. It does not predict the supply of individual amino acids but it does distinguish different relative molar yields of the principal VFA from QFE, SFE and degradable N and so can be modified to predict the amounts and proportions of individual VFA produced by different feed fractions as recommended by AFRC (1987).
(4) 'MENTOR' is already demonstrably better than existing empirical systems based on simple chemical measurements as a predictor of ME_t in grasses. It is equally applicable, in theory, to the prediction of ME_t in clovers, lucerne and balanced compound feeds given improved description of cell-wall carbohydrate. The characterization of feeds in terms of metabolizable protein gives values that are unproven and considerably at variance with ARC (1984), but sensible and in reasonable accord with all other protein systems.

References

AGRICULTURAL RESEARCH COUNCIL (1965). *The Nutrient Requirements of Farm Livestock, No. 2. Ruminants.* Agricultural Research Council, London

AGRICULTURAL RESEARCH COUNCIL (1980). *The Nutrient Requirements of Farm Livestock, No. 2. Ruminants,* 2nd edn. Commonwealth Agriculture Bureaux, Slough

AGRICULTURAL RESEARCH COUNCIL (1984). *The Nutrient Requirements of Ruminant Livestock, Supplement No. 1.* Commonwealth Agriculture Bureaux, Slough

AGRICULTURE AND FOOD RESEARCH COUNCIL (1987). AFRC Technical Committee on Responses to Nutrients. Report No. 1. Characterisation of feedstuffs: Energy. *Nutritive Abstracts & Reviews,* **57,** 507–523

AKIN, D.E. (1986). Chemical and biological structure in plants as related to microbial degradation of forage cell walls. In *Control of Digestion and Metabolism in Ruminants,* pp. 139–157. Ed. Milligan, L.P., Grovum, W.L. and Dobson, A. Prentice-Hall, New Jersey

ALDERMAN, G. (1985). Prediction of the energy value of compound feeds. In *Recent Advances in Animal Nutrition—1985,* pp. 3–52. Ed. Haresign, W. and Cole, D.J.A. Butterworths, London

ALDERMAN, G. (1987). Comparison of rations calculated in the different systems. In *Protein Evaluation of Ruminant Feeds,* pp. 283–297. Ed. Alderman, G. and Jarrige, R. Commission of European Communities, Luxembourg

BALDWIN, R.L., KOONG, L.J. and ULYATT, M.J. (1977). A dynamic model of ruminant digestion for evaluation of factors affecting nutritive value. *Agricultural Systems,* **2,** 255–288

BARBER, W.P., ADAMSON, A.H. and ALTMAN, J.F.B. (1984). New methods of forage evaluation. In *Recent Advances in Animal Nutrition—1984*, pp. 161–176. Ed. Haresign, A. and Cole, D.J.A. Butterworths, London
BEEVER, D.E., BLACK, J.L. and FAICHNEY, G.J. (1981). Simulation of the effects of rumen function on the flow of nutrients from the stomach of sheep. 2. Assessment of computer predictions. *Agricultural Systems*, **6**, 221–241
BEN-GHEDALIA, D. and RUBENSTEIN, A. (1984). The digestion of monosaccharide residues of the cell walls of oat and vetch hays by rumen contents *in vitro*. *Journal of Science and Food Agriculture*, **35**, 1159–1164
BLACK, J.L., BEEVER, D.E., FAICHNEY, G.J., HOWARTH, B.R. and GRAHAM, M.MC. (1981). Simulation of the effects of rumen function on the flow of nutrients from the stomach of sheep. 1. Description of a computer programme. *Agricultural Systems*, **6**, 221–241
BLAXTER, K.L. (1962) *The Energy Metabolism of Ruminants*. Hutchinson, London
BLAXTER, K.L. (1986). An historical perspective of methods for assessing nutrient requirements. *Proceedings of the Nutrition Society*, **45**, 177–183
CHESSON, A. (1986). The evaluation of dietary fibre. In *Feedingstuffs Evaluation, Modern Aspects, Problems, Future Trends*, pp. 18–25. Ed. Livingstone, R.M. FEEDS Publ. No. 1. Aberdeen
CLANCY, M.I. and WILSON, R.K. (1966). A chemical method for predicting digestibility and intake of herbage. *Proceedings Xth International Grassland Congress, (Helsinki)*, pp. 445–453
DEWHURST, R.J., WEBSTER, A.J.F., WAINMAN, F.W. and DEWEY, P.J.S. (1986). Prediction of the true metabolizable energy concentration in forages for ruminants. *Animal Production*, **43**, 183–194
ENGLISH, H.H. and CUMMINGS, J.H. (1984). Simplified methods for the determination of total non-starch polysaccharides by gas-liquid chromatography of constituent sugars as alditol acetates. *Analyst*, **109**, 937–942
ERFE, J.D., SAVER, F.D. and MAHADEVAN, S. (1986). Energy metabolism in rumen microbes. In *Control of Digestion and Metabolism in Ruminants*, pp. 81–99. Ed. Milligan, L.P., Grovum, W.L. and Dobson, A. Prentice-Hall, New Jersey
GILL, M., THORNLEY, J.H.M., BLACK, J.L., OLDHAM, J.D. and BEEVER, D.E. (1984). Simulation of the metabolism of absorbed energy-yielding nutrients in young sheep. *British Journal of Nutrition*, **52**, 621–649
GOERING, H.K. and VAN SOEST, P.J. (1970). *Forage Fibre Analysis Agricultural Handbook, No. 379*. US Department of Agriculture, Washington DC
HENNEBERG, W. and STOHMANN, F. (1860). Beitrage zur Begrundung einer rationelle Futerung der Wiederkauer Braunschweig: C.S. Schweste
INSTITUT NATIONAL DE LA RECHERCHE AGRONOMIQUE (INRA, 1978). *Alimentation des Ruminants*. INRA Publ., Versailles
MCDONALD, T. (1981). A revised model for the estimation of protein degradability in the rumen. *Journal of Agricultural Science, Cambridge*, **96**, 251–252
MINISTRY OF AGRICULTURE, FISHERIES and FOOD (1975). Energy allowances and feeding systems for ruminants. *Technical Bulletin No. 33*. HMSO, London
MINISTRY OF AGRICULTURE, FISHERIES and FOOD (1980). ADAS Advisory Paper No. LGR 21. *Nutrient allowances and composition of feedingstuffs for ruminants*. MAFF, London
MURPHY, M.R., BALDWIN, R.L. and KOONG, L.J. (1982). Estimation of stoichiometric parameters for rumen fermentation of roughage and concentrate diets. *Journal of Animal Science*, **55**, 411–421

NATIONAL RESEARCH COUNCIL (1985). Ruminant nitrogen usage. US National Academy of Science, Washington DC

NEHRING, K. (1969). Investigations on the scientific basis for the use of net energy for fattening as a measure of feeding value. In *Energy Metabolism of Farm Animals*, pp. 5–20. Ed. Blaxter, K.L., Kielanowski, J. and Thorbek, G. Oriel Press, Newcastle-on-Tyne

OLDHAM, J.D. (1987). Testing and implementing the modern systems: UK. In *Protein Metabolism of Ruminant Feeds*, pp. 171–186. Ed. Alderman, G. and Jarrige, R. Luxembourg Commission of European Communities, Brussels

OESKOV, E.R. and MCDONALD, I. (1979). The estimation of protein degradability in the rumen from incubation measurements weighted according to rate of passage. *Journal of Agricultural Science, Cambridge*, **92,** 499–523

OWENS, F.N. (1982). *Protein Requirements for Cattle*. Oklahoma State University, Stillwater, Oklahoma

ROHR, K., LEBZIEN, P., SCHAFFT, H. and SCHULTZ, E. (1986). Prediction of the duodenal flow of non-ammonia nitrogen and amino acid nitrogen in cows. *Livestock Production Science*, **14,** 29–40

RUSSELL, J.B. and HINO, T. (1985). Regulation of lactate production in *Streptococcus bovis*; a spiralling effect that contributes to ruminal acidosis. *Journal of Dairy Science*, **68,** 1712–1721

SMITH, L.W., GOERING, H.K. and GORDON, C.H. (1972). Relationships of forage composition with rates of cell wall digestion and indigestibility of cell walls. *Journal of Dairy Science*, **55,** 1140–1147

STOUTHAMER, A.H. and BETTENHAUSEN, C. (1973). Utilisation of energy for growth and maintenance in continuous and batch cultures of microorganisms. *Biochemica et Biophysica*, **307,** 53–70

TAMMINGA, S. (1982). Energy-protein relationships in ruminant feeding: similarities and differences between rumen fermentation and postruminal utilisation. In *Protein Contribution of Feedstuffs for Ruminants*, pp. 2–17. Ed. Miller, E.L., Pike, I.H. and Van Es, A.J.H. Butterworths, London

TILLEY, J.M.A. and TERRY, R.A. (1963). A two-stage technique for the *in vitro* digestion of forage crops. *Journal of the British Grassland Society*, **18,** 104–111

VAN SOEST, P.J. (1982). *Nutritional Ecology of the Ruminant*. O. & B. Books, Cornvallis, Oregon

VAN SOEST, P.J. (1985). Definition of fibre in animal feeds. In *Recent Advances in Animal Nutrition—1985*, pp. 57–70. Ed. Haresign, W. and Cole, D.J.A. Butterworths, London

WAINMAN, F.W., DEWEY, P.J.S. and BOYNE, A.W. (1975). First report of the Feedingstuffs Evaluation Unit. Rowett Research Institute. Dept. of Agric. & Fisheries for Scotland, Aberdeen

WAINMAN, F.W., DEWEY, P.J.S. and BOYNE, A.W. (1978). Second report of the Feedingstuffs Evaluation Unit. Rowett Research Institute. Dept. of Agric. & Fisheries for Scotland, Aberdeen

WAINMAN, F.W., DEWEY, P.J.S. and BOYNE, A.W. (1981). Third report of the Feedingstuffs Evaluation Unit. Rowett Research Institute. Dept. of Agric. & Fisheries for Scotland, Aberdeen

WAINMAN, F.W., DEWEY, P.J.S. and BREWER, A.C. (1984). Fourth report of the Feedingstuffs Evaluation Unit. Rowett Research Institute. Dept. of Agric. & Fisheries for Scotland, Aberdeen

WALDO, D.R., SMITH, L.W. and COX, E.L. (1972). Model of cellulose disappearance from the rumen. *Journal of Dairy Science*, **55,** 125–129

WATERS, C.J., DEWHURST, R.J. and WEBSTER, A.J.F. (1987). Comparison of systems for estimating protein allowances for dairy cows. *Animal Production*, **44**, 475

WEBSTER, A.J.F. (1987a). Metabolizable protein—the UK approach. In *Protein Evaluation of Ruminant Feeds*, pp. 47–53. Ed. Alderman. G. and Jarrige, R. Commission of European Communities, Luxembourg

WEBSTER, A.J.F. (1987b). *Understanding the Dairy Cow*, pp. 42–47. Blackwell Scientific Publications, London

WEBSTER, A.S.F., KITCHERSIDE, M.A., KIERBY, J. and HALL, P.A. (1984). Evaluation of protein feeds for dairy cows. *Animal Production*, **38**, 548

WEBSTER, A.J.F., SIMMONS, I.P. and KITCHERSIDE, M.A. (1982). Forage protein and the performance and health of the dairy cow. In *Forage Protein in Ruminant Animal Production*, pp. 89–98. Occ. Publ. Br. Soc. Anim. Prod. No. 6

WILSON, P.N. and STRACHAN, P.J. (1980). The contribution of undegraded protein requirements of dairy cows. In *Recent Advances in Animal Nutrition—1980*. Ed. Haresign, W. Butterworths, London

ZINN, R.A. and OWENS, F.N. (1982). Predicting net uptake of non-ammonia N from the small intestine. In *Protein Requirements for Cattle*, pp. 133–140. Ed. Owens, F.N. Oklahoma State University, Stillwater, Oklahoma

V

Nutrition of Alternative Species

12

NUTRIENT REQUIREMENTS OF GAMEBIRDS

J. V. BEER
Game Conservancy Ltd, Fordingbridge, Hampshire, UK

Introduction

The gamebird industry in Britain is substantial and an estimate of the total number of birds hand-reared each summer is 24 million birds (Tapper, personal communication). Four galliform species are involved, the ring-neck or common pheasant, *Phasianus colchicus* which occurs as a variety of races, the red-legged partridge, *Alectoris rufa*, the chukar partridge, *A. chukar* and hybrids, and the grey partridge, *Perdix perdix*. It is estimated that about 86% of those reared are *Phasianus*, 11% *Alectoris* and 3% *Perdix* and the total numbers reared and released has been increasing since the early 1960s (Tapper and Bond, 1987). Indeed there is likely to be an increasing interest in rearing with the current emphasis to take land out of milk or cereal production.

Gamebirds are reared and released by estates into the wild to contribute to winter stocks but rarely direct for the table. During a designated open season, estates and syndicates arrange shooting days, usually on a weekly basis, throughout much of lowland Britain, especially the south east. The harvested birds therefore derive from stock either reared in the wild or hand-reared and progressively adapted to the wild from six to seven weeks of age onwards. Some estates rely entirely on wild-reared birds. A great deal of time, effort and money is spent on establishing a suitable habitat for birds and the character of the British countryside owes much to the production and maintenance of game. Other wildlife and plants also benefit.

Many of the hand-reared birds are produced in quantity by specialized game farms, some of which belong to the Game Farmers Association. Often estates will sell part of their production to offset costs of the shoot. Trading is carried out in a variety of ways involving eggs, day-old chicks, six–eight week old poults, as well as breeders and post-laying-season birds. Custom hatching is fairly common particularly being used by estates who do not want to establish a hatchery, yet who wish to keep control of rearing.

Rearing programmes

Details of many of the possible rearing programmes are covered in a series of Game Conservancy advisory guides (Anon, 1983a, 1983b, 1983c, 1986a, 1986b). A typical pheasant rearing programme covering all stages includes collecting free-living adults (mostly first year birds) from December to February followed by penning for egg production at the end of February or in early March. A breeders diet is fed from this time onwards. The first eggs are laid at the end of March or early April and maximum production is reached at the end of the month or in early May. Peak laying continues for about six weeks declining to zero by the end of July. Few estates, however, retain birds this long and most will return them back to the wild in June where they may lay a further clutch to hatch and rear themselves. A hen can produce about five–six eggs/week at maximum and 30–50 eggs, each weighing about 33 g, in a season. Artificial incubation involves anything from 6 to 12 or 13 weekly settings or two to three monthly settings depending on the production required. Incubation requires 24/25 days.

Day-old chicks are reared in various units, often on grass, ranging in size from 100 chicks to many thousands, the latter usually in 500 chick flocks sectioned off in buildings perhaps adapted from unused barns, etc. Heat is required up to about four weeks when the poults are hardened off. At six to eight weeks of age they are ready to be adapted to the wild in a release pen.

During rearing a starter crumb is fed followed by a rearer crumb or small pellet. Poults on grass supplement their diet to a significant degree with natural food and a small amount of kibbled wheat might be added after a few weeks. In the release pen poults further supplement their diet with other greenfood and seeds. Here they will be changed onto a growers diet plus an increasing proportion of cereals especially wheat, and after a few weeks a poult pellet is often introduced. A short time after release the birds learn to fly in and out of the pen, eventually ranging in the neighbouring fields and spinneys without returning to the pen. By the time they are well feathered with an adult plumage (four months) they will be eating natural foods supplemented with cereals. However, it is now commonplace to provide a small amount of a maintenance pellet during the winter, partly to boost trace nutrients as well as to discourage the birds from wandering off the estate. Partridges are reared and released in a similar manner but there are differences in detail. For instance, release is done in much smaller groups.

Game-farm stock is reared in a similar manner but is more likely to be overwintered in large fixed pens; none are released except where a game farmer may be running a shoot for a client.

Gamebird nutrition

Gamebirds are omnivorous and in the wild take a wide range of vegetable and animal foods depending on availability (Dalke, 1937). Animal foods, generally insects, are very important during the first two weeks of life (Hill, 1985). Early attempts at hand-rearing often involved a chicken feed supplemented with such extras as boiled eggs, rabbit, ant eggs, fish meal or milk powder and even custard in Victorian days (Walsingham and Payne Gallwey, 1889). Post-war complete commercial game feeds differed enormously, the protein content varying from 12 to 25% (Anon, 1954).

A survey of available information shows that while some research has been published in detail, other work has been published only as dietary ingredients. In America, Ewing (1963) reviewed the nutrition of gamebirds and Scott (1978) covered 25 years of research into gamebird nutrition carried out by his group. Streib, Streib and Fletcher (1973) and Summers and Leeson (1985) in Canada listed diets and feeding regimes while Leclercq et al. (1987) listed the components of diets based on work in France. Apart from the feed components listed by Burdett, Woodward and Wenham (1978) there are few British publications which include gamebirds. The Agricultural Research Council (1975) published data for poultry and gamebirds and a conference on 'Nutrient Requirements of Poultry and Nutritional Research' produced just one comment, on gamebird vitamin requirements (Whitehead, 1986). A list of vitamin requirements is published in the commercial literature (Anon, 1986c).

PROTEIN

It is generally thought that gamebirds are nearer to the turkey than the chicken in their overall nutritional requirements. *Table 12.1* gives a range of crude protein content and dietary energy levels of pheasant and partridge diets. Starter crumbs can be as low as 22–23% crude protein when metabolizable energy (ME) is also low but the consensus is to use 28–30% protein with energy levels near to 12.55 MJ ME/kg. This is fed for two to four weeks followed by a rearer diet with less protein (20–25%) and a lower energy level (about 11.72–12.13 MJ ME/kg) until about five to ten weeks. A growers diet is next fed for several weeks with a gradually increasing proportion of wheat; the protein is normally below 20%. This can be fed as early as six weeks until feathering is virtually complete at four months. A poult pellet with around 15% protein is often fed once the birds are well established in the release pen. Maintenance or winter pellets with 10–15% protein are often fed periodically during the winter. A breeder diet does not need a high protein level and usually has about 20% although some workers suggest diets with crude protein as low as 15.1–12.5%.

AMINO ACIDS

The requirements of gamebirds for amino acids tend to be high (*Table 12.2*) but Leclercq et al. (1987) quoted lower figures for partridge than for pheasant. Since reared gamebirds released to the wild must be able to fly and survive, their plumage must be well developed at an early stage for good thermal insulation, water repellency and flying (Scott, 1978; Deschutter and Leeson, 1986). In addition, feather pecking is often a problem and sufficient protein/amino acids must be available for both adequate feather production and replacement—in particular cystine and methionine must be high. Scott, Holm and Reynolds (1963) reported that 26.5% protein diets should be supplemented with 0.1% methionine.

FATS AND FIBRE

The amount of fat present in feeds ranges from 2 to 3.9% during rearing (Summers and Leeson, 1985). Essential fatty acids range from 0.6 to 1.2%. Crude fibre ranges from 3.0% (Summers and Leeson, 1985) to as high as 9% in winter feeds (Wöhlbier, 1974).

Table 12.1 CRUDE PROTEIN (%) AND METABOLIZABLE ENERGY (MJ/kg) IN PHEASANT AND PARTRIDGE DIETS

	Pheasant										Partridge
Diet	1	2	3	4	5	6	7	8	9	10	11
Starter											
% Protein	21–30	27.25	27	22–25	23.4	30	26.5–30.0	23.4	30	29.1–29.4	23.1–28.7	17.6–20.4
ME				9.62–10.46	11.38	11.72	12.55	11.38	11.72	11.53–11.89	10.46–12.97	10.88–12.55
Rearer											
% Protein				18–22			24			23.3–25.5		14.0–16.0
ME				9.62–10.46			12.55			12.00–13.36		10.46–12.13
Grower											
% Protein			26		19.8	24	20	19.8	24	18.5–18.7	14.8–17.2	
ME					11.13	11.97		11.13	12.17	12.43–12.66	10.46–12.12	
Poult											
% Protein											13.0–15.0	
ME											10.46–12.13	
Winter											
% Protein				13–15			12					
ME				12.13–13.39								
Breeder											
% Protein			20		15.1	19	15.5–20.0	15.1	19	16.9–17.0	12.5–14.5	14.7–17.0
ME					10.75	11.51		10.75		11.59–11.84	10.46–12.13	10.88–12.55

Sources:
1 Norris et al. (1936)   2 Scoglund (1940)   3 Stanz (1952)   4 Wöhlbier (1974)
5 Woodard et al. (1977)   6 Burdett et al. (1978)   7 Scott (1978)   8 Woodard et al. (1978)
9 Jee and Wilson (1981)   10 Summers and Leeson (1985)   11 Leclercq et al. (1987)

Table 12.2 AMINO ACID CONTENT OF PHEASANT DIETS (%)

Amino acid	Starter			Rearer			Grower		Poult	Breeder		
	1	2	3	1	2	3	1	2	1	1	2	3
Lysine	1.4	1.75	1.75	1.3		0.8	0.98	0.9	0.7	0.72	0.9	0.8
Methionine	0.5	0.8	0.8	0.5		0.6		0.4	0.27	0.31	0.4	0.4
Methionine + cystine	1.0	1.1		0.5			0.4	0.6	0.54	0.55	0.68	
Tryptophan	0.22	0.42		0.32			0.26	0.15	0.13	0.15	0.27	
Threonine	0.85	1.1		0.9			0.75	0.5	0.41	0.48	0.7	

Sources:
1 Leclercq *et al.* (1987)
2 Summers and Leeson (1985)
3 Burdett *et al.* (1978)

ENERGY

The range of energy levels quoted in *Table 12.1* covers 10.46–12.97 MJ ME/kg but the upper part of the range is preferred since high protein diets may be used (Leclercq *et al.*, 1987).

MINERALS

The calcium level needed by the growing bird is 1% and available phosphorus 0.49% (Scott, 1978; Reynnells and Flegal, 1979). The breeder requires 2.8 and 0.34% respectively. The requirements for manganese and zinc are important and the totals should be 95 ppm and 62 ppm respectively (Scott, 1978). Salt should be a maximum of 0.35–0.40% and sodium a minimum of 0.12% (Burdett, Woodward and Wenham, 1978).

VITAMINS

Vitamin requirements for gamebirds tend to be higher than for poultry and are listed by Scott (1978), Summers and Leeson (1985), Burdett, Woodward and Wenham (1978), Leclercq *et al.* (1987) and Anon. (1986c). Where mixes are not specifically available for gamebirds turkey mixes have often been suggested. Because gamebirds are reared and maintained under a variety of conditions it is prudent to use the high levels. For example, the absolute requirement for Vitamin A under ideal conditions for pheasant chicks is much lower than the normal amounts provided in commercial feeds (Scott, 1978).

Scott (1978) discusses the importance of vitamins A, D, K, B_2, niacin and choline for gamebird growth and hatchability. There are indications that high levels of biotin should be used where wheat is present in large amounts (Anon, 1986c). Vitamin C levels may be one factor associated with the fracture of long bones in two and a half week old grey partridge chicks reared indoors on shavings (Beer and Jenkinson, 1982). This condition has not been found in grey partridge reared out-of-doors nor in the pheasant or redleg, however reared.

Growth rates, feed consumption and feed efficiency

It is not possible to give precise body weights because of the varied genetic makeup of gamebirds and conditions of rearing, but a guide for young mixed sexes and adult cocks and hens is given in *Table 12.3* (Beer, 1988). The cock bird is significantly heavier than the hen.

The quantity of crumbs, pellets and grain consumed by hand-reared pheasants is given in *Table 12.4*. The amounts consumed by the red-legged partridge is about one-half to one-third, and the grey partridge one-third to one-quarter of the pheasant. Summers and Leeson (1985) quote feed conversion ratios for pheasants reared in Canada which ranged from 1.71 at two weeks to 4.14 at 18 weeks for cocks, and 1.71 to 5.40 for hens. No figures are available for partridges.

Table 12.3 GUIDE TO MEAN WEIGHTS OF THREE GAMEBIRD SPECIES

Age	Pheasant	Weight (g) Red-legged partridge	Grey partridge
Day old	20	12	7
1 week	50	17	11
2 weeks	80	30	18
3 weeks	120	45	28
4 weeks	200	75	45
5 weeks	300	100	62
6 weeks	380	140	86
7 weeks	450	190	115
8 weeks	550	230	140
9 weeks	650	280	170
10 weeks	750	330	210
12 weeks	900	400	250
17 weeks	1000	490	320
Adult male	1300	540	400
Adult female	1050	460	370

Beer (1988)

Table 12.4 CONSUMPTION OF CRUMBS, PELLETS AND GRAIN BY 100 PHEASANTS

Age (weeks)	1	3	5	7	9	11	13	15	17	19	Breeders
Weekly consumption (kg)	7	18	28	34	41	47	48	45	44	42	56
Total consumption (kg)	7	38	88	153	232	323	422	515	604	690	975 (March-June)

Discussion

The amount of published work on gamebird nutrition is much less than for poultry and since the genetic variation is so large the values for dietary components cannot be exact.

There is increasing evidence that hand-reared pheasants do not survive in the wild as well as wild-reared birds (Hill and Robertson, 1986). Scott (1978) noted that the ability of young pheasants to withstand the stress of cold, drenching rain and resistance to general stress improved by increasing protein from 28 to 34%. Good feathering is needed to combat heat loss but a poorly developed or damaged plumage in hand-reared gamebirds is not uncommon. Deschutter and Leeson (1986) in their review of growth and development of feathers in poultry consider that the levels of methionine and cystine must relate to the growth and rate of production of feathers and their replacement and not just meat. Woodard, Vohra and Snyder (1977) found that feathering was better when the protein level was at least 25% for the first five weeks. Zinc is important in feathering and Scott (1978) indicated that cereal diets for pheasants need to be supplemented to avoid poor development. If the mycotoxin T-2 is present in chick diets it depresses the production of feathers (Wyatt, Hamilton and Burmeister, 1975). The reduction in costs when using lower protein feeds is offset by higher mortality (Woodward, Vohra and Snyder, 1977).

Since released birds undergo various environmental stresses and may not feed adequately despite provision of compound feeds and cereals alongside the natural foodstuffs growing in the pen, losses from these causes and predation may be significant. Thomas (1986) suggested that shortly before release, the energy content of the feeds should be increased to ensure extra internal energy supplies while the birds are changing to the new regime in the wild. Hand-reared red grouse (*Lagopus lagopus scoticus*) are not suitable for release because they show a less well developed gut than a truly wild bird, a feature associated with the feeding of compounded easily digestible feeds, rather than the high fibre, lower digestible heather shoots taken by the wild birds (Moss and Hanssen, 1980). Thomas (1986) fed pheasant chicks an experimental diet containing inert fibrous filler. They grew more slowly than the controls but did eventually reach a normal size and again had larger digestive systems. However, in one small experiment involving the release of more than 300 experimental and control pheasant poults, mortality within two weeks from predation and starvation was high in both groups. He considers that foraging and predator avoidance behaviour should be studied as well as early nutrition and gut development.

The modern tendency to rear intensively various gamebirds as rapidly and to as large a size as possible for release is not likely to contribute to optimum survival. What is needed is a vigorous, lean, hardy bird with a well developed plumage, whose gut is able quickly to make maximum use of natural foods when released to allow the bird to survive stressful situations and to fly well. Feeds should be compounded to help with this aim and also to take account of species differences where relevant.

Acknowledgements

I am most grateful for the help provided by Criddle Peters Feeds Ltd, Heygate and Sons Ltd, Pauls Agriculture Ltd, Roche Products Ltd, Sportsman Game Feeds and Spratt's Game Foods. Also I thank Drs S. Tapper and P. Robertson who commented on the drafts.

References

AGRICULTURAL RESEARCH COUNCIL (1975). *Nutrient Requirement of Farm Livestock, No. 1. Poultry*. Agricultural Research Council, London

ANON (1954). *Annual Report ICI Game Research Station*. Fordingbridge, Hampshire

ANON (1983a). *Pheasant Rearing and Releasing*. Game Conservancy, Fordingbridge, Hampshire

ANON (1983b). *Egg Production and Incubation*. Game Conservancy, Fordingbridge, Hampshire

ANON (1983c). *Red-legged Partridges*. Game Conservancy, Fordingbridge, Hampshire

ANON (1986a). *The Grey Partridge*. Game Conservancy, Fordingbridge, Hampshire

ANON (1986b). *Game in Winter, Feeding and Management*. Game Conservancy, Fordingbridge, Hampshire

ANON (1986c). Roche Vitec 2. Roche, Welwyn Garden City, Hertfordshire

BEER, J.V. (1986). *Annual Review The Game Conservancy*, **17**, 140–143

BEER, J.V. (1988). *Diseases of Gamebirds and Wildfowl*. Game Conservancy, Fordingbridge, Hampshire (in press)

BEER, J.V. and JENKINSON, G. (1982). *Annual Review The Game Conservancy*, **13**, 112–115
BURDETT, B., WOODWARD, P. and WENHAM, T. (1978). *Poultry World*, **131** (43), 17–38
DALKE, P.L. (1937). *Ecology*, **18** (2), 199–213
DESCHUTTER, A. and LEESON, S. (1986). *World Poultry Science Journal*, **42**, 259–267
EWING, W.R. (1963). *Poultry Nutrition*, 5th Edition. The Pay Ewing Company, Pasadena, California
HANSSEN, I., GRAV, H.J., STEEN, J.B. and LYSNES, H. (1979). *Journal of Nutrition*, **109**, 2260–2276
HILL, D. (1985). *Annual Review The Game Conservancy*, **16**, 41–46
HILL, D. and ROBERTSON, P. (1986). *Annual Review The Game Conservancy*, **17**, 76–84
JEE, D. and WILSON, S. (1981). *Poultry World*, **133** (47), 11–30
LECLERCQ, B., BLUM, J.C., SAUVEUR, B. and STEVENS, P. (1987). In *Feeding of Non-Ruminant Livestock*, pp.116–119. Ed. Wiseman, J. Butterworths, London
MOSS, R. and HANSSEN, I. (1980). *Nutrition Abstracts and Reviews—Series B*, **50**, 555–567
NORRIS, L.C., ELMORE, L.J., RINGROSE, R.C. and BUMP, G. (1936). *Poultry Science*, **15**, 454–459
REYNNELLS, R.D. and FLEGAL, C.J. (1979). *Poultry Science*, **58**, 1097–1098
SCOTT, M.L. (1978). *World Pheasant Association Journal*, **3**, 31–45
SCOTT, M.L., HOLM, E.R. and REYNOLDS, R.E. (1963). *Poultry Science*, **42**, 676
SKOGLUND, W.C. (1940). *Bulletin 389*. Pennsylvania State College, School of Agriculture and Experimental Station, Pennsylvania
STANZ, H.E. (1952). *Technical Wildlife Bulletin No. 3*. Wisconsin Conservation Department, Madison
STREIB, A., STREIB, D. and FLETCHER, D.A. (1973). *Pheasants*, Pub. 1514, Canada Department of Agriculture, Ottawa
SUMMERS, J.D. and LEESON, S. (1985). *Poultry Nutrition Handbook*. University of Guelph, Ontario
TAPPER, S. and BOND, P. (1987). *Annual Review The Game Conservancy*, **18**, 167–173
THOMAS, V.G. (1986). *World Pheasant Association Journal*, **11**, 67–75
WALSINGHAM and PAYNE GALLWEY, R. (1889). *Shooting*, 3rd Edition, pp. 246–250. Longmans Green, London
WHITEHEAD, C.C. (1986). In *Nutrient Requirements of Poultry and Nutritional Research*, pp. 173–189. Eds. Fisher, C. and Boorman, K.N. Butterworths, London
WÖHLBIER, W. (1974). *The Supplementary Feeding of Game Animals and Fowl*. Roche, Basle
WOODARD, A.E., ERNST, R.A., VOHRA, P., NELSON, L. and PRICE, F.C. (1978). *Raising Game Birds*, Leaflet 21046. University of California, Berkeley
WOODARD, A.E., VOHRA, P. and SNYDER, R.L. (1977). *Poultry Science*, **56**, 1492–1500
WYATT, R.D., HAMILTON, P.B. and BURMEISTER, H.R. (1975). *Poultry Science*, **54**, 1042–1045

13

NUTRITION OF THE LEISURE HORSE

D. L. FRAPE
British Horse Feeds Ltd, Rugby, Warks, UK

Defining the task

Important response characteristics of the leisure horse include a long working life, freedom from debilitating metabolic and transmissible diseases, good temperament and amenability, willingness to be exercised, absence of premature fatigue, adequate and sustainable speed, rapid recovery from hard exercise, absence of excessive sweating and absence of vices. Each of these is influenced by diet in varying degrees. There is nevertheless no good evidence as to what is the optimum diet for the achievement of any one of them. Nor is it possible to ascribe with any precision at all the contribution of trends in diet and feed management practices to secular changes in performance. By contrast the role of diet in a strictly limited group of clinically observable debilitating conditions is widely appreciated. These include forms of respiratory hypersensitivity (McPherson *et al.*, 1979a), rickets and related bony abnormalities, enterotoxaemia (Carroll *et al.*, 1987) and endotoxaemia with associated laminitis (Garner *et al.*, 1978). Evidence of requirements for nutrients is derived from balance studies, in a few cases from enzyme measurements from qualitative and gross responses and in many cases by deductions from evidence in other domestic species.

An adequate description of the dietary needs of the leisure horse may be predicated only after an acceptable and sufficiently precise definition of that horse, and its dietary nutrient requirements, have been established. A quantitative assessment of those requirements may be based on both a deductive process and by reference to empirical evidence; and they have meaning only in relation to specified objectives. For the purpose of the discussion the horse is defined as an adult non-growing animal that is worked three to four days per week in the summer and one to two days per week in the winter, with the daily exercise amounting to a maximum of $1\frac{1}{2}$–2 h. On occasion as much as 8 h of work in a day may be required but this would be followed by several days with very little. Fitness will not be discussed but it should be self-evident that 8 h work would not be an inevitable outcome of the provision of the feed energy resource.

Many horse owners would wish that nutrient requirements be addressed in terms of units/kg of feed. This presents something of a dilemma. There is inadequate space available here to provide a convincing argument that for several important nutrients minimum requirements only have meaning where they are given as amounts/day for a horse of a given weight. Energy expenditure can vary as much as three to three-and-a-half-fold daily and one might speculate that, at the very least, this would influence the daily requirement of nutrients the metabolism of which is closely allied to energy utilization. That argument is partly abrogated by the facts that it may take two to three days to refurbish the energy reserves expended in one day and the requirements for several nutrients of considerable economic worth seem to be quite unrelated to energy expenditure. These include calcium, retinol and lysine.

There is another problem of a quite different dimension and this is that compounded feedstuffs very rarely represent the complete diet. Many horses in addition receive oats and bran, or similar, and nearly all receive a forage providing bulk and long fibre.

The current situation as viewed from the perspective of the feeder is that he, or she, is confronted by a welter of commercial products, both as compounds and as supplements. There is clearly some need for classification of products by those manufacturers, who do not do so, in a fashion similar to that adopted for farm animals. There is also no obvious means by which the feeder may distinguish between reliable products and others, and between worthwhile supplements and those of doubtful composition, but that offer great expectations. Presumably on the assumption by the feeder that there is safety in numbers, or greater satisfaction for the horse in variety, alarming mixtures are frequently formed, without good reason, creating diets of abnormal chemical composition. Education must lead to the posing of more critical questions than hitherto, which in turn should spawn a more rational approach by all concerned.

The solution

FEED INTAKE

The interests of the client and the horse are probably best served by the compounder formulating products of clearly defined nutrient composition. The problems outlined above are partly resolved by the formulation of products of differing energy densities and recommending daily rates of feeding of each product for various amounts of exercise by horses and ponies of different sizes. This assumes that the body weight is known, or can be estimated. Various systems have been proposed for predicting body weight—one of the simplest, but giving inaccurate estimates that are biased for many breeds, assumes it is directly related to wither's height. Recently Carroll and Huntington (1988) derived a general relationship between body weight, kg (W) and two linear measurements from 372 horses of various breeds in Australia and this is presented in Equation (13.1).

$$W = g^2 \times l \times 11\,877^{-1} \tag{13.1}$$

where g and l respectively are heart girth and length from the point of the shoulder to the tuber ischii in cm. This relationship also fits well for groups of English Thoroughbreds and Welsh ponies.

Recommendations of daily feeding rates of products of known energy density can be only a guide as undoubtedly individual animals differ in the extent to which feed energy is diverted to fat deposition and presumably the extent to which that fat is called on as a major source of fuel. In other words the amount of daily exercise individual horses are willing to undertake is not the same. Furthermore, the appetite of individuals varies remarkably and it is desirable to satisfy that hunger, preferably

Table 13.1 SPECIES COMPARISON OF STOMACH AND CAECUM + COLON CAPACITY IN RELATION TO BODY WEIGHT ($W^{0.75}$)

		Gut capacity ($1 \times 10^{-3}/kg^{0.75}$)		
	Liveweight (kg)	Stomach	Caecum + colon	Stomach + caecum + colon
Horse	545	74.5	595[a]	669.5
Pig	73	71.3	358	429.3
Sheep	36	772	191	963
Pig	9	61.6	135	196.6
Rabbit	2.36	30.5	104	134.5
Guinea pig	0.58	71.1	75.4	146.5
Rat	0.30	16.9	36.3	53.2
Hamster	0.222	5.4	37.1	42.5
Gerbil	0.068	16.3	45.8[b]	62.1
Mouse	0.030	20.8	34.7	55.5

(After Frape, 1984)
[a] Excluding dorsal colon
[b] Caecum only

by additional bulky feeds where the appetite is considerable. Bulkiness may effectively limit dry matter intake even where the bulky feed is highly digestible as occurs with hydroponic grass. Cuddeford (1987, personal communication) found that hydroponic grass possesses a DE of 15 MJ/kg DM yet horses given 4 kg hay would voluntarily consume forage (grass + hay on an 88% DM basis) at a daily rate equivalent to only 1.7–1.9% of body weight (up to 60 kg hydroponic grass daily). This low level of consumption may reflect minimal exercise, and voluntary restriction of consumption is deserving of approbation in the husbandry of leisure horses. *Ad libitum* consumption of the hydroponic grass in mat form introduces wastage of material according to Cuddeford. Although wastage is overcome by shredding, additional labour is incurred. For the average pleasure horse, voluntary food intake increases with work rate within the approximate bounds of 0–2 h work of moderate intensity/day (Orton, Hume and Leng, 1985a,b).

Capacity of herbivorous and omnivorous animals for feedstuffs and the relative efficiency of utilization of their structural carbohydrates are related to the volume of the gastrointestinal tract in relation to body size (*Table 13.1*), the anatomical position of any region primarily devoted to microbial fermentation, the rate of passage of feed residues, the species distribution of microbial flora and physiological mechanisms that influence the composition of their media (Frape, 1984). The horse is generally less efficient than are either cattle or sheep in the utilization of dietary fibre (*Table 13.2*) and hence most forages are accorded digestible energy values 15–20% lower than those for ruminants (Hintz, 1969; Hintz, Schryver and Stevens, 1978).

ENERGY NEEDS

Although energy may not be the 'hook' upon which to hang the dietary requirements for most nutrients not degraded by horses as energy sources, energy is the major dietary cost and therefore deserves precedence in discussion over other entities. The digestible energy requirements of horses of three weights for maintenance and work of various categories are depicted in *Figures 13.1* and *13.2* and given in *Table 13.3*. It

Table 13.2 APPARENT FIBRE DIGESTIBILITY (%) OF A RANGE OF ROUGHAGES IN CATTLE AND HORSES

	Digestibility (%)	
	Crude fibre	Cellulose
	Ochard grass hay	
Steers	53.4	64.3
Geldings	43.1	52.1
	Dehydrated timothy	
Steers	61.5	67.2
Geldings	43.9	48.3
	Dehydrated brome grass	
Steers	59.9	63.7
Geldings	34.5	37.8
	Oat straw	
Heifers	59	—
Horses	51	—
	Wheat straw	
Heifers	58	—
Horses	59	—

(After Hintz, 1969; Noot and Gilbreath, 1970)

might be assumed that the oxygen consumption above maintenance for travel on level ground at any speed for a preset distance is an approximate constant. However, this is not so, the oxygen consumption, and therefore energy utilization, rises at an increasing rate as speed increases (*Figure 13.2*) (Pagan and Hintz, 1986b). This information is used together with that of Pagan and Hintz (1986a) in calculating the digestible energy requirements per day of horses of three live weights subjected to 4, or 14 h of work/week, or none at all, at three average speeds (*Figure 13.2*). The division of the digestible energy requirements by the assumed daily appetite for energy containing feed yields the necessary energy density of that feed to achieve energy balance. The appetite here is assumed to be that associated with twice daily feeding. Certainly many professionally maintained Thoroughbreds weighing approximately 500 kg may consume 15 kg of air dried feed daily when receiving four meals/day and nearly *ad libitum* diurnal access to concentrates.

Feeding behaviour and appetite for a meal depend upon gastrointestinal (Ralston and Baile, 1982b; Ralston and Baile, 1983), large intestinal (Ralston, Freeman and Baile, 1983) and intermediary (Ralston and Baile, 1982a) metabolic events. 'Capacity' for feed may be considered synonymous with appetite, or voluntary intake, for the purpose of this discussion. *Table 13.4* assumes that the voluntary consumption of air dried feed by a leisure horse results from two meals daily and is somewhat less than that of the Thoroughbred in training, but in accord with evidence of normal rates of consumption. Capacity and energy requirement are the common factors of energy density given in *Tables 13.4* and *13.5*.

If one assumes that forages are required in the diet in reasonable quantities for more reasons than simply as a source of nutrients, then it is necessary to be aware of the range of energy densities in forages and, secondly, to calculate the impact that forages have on the required energy densities of compounded feeds consumed with them. The energy contents of common forages on an 88% DM basis range from 5.5–6.0 MJ/kg for wheat and barley straw at one extreme to over 13 MJ/kg for lush, growing forages at the other extreme. The capacity of horses for feed energy tends to

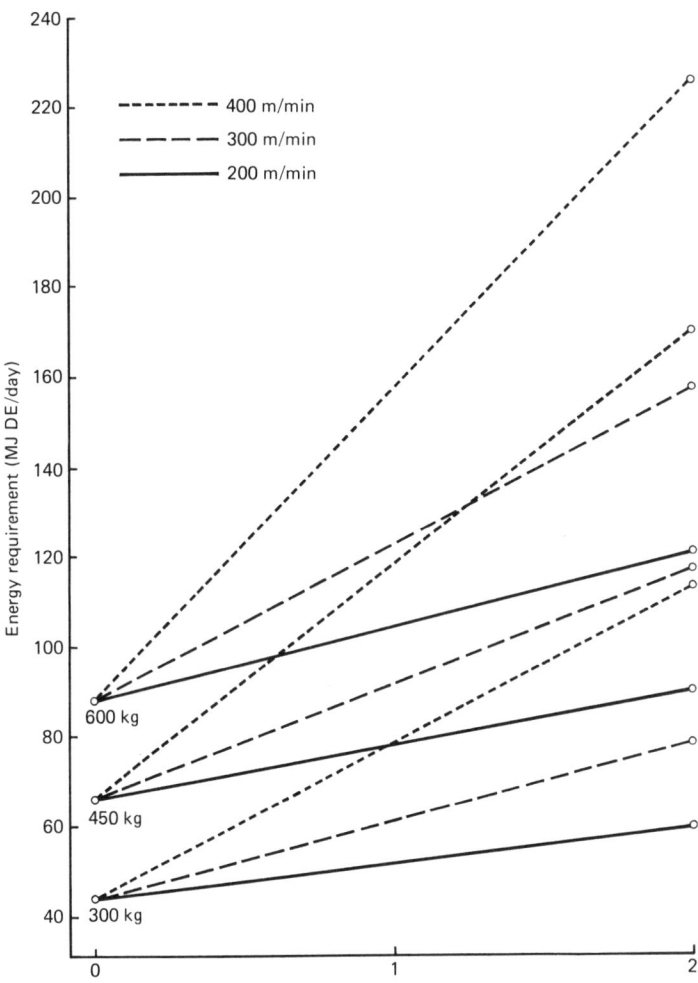

Figure 13.1 Daily energy requirements of horses of three body weights (300, 450 or 600 kg) for maintenance plus exercise at three speeds (200, 300 or 400 m/min) of varying durations (After Pagan and Mintz, 1986b)

Table 13.3 DIGESTIBLE ENERGY REQUIREMENTS (MJ DE/DAY) OF HORSES OF DIFFERENT BODY WEIGHTS FOR MAINTENANCE PLUS EXERCISE

Exercise level		Body weight (kg)[a]		
Duration (h/day)	Speed (m/min)	300	450	600
		DE requirements (MJ/day)		
0	—	44.1	66.2	88.3
0.57	200	48.8	73.1	97.5
2.0	200	60.3	90.5	120.7
0.57	300	54.0	80.9	107.9
2.0	300	78.5	117.8	157.1
0.57	400	63.9	95.9	127.8
2.0	400	113.4	170.0	226.7

[a]Including rider

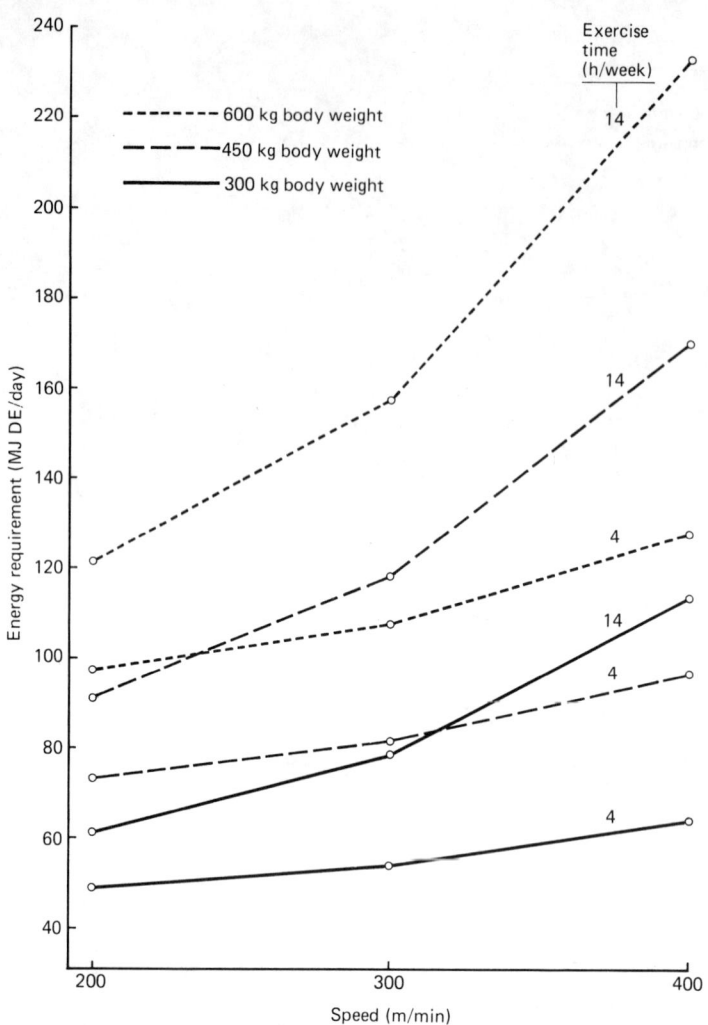

Figure 13.2 Daily energy requirements of horses of three body weights (300, 450 or 600 kg) for maintenance plus exercise of two durations (4 or 14 h/week) at various speeds (After Pagan and Hintz 1986b)

increase with the digestibility of that feedstuff (Frape, 1984), so that appetite for energy is assumed to increase with both moderate exercise (Orton, Hume and Leng, 1985a,b) and increased feed quality. These factors complicate the calculations as *Tables 13.4* and *13.5* show. *Table 13.5* gives the amounts of forage and compound feedstuff of three different energy densities required by a 600 kg horse to just satisfy its appetite when it accomplishes different amounts of daily exercise. An extreme range of both feed energy density and energy expenditure is assumed in order to cover all possible contingencies. It should be appreciated, however, that the highest assumed level of activity could not be undertaken day in and day out, because of exhaustion. This implies that the greatest energy expenditure could not be met on a daily basis by any mixture of common feedstuffs. Hence several rest days must follow days of extreme exertion.

Table 13.4 DIGESTIBLE ENERGY DENSITY OF DIETS REQUIRED TO MEET THE ENERGY EXPENDITURE RESULTING FROM DIFFERENT AMOUNTS OF EXERCISE BY HORSES HAVING AVERAGE 'APPETITES'. (VALUES IN BRACKETS ARE 'APPETITES' EXPRESSED IN kg FEED/DAY OF 88% DRY MATTER)

Exercise level		Body weight (kg)[a]		
Duration (h/day)	Speed (m/min)	300	450	600
		Energy density (MJ DE/kg feed)		
0	—	6.0 (7.3)	6.5 (9.2)	7.5 (10.6)
0.57	200	6.4 (7.3)	6.8 (9.2)	7.9 (10.6)
2.00	200	7.5 (7.3)	8.0 (9.2)	9.3 (10.6)
0.57	300	7.1 (7.6)	7.6 (10.7)	8.7 (12.4)
2.00	300	9.8 (7.6)	10.5 (10.7)	12.1 (12.4)
0.57	400	8.4 (8.0)	9.0 (12.3)	10.3 (14.3)
2.00	400	14.2 (8.0)	15.0 (12.3)	17.4 (14.3)

[a] Including rider

PROTEIN NEEDS

Not only do forages differ widely in energy density but they differ even more widely in crude protein content. High quality growing forages may contain as much as 180 g crude protein/kg DM, whereas poor quality hay and cereal straws both contain less than 40 g crude protein/kg DM. Moreover, some high fibre by-products that may be used to satisfy the appetite of resting horses contain less than 30 g crude protein/kg DM.

Compounded feedstuffs are generally required to supplement a forage. It is unlikely that the daily energy demands will be exactly met by any compound feedingstuff supplementing a forage of unknown quality. The deposition and mobilization of depot fat provides a natural buffer to these exigent requirements. Indeed depot fat serves as a major and direct energy source to extended work (Carlson, Fröberg and Persson, 1965; Nimmo and Snow, 1983). This in no way implies that horses should be fattened for 'battle'. Quite the reverse, as such a manoeuvre could curb the facility for fat mobilization in time of need (Frape, 1986; 1988). Similarly the diversity in protein content of forages inevitably means that with the best endeavours the minimum protein requirements of the horse are unlikely to be met exactly on a daily basis.

The daily minimum protein requirement of the adult horse for maintenance is low, approximately equivalent to 2.7 g digestible protein/kg $W^{0.75}$, where W is body weight (Slade, Robinson and Casey, 1970; Hintz and Schryver, 1972; Prior et al., 1974; Quinn, 1975), and this estimate is probably uninfluenced by moderate exercise (Orton, Hume and Leng, 1985a). This is not the place to dissert upon the meaning of a minimum requirement, but it is well established that for most nutrients the efficiency of utilization seems to increase with decreasing intakes so that the minimum, or negative balance, is approached asymptotically. Thus, under maintenance conditions it would require a large number of animals to detect small shifts in the requirement for nutrients other than energy sources, resulting from the imposition of light or moderate exercise. Success is more elusive when natural conditions are simulated. For example, it is quite difficult to induce a sodium deficiency in a horse given a natural diet (Meyer et al., 1984). On the other hand, hard exercise would soon precipitate such a deficiency in an acute and devastating form.

The data in *Table 13.6*, based on one set of calculations, indicate that the

Table 13.5 APPROXIMATE FORAGE AND COMPOUND REQUIREMENTS (kg/DAY of 88% DRY MATTER) THAT JUST SATISFY THE APPETITE OF A 600 kg HORSE

Appetite (kg/day)		10.6		12.4		14.3	
DE (MJ/kg)		6.0	8.8	8.5	10.0	10.0	11.5
		Forage	Compound	Forage	Compound	Forage	Compound
Exercise level				Requirements (kg/day)			
Duration (h/day)	Speed (m/min)						
0	—	1.9					
0.57	200	0.4					
2.0	200		8.7				
0.57	300		10.2				
2.0	300			9.9	2.5		
0.57	400			10.7	1.7	9.5	4.8
2.0	400					7.15[a]	7.15[a]

[a]None of the feedstuffs would satisfy the energy expenditure—a 50:50 mix of the high energy forage and compound would meet 68% of the energy demand

Table 13.6 THE RELATIONSHIP AMONGST THE RANGE OF ESTIMATED MINIMUM CRUDE PROTEIN REQUIREMENTS, FORAGE QUALITY, WORK INTENSITY AND THE CALCULATED DESIRABLE MINIMUM PROTEIN CONTENT OF A COMPOUNDED FEEDSTUFF FOR HORSES OF VARIOUS BODY WEIGHTS BUT WITH A CONSTANT 'APPETITE'

			Required protein content of compounded feed (g/kg)			
			Light work (hay 70% of diet)		Heavy work (hay 40% of diet)	
Body weight (kg)	Minimum protein[a] requirement (g/day)	Food intake (88% dry matter) (kg/day)	Hay protein (g/kg)			
			40	80	40	80
200	300–460	5.0	107–213	13–120	73–127	47–100
400	520–850	9.5	89–205	4–112	65–122	38–96
500	630–1080	12.0	82–207	nil–113	61–123	34–97
600	740–1230	13.5	89–210	nil–117	65–125	38–99

[a]Containing 3.5–3.8 g lysine/kg

theoretical minimum desirable crude protein content of a compounded feedstuff used to supplement two typical hay samples would range from nil to 213 g/kg. Moderate excesses in the consumption of a variety of nutrients at a constant energy intake have little practical untoward effect on the leisure horse, whereas they would do so for the competition horse. Of course, the excess measured in multiples of the minimum dietary requirement that causes toxic complications varies widely with the nutrient. Two characteristics of feedstuffs of practical significance here are protein and calcium, and daily intakes up to double, or even treble, the requirement have no adverse effect in active horses with a healthy renal system. From experience, however, this assertion is assailed by the observation of acute complications apparently arising from such excesses in the occasional horse otherwise visibly enjoying good health.

Diet-related ailments

Major diet-related problems of horses include laminitis, endotoxaemia, enterotoxaemia, some other causes of colic and chronic obstructive pulmonary disease (COPD). Certainly the likelihood of encountering the first three of these conditions is lessened by ensuring that high energy feedstuffs are offered in small quantities, and this may entail providing more than two feeds/day, and secondly of meeting an increased energy requirement by a gradual change in the daily intake of a high energy feedstuff (Garner *et al.*, 1978, Carroll *et al.*, 1987). The prevailing view that laminitis in horses is commonly caused by excessive dietary protein not acting as an energy source is unsupported by investigation (Yoakam, Kirkham and Beeson, 1978).

DIETARY FIBRE

Respiratory hypersensitivity, expressed as COPD, is contained in sensitive animals by improving the ventilation of the stable (McPherson *et al.*, 1979b), and this is rarely satisfactory, and by reducing the dustiness of forage and bedding. To this end, long hay may be replaced by high fibre compounded feedstuffs, or by long fibre hay cobs, or indeed by silage, haylage or hydroponic barley. The disadvantage of high fibre compounded feedstuffs is that the fibre is typically in a ground form. Despite the pulverization of hay by the molar and premolar teeth, long fibre seems to have beneficial effects in both higher digestibility (Schurg *et al.*, 1978) and greater contentment (Haenlein, Holdren and Yoon, 1966; Haenlein, 1969). The former of these two advantages is, from a practical point of view, of little significance, but the latter is generally considered to be important. The question of contentment touches on another subject of 'heating' feeds. These are feedstuffs said to cause difficulty in the management of horses, and the effects are probably in large measure associated with rapid hindgut fermentation and a precipitate postprandial glycaemic index curve. A mixed diet containing long fibre of only moderate digestibility seems here to provide some advantages. It is relevant to point out that very lush forage and high starch diets both can cause metabolic upsets and should be used by the amateur horseman with considerable care. These problems relate to the number of chewing movements required in the consumption of feedstuffs of different physical qualities. It has been demonstrated, for example, in German work (Meyer, Ahlswede and Reinhardt, 1975) that three and a half times as many chewing movements are required in the consumption of long hay as are required in the consumption of an equal weight of concentrated feedstuffs. The delay that results in stomach filling prevents an undesirable rise in stomach pH and similarly reduces the extent of fall in caecal pH (Meyer, Ahlswede and Pferdekamp, 1980; Meyer, 1983; Frape, 1986). Both of these changes must be avoided if acute digestive upsets are to be prevented.

MINERALS

Considerable interest still remains in calcium, phosphorus and magnesium nutrition of horses; but there has been no recent unexpected finding in adult horses, apart from preliminary observations linking a form of poor quality eroded hoof horn with inadequate dietary calcium (Kempson, 1987). The requirements for sodium, potassium and chloride are considerable in horses subjected to endurance work in hot weather (Carlson, 1983). However, the additional requirements for these electrolytes represent an acute need that may not be met through building up any reserve by way of the daily ration.

Responses of horses under 'field' conditions to several trace elements, particularly that of selenium, but also those of zinc, manganese and possibly copper have been recorded. The selenium intake of many stud animals (Basler and Holtan, 1981) and many riding horses (Blackmore *et al.*, 1982) has been shown to be inadequate for maximum performance. On the other hand metabolic changes, that are measurable and that result from natural variation in the provision of these nutrients by forages, probably have little impact on the performance of leisure horses.

VITAMINS

Interest still attaches to dietary requirements for several vitamins despite a paucity of appropriate recent experimental work. In Britain stabled horses receiving traditional feedstuffs have a need for supplementary sources of vitamins A and D. The introduction of some exotic feeds and by-products to compounded products requires that careful attention is paid to their composition, in respect of both major nutrients and potential contaminants. Their use can invoke a need for several vitamins. For example, the barn drying of forage affects the dietary need for calciferol; the ensilage of forage increases the supplementary demand for alpha-tocopherol; and the substitution of root vegetables or by-products for cereals and their by-products alters the dietary supply of several water-soluble vitamins. Many fibre sources in general use are practically devoid of micronutrients, but with supplementation they are of considerable use, even though recognition of this may influence the perception of their monetary worth.

Field evidence (Smith and Wright, 1984) indicates that, at the latitude of the British Isles, natural sources of vitamin D may not be adequate in the achievement of 'normal' blood concentrations of the monohydroxy metabolites. As the Ca:P ratio of diets varies over a considerable range this observation has practical importance. There would seem to be a very low dietary requirement for menaphthone, but the need of horses for the two other major fat-soluble vitamins—retinol and alpha-tocopherol—has not been established. The horse is inefficient in the utilization of beta-carotene as a precursor of retinol. Whether beta-carotene itself contributes anything to the common weal of equids is unknown. There is little concordance in the estimated, or implied, requirements for α-tocopherol. Evidence indicates a range from 11 i.u. (Stowe, 1968) to 225 i.u. (Ronéus et al., 1985)/kg diet. The effect of vitamin supplementation on the extent of tissue saturation of certain vitamins or their metabolites has also been demonstrated for some water-soluble vitamins. This may indicate a need of hard-worked adult animals for supplementary folic acid and cyanocobalamin (Frape, 1986). Whether dietary sources of ascorbic acid exert any metabolic effect, or indeed whether ascorbic acid is absorbed from the intestines of the horse in measurable quantities, is the subject of debate (Löscher, Jaeschke and Keller, 1984; Snow, Gash and Cornelius, 1987). Thus there is no semblance of precision in the estimates of requirements for dietary vitamins. Demonstrated responses of horses under natural conditions to supplementary vitamins are exceptional and almost curiosities. However, improved hoof wall quality has been demonstrated with biotin supplementation of adult horses displaying one type of crumbly hoof (Comben, Clark and Sutherland, 1984).

Formulating the diet

It has been asserted that to meet a range of daily energy needs, or more precisely the range of average requirements over a few days whilst hunger is satisfied, necessitates the formulation of compounded feedstuffs of differing energy densities. For this to be accomplished not only should digestible sources of energy and protein be reserved but also reliable and palatable sources of edible fibre should be available. Many sources of these that have proved suitable are listed in *Table 13.7* and proposed dietary nutrient concentrations are given in *Table 13.8*. Excepting finer organoleptic considerations, typical formulations that could result from the foregoing are suggested in *Table 13.9*.

Feedingstuffs used in the formulation of diets for horses should be of good and dependable physical quality and should not be a vector of enteric infection—commonly caused by salmonella bacteria, but also by nematode, cestode and trematode parasites. Acceptability of feedstuffs involves both their smell, physical

Table 13.7 RAW MATERIALS OF PRACTICAL VALUE IN HORSE FEEDS IN APPROXIMATELY DESCENDING ORDER OF 'VALUE'[a] WITHIN CLASS

Cereals
Oats
Maize
Barley
Millet
Grain sorghum
Wheat
Rice

Protein concentrates
Single cell proteins
Soya
Peas
Brewers yeast
00 rapeseed meal
Low gossypol cottonseed meal
Extr. sunflower seed meal
Field beans
Maize gluten feed/meal
White fishmeal
High protein grassmeal/lucerne
Hydroponic grass/barley[b]
Dried skim-milk
Exp. linseed

By-products
Wheat bran/wheatfeed
Sugar beet pulp
Grain distillers, dark grains
Malt distillers, dark grains
Distillers dried solubles
Malt culms
Extr. rice bran

Fibre sources
NIS
Straw cubes
Oatfeed
Citrus pulp
Olive pulp
Rice husk pellets

Bulky feeds
(consumption rate controllers)
Hay cobs
Grass hay
Clover hay
Lucerne hay
Sugar beet pulp
Haylage
Hydroponic grass/barley

Sundries
Fat sources—as for poultry
Cane molasses
Beet molasses
Locust beans

[a]Excluding cost
[b]Perhaps more appropriately an energy concentrate

Table 13.8 PROVISIONAL DIETARY REQUIREMENTS OF A 500 kg HORSE FOR SOME NUTRIENTS. (DATA IN BRACKETS ARE SPECULATIVE)

	Per kg diet (88% dry matter)		Per kg bodyweight daily	
	Maintenance	Working	Maintenance	Working
Crude protein (g)	80[a]	70[a]	1.5	1.6
Calcium (g)[b]	2.7–3.2		0.06	0.06
Phosphorus (g)[b]	2.0–2.5		0.035	0.035
Sodium (mg)	80 min		1.6 min	up to 260
Potassium (mg)	1900 min		30 min	up to 130
Chloride (mg)	175[c] min		3.2[c]–120	up to 570
Magnesium (mg)	1000		13	up to 25
Iron (mg)	50		1	1
Selenium (mg)	0.2		0.004	0.004
Vitamin A (i.u.)	6000		100	(150)
Vitamin D_2 or D_3 (i.u.)	1000		16	16
Vitamin E (i.u.)	25	(25–225)	0.5	(0.5–5.0)
Ascorbic acid (mg)		(100)		(3)
Folic acid (mg)		(1–2)		(0.05)
Biotin (µg)	100		1.6	(1.6)
Vitamin B_{12} (µg)	0	(5–10)	0	(0.2)

Water requirement: 16 litre minimum/horse daily for maintenance; 20–60 litre daily during work

After Frape (1988)
[a]Protein containing 38 g lysine/kg
[b]Ca:P ratio between 1:1 and 2:1
[c]Assuming no sweat loss

Table 13.9 UNROUNDED PRINTOUTS OF THE COMPOSITION OF THREE HORSE COMPOUNDS OF HIGH, OR LOW, FIBRE AND HIGH, OR LOW, PROTEIN CONTENTS

	High fibre, low protein	Low fibre, low protein	Low fibre, high protein
	Weight (kg) of individual ingredients		
Lucerne meal, 16	—	—	100.0
Barley	—	307.1	351.5
Sugar beet pulp	100.0	102.4	—
NIS	200.0	—	—
Soya extn 44	90.0	30.7	150.0
Wheatfeed	252.1	307.1	250.0
Dried grass 16	100.7	102.4	—
Oatfeed	100.0	—	45.0
HEF	26.0	20.5	10.0
Molasses, cane	88.2	71.6	50.0
Limestone	4.0	6.1	7.5
Salt	4.0	6.1	6.0
Vits, trace minerals	20.0	20.4	20.0
Dried brewers yeast	—	10.2	—
Borrebond	15.0	15.4	10.0
Total	1000	1000	1000
Calculated composition			
Oil (g/kg)	44	41	30
Crude protein (g/kg)	117	123	160
Crude fibre (g/kg)	164	85	87
Calcium (g/kg)	7.5	7.4	7.7
Total phosphorus (g/kg)	4.6	5.6	5.7
Total salt (g/kg)	7.2	9.4	7.7
Starch (g/kg)	176	288	265
DE, horse (MJ/kg)	9.9	11.2	11.2
Methionine (g/kg)	1.7	1.9	2.3
Lysine (g/kg)	5.8	5.5	7.8

texture, particle size and the amount of chewing required prior to their deglutition. It is equally important to meet nutrient needs and to avoid diet related metabolic upsets. Thus the formulation of the diet, its physical characteristics and feed management are all important in achieving health in horses that are capable of carrying out the exercise demanded in a controllable manner by riders with a range of skills and abilities.

References

BASLER, S.E. and HOLTAN, D.W. (1981). *Proceedings of the Western Section of the American Society of Animal Science*, **32**, 399–400

BLACKMORE, D.J., CAMPBELL, C., DANT, C., HOLDEN, J.E. and KENT, J.E. (1982). *Equine Veterinary Journal*, **14**, 139–143

CARLSON, G.P. (1983). In *Equine Exercise Physiology*, pp. 291–309. Ed. Snow, D.H., Persson, S.G.B. and Rose, R.J. Granta Editions, Cambridge

CARLSON, L.A., FRÖBERG, S. and PERSSON, S. (1965). *Acta Physiologica Scandinavia*, **63**, 434–441

CARROLL, C.L., HAZARD, G., COLOE, P.J. and HOOPER, P.T. (1987). *Equine Veterinary Journal*, **19**, 344–346
CARROLL, C.L. and HUNTINGTON, P.J. (1988). *Equine Veterinary Journal*, **20**, 41–45
COMBEN, N., CLARK, R.J. and SUTHERLAND, D.J.B. (1984). *Veterinary Record*, **115**, 642–645
CUDDEFORD, D. (1987). Personal communication
FRAPE, D.L. (1984). In *Straw and other Fibrous By-Products as Feed*, pp. 487–532. Ed. Sundstøl and Owen, Elsevier, Amsterdam
FRAPE, D.L. (1986). *Equine Nutrition and Feeding*. Longmans, London.
FRAPE, D.L. (1988). *Equine Veterinary Journal* (in press)
GARNER, H.E., MOORE, J.N., JOHNSON, J.H., CLARK, L., AMEND, J.F., TRITSCHLER, L.G., COFFMAN, J.R., SPROUSE, R.F., HUTCHESON, D.P. and SALEM, C.A. (1978). *Equine Veterinary Journal*, **10**, 249–257
HAENLEIN, G.F.W. (1969). *Feedstuffs*, **41** (No. 6), 19–20
HAENLEIN, G.F.W., HOLDREN, R.D. and YOON, Y.M. (1966). *Journal of Animal Science*, **25**, 740
HINTZ, H.F. (1969). *Veterinarian*, **6**, 45–51
HINTZ, H.F. and SCHRYVER, H.F. (1972). *Journal of Animal Science*, **34**, 592–97
HINTZ, H.F., SCHRYVER, H.F. and STEVENS, C.E. (1978). *Journal of Animal Science*, **46**, 1803–7
KEMPSON, S.A. (1987). *Veterinary Record*, **120**, 568–570
LÖSCHER, W., JAESCHKE, G. and KELLER, H. (1984). *Equine Veterinary Journal*, **16**, 59–65
MCPHERSON, E.A., LAWSON, G.H.K., MURPHY, J.R., NICHOLSON, J.M., BREEZE, R.G. and PIRIE, H.M. (1979a). *Equine Veterinary Journal*, **11**, 159–166
MCPHERSON, E.A., LAWSON, G.H.K., MURPHY, J.R., NICHOLSON, J.M., BREEZE, R.G. and PIRIE, H.M. (1979b). *Equine Veterinary Journal*, **11**, 167–171
MEYER, H. (1983). *Proceedings Horse Nutrition Symposium*, Uppsala Conference, pp. 95–109
MEYER, H., AHLSWEDE, L. and PFERDEKAMP, M. (1980). *Deutsch Tierärztlich Wochenschrift*, **87**, 43–47
MEYER, H., AHLSWEDE, L. and REINHARDT, H.J. (1975). *Deutsch Tierärztlich Wochenschrift*, **82**, 54–58
MEYER, H., SCHMIDT, M., LINDNER, A. and PFERDEKAMP, M. (1984). *Tierernahrung und Futtermittelkunde*, **51**, 182–196
NIMMO, M.A. and SNOW, D.H. (1983) in *Equine Exercise Physiology*, pp. 237–244. Ed. Snow, D.H., Persson, S.G.B. and Rose, R.J. Granta Editions, Cambridge
NOOT, G.W.V. and GILBREATH, E.B. (1970). *Journal of Animal Science*, **31**, 351–5
ORTON, R.K., HUME, I.D. and LENG, R.A. (1985a). *Equine Veterinary Journal*, **17**, 381–385
ORTON, R.K., HUME, I.D. and LENG, R.A. (1985b). *Equine Veterinary Journal*, **17**, 386–390
PAGAN, J.D. and HINTZ, H.F. (1986a). *Journal of Animal Science*, **63**, 815–821
PAGAN, J.D. and HINTZ, H.F. (1986b). *Journal of Animal Science*, **63**, 822–830
PRIOR, R.L., HINTZ, H.F., LOWE, J.E. and VISEK, W.J. (1974). *Journal of Animal Science*, **38**, 565–571
QUINN, C.R. (1975). PhD Dissertation, Colorado State University
RALSTON, S.L. and BAILE, C.A. (1982a). *Journal of Animal Science*, **54**, 1132–1137
RALSTON, S.L. and BAILE, C.A. (1982b). *Journal of Animal Science*, **55**, 243–253
RALSTON, S.L. and BAILE, C.A. (1983). *Journal of Animal Science*, **56**, 302–308

RALSTON, S.L., FREEMAN, D.E. and BAILE, C.A. (1983). *Journal of Animal Science*, **57**, 815–825

RONÉUS, B.O., HAKKARAINEN, R.V.J., LINDHOLM, C.A. and TYÖPPÖNEN, J.T. (1985). *Equine Veterinary Journal*, **18**, 50–58

SCHURG, W.A., PULSE, R.E., HOLTAN, D.W. and OLDFIELD, J.E. (1978). *Journal of Animal Science*, **47**, 1287–1291

SLADE, L.M., ROBINSON, D.W. and CASEY, K.E. (1970). *Journal of Animal Science*, **46**, 983–991

SMITH, B.S.W. and WRIGHT, H. (1984). *Veterinary Record*, **115**, 579

SNOW, D.H., GASH, S.P. and CORNELIUS, J. (1988). *Equine Veterinary Journal* (in press)

STOWE, H.D. (1968). *American Journal of Clinical Nutrition*, **21**, 135–143

YOAKAM, S.C., KIRKHAM, W.W. and BEESON, W.M. (1978). *Journal of Animal Science*, **46**, 983–991

14

NUTRITION OF THE DOG

J. CORBIN
Department of Animal Sciences, University of Illinois, Urbana, Illinois, USA

The total American dog population is estimated to be 52 million, only recently surpassed in number by its 56 million cats. These dogs consume an estimated 3 381 500 tons of dog food sold through food markets plus an additional 800 000 tons sold through feed outlets. Total dog food sales in US dollars are $3.599 billion and cat sales are $2.356 billion (*Table 14.1*) for a total of $5.955 billion.

Extensive nutritional studies of dogs have been conducted within the pet food industry, but with a minimum release of data, which are generally considered proprietary. The rapid escalation of pet numbers during the past three decades has not generated support and research in public institutions comparable with that of poultry and livestock, thus a relative shortage of published nutritional data developed. The assignment of human-like qualities to pets has removed dogs from the usual economic status of most animals and placed them in an anthropomorphic status where expenditure is based less and less on farm animal economics. Instead, dogs have been elevated to a role considerably above farm animals and deserving the best nutritional products as perceived by their owners. Extensive advertising by producers of commercial pet foods has promoted the concept that dogs merit food equal to or superior to that consumed by humans. Most dog owners acknowledge that their dogs fed good commercial diets receive a more carefully balanced diet than is consumed by America's youngsters.

The role of dogs in the family and the perception of the availability of quality dog food have created a wide variation in available products. America's 3200 brands of commercial pet foods have a wide range of guaranteed levels of ingredients: protein 150 to 300 g/kg, fat 50 to 300 g/kg, fibre 15 to 80 g/kg, as well as enormous variation in calcium phosphorus, and other nutrients. One kilogram of 'complete' dog food may have a price high of 35 times the low range on a dry matter basis. Advertising appears far more influential than economics in selecting dog foods. Such buying habits—based on owners' concepts of price, ingredient inclusion, shape, appearance, texture, colour, aroma, acceptance by the dog, droppings condition, plus advertising and packaging—have impeded research progress into the basic nutritional requirements of dogs.

Nutritional research and publications

Canine nutritional data published throughout the world were scrutinized and combined with safety margins and tempered experience and published in the USA under the auspices of the National Research Council (NRC, 1974; 1985). NRC (1985)

Table 14.1 US DOG AND CAT FOOD SALES (52 WEEKS ENDING 19 JUNE 1987)

	Tons (as sold)	%	% Dry matter	Tons (dry-matter basis)	%
Dog food					
Dry	2 276 000	67.3	90	2 048 400	84.0
Canned	849 000	25.1	22	186 780	7.6
Semi-moist	124 500	3.7	70	87 150	3.6
Snacks-treats	132 000	3.9	90	118 000	4.8
	3 381 500	100.0		2 440 330	100.0
Cat food					
Dry	529 500	42.1	90	475 550	70.0
Canned	638 000	50.7	22	140 360	20.7
Semi-moist	90 000	7.2	70	63 000	9.3
	1 257 500	100.0		678 910	100.0
Total SAMI[a]	4 639 000				
'Feed mills'[b]	927 000				
Total	5 566 000				

	Dollars (millions)
Dog food	
Dry	$1754
Canned	828
Semi-moist	191
Snacks-Treats	357
Total SAMI	3130
'Feed mills'[b]	469
Total dog	3599
Cat food	
Dry	$ 776
Canned	1133
Semi-moist	233
Total SAMI	2142
'Feed sources'[b]	214
Total	$2356
Total dog and cat	$5955

[a]Source: Based on SAMI report of warehouse withdrawals at Pet Food Institute, 17 September 1987, Denver, Colorado.
[b]Based on USDA Survey reported in *Feedstuffs* **59**(17), **69**, 1987 (27 April, 1987) with 25% of specialty feed estimated to be *800 000* tons of dog food and *127 000* tons of cat food with 60% of the value of SAMI (1986) per pound. 'Feed sources' include sales through feed stores, pet stores, direct to some large retailers, plus veterinarians and sales direct to larger kennels and research facilities

records minimum dietary nutrient requirements and is based primarily on results obtained with growing puppies. These publications are used as the broad technical basis for the formulation and manufacture of dog foods. However the NRC (1985) data are based primarily on nutrient levels obtained with purified diets, with amino acids, vitamins and minerals supplied in crystalline form. Since commercial foods consist mainly of feedstuff ingredients, the more readily adaptable NRC (1974) version with its acknowledged safety factors, is still the reference of choice of the American dog food industry. An immense amount of additional comparative information has been obtained during the past few years. Some of those data are referenced here.

Vitamin requirements

Hazewinkel et al. (1987a) kept two groups of puppies indoors and raised them on a diet meeting the recommendations of the NRC (1974) without the inclusion of dietary vitamin D. One group was given daily exposure to low-level artificial UV-B in a manner shown to be effective in producing cholecalciferol in other species. The other group was exposed to high-energy radiation. Blood chemistry at regular intervals measured vitamin D metabolites (i.e. 25 (OH) vitamin D, 24,25 $(OH)_2$ vitamin D and 1,25 $(OH)_2$ vitamin D). Histologic and radiographic measurements were also taken. Circulating concentrations of both 25 (OH) vitamin D and 24,25 $(OH)_2$ vitamin D decreased in both groups to subnormal levels and only 1,25 $(OH)_2$ vitamin D increased during the period of high-energy radiation to attain low to normal levels. Both groups of puppies developed histological and radiographic signs of rickets. This study corroborates the suggestion by Wheatley and Sher (1961) that without 7-dehydrocholesterol in the skin of dogs, there is no synthesis of vitamin D.

Apparently hypocalcaemia tetany in association with rickets is still a problem with dogs. Lavelle (1987) described the clinical history and signs and pathological findings in a three-month-old collie and a litter of ten-week-old greyhounds which had hypocalcaemic tetany in association with rickets. They had been reared inside, which apparently makes little difference since dog skin contains no 7-dehydrocholesterol (Wheatley and Sher, 1961; Hazewinkel et al., 1987b). Prior to presentation, the collie was treated for hypocalcaemia and died shortly after admission. The greyhounds were treated for poisoning (the type of poisoning was not specified) with one making a recovery and the other two dying overnight. When this author fed English pointers on lean meat with no added vitamin D and a wide Ca:P ratio, hypocalcaemia was encountered with thin cortical bone and wide flaring growth plates in the metaphyseal region, but no acute death losses occurred.

The effects of 1,25-dihydroxyvitamin D [1,25-$(OH)_2$ vitamin D] alone or with parathyroid hormone (PTH) were evaluated in long-term studies in dogs (Malluche et al., 1986). Histomorphometric evaluation of static parameters of bone after eight months of experimental observations indicated that deficiency in 1,25-$(OH)_2$ vitamin D and PTH resulted in decreased number and activity of bone-forming and resorbing cells. Adding 1,25-$(OH)_2$ vitamin D only increased the activity but not the number of bone cells while adding PTH only increased the number but not the activity of bone cells.

A deficiency of vitamin E may result in blindness in dogs, similar to senile macular degeneration of humans (Loew, 1987). Dogs deficient in vitamin E also had eye lesions and produced night blindness followed by total blindness. This may be aggravated by hypervitaminosis A when associated with a deficiency of vitamin E, and reflects an increase of free-state vitamin A. This high concentration of free vitamin A begins to destroy cell membranes. Loew (1987) reported that dogs deficient in both vitamins A and D may encounter less severe damage than with diets supplying hypervitaminosis A.

A possible vitamin C requirement in racing sled dogs trained on a high-fat diet has been advocated (Donoghue, Kronfeld and Banta, 1987) based on four groups of nine dogs fed the same basal diet supplemented with 0, 500, 1000 and 2000 mg/day of ascorbic acid. Only the high dose had a significant effect on the serum ascorbate concentration and, 0.24 mg/KJ ME appears to confer an advantage to sled dogs in regard to adrenal responsiveness and, perhaps, fat oxidation in muscle.

Digestion and metabolism

The influence of various food components on the intestinal flora of the dog was investigated by Amtsberg (1987), who determined the qualitative and quantitative bacterial populations of the ileum, colon and faeces of adult dogs fed different ingredients. Cattle lungs and raw potato starch were used to produce nutritional diarrhoea in dogs. Cattle lungs produced significant alterations in microbial flora.

The digestibility of various carbohydrates was documented by Schünemann, Mühium and Meyer (1987) with fistulated and intact dogs. Sucrose was digested almost totally in the ileum (99.7%). Lactose digestion in the small intestine varied from none in one dog to about 60% in another, but colonal fermentation of lactose by bacteria boosted total digestibility of approximately 99%.

Carbohydrates with low digestibility, e.g. potato starch and lactose, reduced digestibility of meatmeal (Meyer et al., 1987). Orally administered antibiotics had little influence on total and partial digestibility. The total endogenous N-secretion up to the end of the small intestine was about 40 mg/kg bodyweight/day (calculated from rations with low N-content or high protein digestibility).

Flatulence has been highly associated with some feedstuffs and chyme from fistulated dogs was incubated under standardized conditions with a buffer with pH 7 or 8 for 5 h at 38°C. Soyabean meal with tapioca starch produced almost three times as much gas production as soyabean meal plus rice (Klocke et al., 1987). Hydrogen production, as detected in the breath of dogs fed different rations, varied enormously, with soyabean meal and tapioca starch producing four times as much hydrogen as soyabean and rice (Schünemann and Ingwersen, 1987).

Energy turnover of beagles at maintenance, gestation, and lactation was estimated using four energy levels, and energy requirements of gestating and lactating dogs were determined on the basis of basal metabolic rates and heat production value using typical published conversion factors (Männer, Bronsch and Wagner, 1987). Energy requirements of adult dogs are being measured (Booles and Rainbird, 1987) by direct calorimetry in a whole body direct calorimeter. A sled-dog study on more than 100 dogs (Grandjean, Paragon and Grandjean, 1987) indicates that sled dogs have particular nutritional requirements due to their digestive and metabolic characteristics and the physiological demands of their efforts.

Growth and energy requirements of several large breeds of dogs have been found to be closely related to body weight raised to some power, W^x, where W equals weight in kg and x is an exponent calculated from experimental data (Rainbird, 1987). This helps relate requirements for energy, protein, and minerals on the basis of available information with the help of a factorial method (Leibetseder, 1987). Requirements for trace elements and vitamins can be estimated from dosage–response relationships (Leibetseder, 1987). Calculations are based on energy exchange and endogenous losses in the basal state, work and quantity, composition and rate of synthesis of products, including hair, fetal deposition, reproductive organs, milk, body mass while growing, plus digestibility and net absorption of nutrients and efficiency of utilization.

Proteins and amino acids

Extensive research at the University of Illinois has provided published information on which to base the amino acid requirements of growing puppies (Milner, 1979; Milner, 1984; Milner, 1981; Burns and Milner, 1981; Burns and Milner, 1982; Czarnecki and

Baker, 1984; Czarnecki and Baker, 1982; Hirakawa and Baker, 1985; Hirakawa, 1986). Most of these studies utilized crystalline amino acids fed initially in agar-gel form to facilitate feed consumption. Since added moisture could hasten Maillard-type reactions and the formation of other crosslinked products, sucrose and maize starch were later included as ingredients and the diet was consumed without added moisture.

Information on the requirements for specific amino acids for individual breeds of dogs is not plentiful. Labradors have a higher requirement for sulphur amino acids than beagles (Blaza et al., 1982). Recent studies in our laboratory by Hirakawa and Baker (1985) demonstrated that the sulphur amino acid requirement of pointer puppies is also different from those reported for either beagles or labradors by Blaza et al. (1982). These are differences between breeds relatively close in weight. The requirements of other breeds including chihuahuas with adult weights of perhaps 1 kg to St Bernards weighing more than 100 kg need to be investigated. Also, body composition differences as a function of age, sex and breed have not been established which makes it difficult to predict how one breed may differ from another in its requirement for an amino acid.

Four types of diets for puppies

Comparative utilization of four diets by English pointer puppies was evaluated by Hirakawa (1986). These included a chemically defined crystalline amino acid diet (147.8 g/kg crude protein, 17.15 MJ ME/kg), a methionine-fortified casein diet (223.6 g/kg crude protein, 15.31 MJ ME/kg), a maize-soya-meat and bonemeal food (206.6 g/kg crude protein, 13.54 MJ ME/kg) and a commercial puupy food (274.7 g/kg crude protein, 14.41 MJ ME/kg). Rate of gain, feed efficiency and metabolizable energy utilization were greatest in puppies consuming the casein diet. Nitrogen utilization was most efficient in puppies fed the crystalline amino acid diet. Comparisons of this type are needed since little information is available correlating the efficiency of free amino acid diets with intact-protein diets for growing puppies. However, more comparative studies are needed to provide parameters for evaluation predictions of practical-intact ingredients formulated into normal commercial diets.

Sulphur amino acid requirements of the growing puppy

Pointer puppies were fed crystalline amino acid diets (17.15 MJ ME/kg) by Hirakawa (1986) in a series of experiments to determine the sulphur amino acid (SAA) requirement of English pointer puppies and to evaluate the optimal cystine–methionine ratio for weight gains. The diet without SAA included 155.9 g/kg of amino acid mixture and is shown in *Table 14.2*. Total sulphur amino acids, half L-methionine and half L-cystine, were fed at levels from 2.5 g/kg to 5.5 g/kg of the diet (*Table 14.3*). The refinement series included SAA levels in incremental increases from 3.5 g/kg through to 6.5 g/kg diet (*Table 14.4*) and indicated that 4.5 g/kg in the diet gave the maximum growth response.

The replacement value of cystine for methionine is limited as indicated by the results based on weight gain and gain/feed data (*Table 14.5*). Excess L-cystine was also found to antagonize methionine utilization and depress growth (*Table 14.6*). Since a scarcity of information exists on the effects of feeding excess or imbalances of amino acids to dogs, the tendency to add methionine indiscriminately to diets for dogs is

Table 14.2 COMPOSITION OF PURIFIED BASAL DIET AND CRYSTALLINE AMINO ACID MIXTURE[a]

Ingredient	(g/kg diet)	Amino acid mixture	(g/155.9 g)
Pregelatinized maize starch[b]	to 1000.0	L-Arginine.HCl	11.5
Sucrose	290.0	L-Histidine.HCl.H$_2$O	4.5
Amino acid mixture	155.9	L-Lysine.HCl	11.4
Corn oil	150.0	L-Phenylalanine	5.0
Mineral mixture[c]	53.7	L-Tyrosine	4.5
Solka floc	30.0	L-Tryptophan	1.6
NaHCO$_3$	15.0	L-Methionine	—
Choline.Cl	3.0	L-Cystine	—
Vitamin mixture[d]	3.0	L-Threonine	6.5
MgSO$_4$	0.3	L-Leucine	10.0
ZnCO$_3$	0.2	L-Isoleucine	6.0
DL-α-tocopheryl acetate (50 mg/kg)	+	L-Valine	6.9
Ethoxyquin (125 mg/kg)	+	Glycine	16.0
		L-Proline	8.0
		L-Glutamic acid	24.0
		L-Alanine	16.0
		L-Asparagine.H$_2$O	16.0
		L-Serine	8.0
		Total	155.9

From Hirakawa (1986)
[a] Diet contained 17.15 MJ ME/kg and 142.6 g/kg crude protein (N × 6.25)
[b] Pre-Jel, The Hubinger Company, Keokuk, Iowa
[c] Mineral mix provided/kg of diet: CaCO$_3$, 3.0 g; Ca$_3$(PO$_4$)$_2$, 28.0 g; K$_2$HPO$_4$, 9.0 g; NaCl, 8.8 g; MgSO$_4$. 7H$_2$O, 3.5 g; MnSO$_4$. H$_2$O, 0.65 g; ferric citrate, 0.50 g; ZnCO$_3$, 0.10 g; CuSO$_4$. 5H$_2$O, 20.03 mg; H$_3$BO$_3$, 90.2 mg; Na$_2$MoO$_4$: 2H$_2$O, 90.2 mg; KI, 40.6 mg; Na$_2$SeO$_3$, 0.215 mg
[d] Vitamin mix provided/kg of diet: thiamin HCl, 150 mg; niacin, 150 mg; riboflavin, 24 mg; Ca-panthothenate, 30 mg; vitamin B$_{12}$, 0.03 mg; pyridoxine. HCl, 9.0 mg; biotin, 0.9 mg; folic acid, 6 mg; inositol, 150 mg; para-aminobenzoic acid, 3 mg; menadione, 7.5 g; ascorbic acid, 375 mg; retinyl acetate (250 000 i.u./g), 15 000 i.u.; cholecalciferol (200 000 i.u./g), 900 i.u.

Table 14.3 SULPHUR AMINO ACID REQUIREMENTS OF THE GROWING PUPPY[a]

	g/kg Diet		Gain[b]	Gain: feed[c]
L-Methionine	L-Cystine	TSAA	(g/day)	(g/kg)
1.25	1.25	2.5	22.1	192
1.75	1.75	3.5	42.9	336
2.00	2.00	4.0	54.9	429
2.25	2.25	4.5	64.5	481
2.75	2.75	5.5	76.3	517
Pooled SEM			2.8	14

From Hirakawa (1986)
[a] Results represent mean values of four observations in a 21-day randomized complete block design experiment. The 20 puppies were 28 days of age at the initiation of the experiment with an average initial weight of 1283 g
[b] Initial weight was used as a covariant in the analysis. Linear effect significant ($P < 0.005$)
[c] Quadratic effect significant ($P < 0.005$)

Table 14.4 SULPHUR AMINO ACID REQUIREMENTS OF THE GROWING PUPPY[a]

	g/kg Diet		Gain[b]	Gain: feed[b]
L-*Methionine*	L-*Cystine*	TSAA	(g/day)	(g/kg)
1.75	1.75	3.5	48.4	387
2.00	2.00	4.0	71.6	429
2.25	2.25	4.5	93.4	528
2.75	2.75	5.5	97.5	538
3.25	3.25	6.5	89.2	543
Pooled SEM			6.3	17

From Hirakawa (1986)
[a] Results represent mean values of three observations in a 28-day randomized complete block design experiment. The 15 puppies comprised nine males and six females from three litters. Puppies were 28 days of age at the initiation of the experiment with an average initial weight of 1444 g
[b] Initial weight was used as a covariant for analysis of weight gain and gain:feed. Quadratic effect significant ($P < 0.005$)

Table 14.5 REPLACEMENT VALUE OF CYSTINE FOR METHIONINE[a]

	g/kg Diet	Cystine	Gain[b]	Gain: feed
L-*Methionine*	L-*Cystine*	(% SAA)	(g/day)	(g/kg)
4.500	0	0	66.7[x]	537[x]
3.375	1.125	25	68.9[x]	530[x]
2.250	2.250	50	55.8[x]	479[x]
1.125	3.375	75	2.9[y]	51[y]
Pooled s.e.m.			7.5	20

From Hirakawa (1986)
[a] Results represent mean values of three observations in a 21-day randomized complete block design experiment. The 12 puppies represent five males and seven females from two litters. Puppies were 28 days of age at the initiation of the experiment with an average initial weight of 1396 g
[b] Means with different letters are significantly different ($P < 0.001$)

Table 14.6 RESPONSE OF PUPPIES FED DIETS CONTAINING VARYING RATIOS OF METHIONINE TO CYSTINE[a]

	g/kg Diet		Cystine	Gain[b]	Gain: feed[b]
L-*Methionine*	L-*Cystine*	TSAA	% SAA	(g/day)	(g/kg)
2.250	2.250	4.5	50	78.7[x]	420[x]
1.125	3.375	4.5	75	14.3[y]	114[y]
1.125	1.125	2.25	50	21.2[y]	158[y]
Pooled SEM				8.9	28

From Hirakawa (1986)
[a] Results represent mean values of five observations in a 21-day randomized complete block design experiment. Nine puppies were 56 days of age and six puppies were 28 days of age at the initiation of the experiment with an average initial weight of 3072 g and 1629 g, respectively
[b] Means with different superscript letters are significantly different for weight gain ($P < 0.005$) and gain:feed ($P < 0.001$)

alarming. An excess of methionine tends to be toxic. Excesses of other amino acids, including lysine and cystine have been shown to produce toxicity in puppies. Pointer puppies that consumed 40 g/kg excess lysine, which antagonized arginine, had reduced weight gain (Czarnecki, Hirakawa and Baker, 1985). Lack of specific information on the role of amino acid excesses and antagonisms or imbalances, destruction during food processing and storage, and digestibility and period of availability for natural ingredients adds to the complexity of evaluating dietary amino acid requirements. This is additionally complicated by the effects of other nutrients. Ontko, Wuthier and Phillips (1957) demonstrated that increased energy levels in the diet affects the protein requirement of puppies. Fibre levels also influence nutrient utilization of rations consumed by adult dogs and the utilization varies with the fibre composition (Burrows et al., 1982; Moore et al., 1980; Allen et al., 1981).

Zinc, calcium and iron

A zinc-deficiency syndrome was produced in dogs by Robertson and Burns (1963) by adding 20 g/kg calcium carbonate to a diet containing 3 g/kg calcium and 33 mg/kg of zinc. The syndrome was accompanied by emaciation, emesis, conjunctivitis, keratitis, and growth retardation. Calcium deposits were observed in the renal tubules.

Sanecki, Corbin and Forbes (1982) fed six-week-old puppies a maize–soya based zinc-deficient diet (Table 14.7) for five weeks and then transferred them to the same basic diet plus zinc carbonate at 200 mg/kg diet. Littermates were offered the same basic diet plus zinc carbonate at 200 mg/kg diet for five weeks and then transferred to the basal diet without added zinc for a period of an additional six weeks. Puppies on

Table 14.7 CONSTITUENTS OF CONTROL (ZINC ADEQUATE) DIET

Ingredients	(g/kg)
Maize (kibbled), extruded	426.5
Soya (cakelets), extruded	437.5
Corn oil	30.0
Vitamin premix in dextrose[a]	5.8
Choline chloride	4.2
Trace minerals mix in dextrose[b]	5.0
Magnesium oxide	5.0
Sodium chloride	10.0
Potassium chloride	10.0
Tricalcium phosphate	64.5
DL-Methionine	1.5

[a]Provides in the diet (mg/kg): vitamin A palmitate (250 000 i.u./g) 20, vitamin D_2 (850 000 i.u./g) 0.6, menadione sodium bisulphate 20, D-α-tocopherol succinate 50, D-biotin 2, thiamine HCl 20, riboflavin 10, pyridoxine HCl 20, DL-Ca pantothenate 36, folic acid 7.5; nicotinic acid 100, vitamin B_{12} in mannitol 22, ascorbic acid 20, para-aminobenzoic acid 6.
[b]Provides in the diet (mg/kg): potassium iodide 7, cobalt sulphate 7 H_2O 1, cupric sulphate 5 H_2O 30, ferrous sulphate 7H_2O 300 sodium molybdate 2H_2O 2, boric acid 10, sodium fluoride 2, sodium selenite 5H_2O 1, manganese sulphate 50, zinc carbonate (omitted from experimental [zinc deficient] diet) 200
The total diet contained: metabolizable energy 13.13 MJ/kg; Ca 26.4 g/kg; phosphorus 141 g/kg; protein 260 g/kg; zinc content in control diet was 120–130 ppm; zinc content in the deficient diet was 20–35 ppm

the zinc-deficient diet developed lesions of parakeratosis, mild hyperkeratosis, alterations in germinal epithelium, erosions, ulcerations, vesiculation, alopecia and inflammation of the skin. These changes were prominent in the skin of dependent regions, in areas of stretch and friction, and external contact. These severe epithelial lesions developed by week five. With the change of diet to the basal with zinc-addition, the epithelial lesions were reversible with complete remission of external lesions after five weeks in the diet. These changes are quantified in *Table 14.8*. In an additional series, Sanecki, Corbin and Forbes (1985) fed four English pointer puppies a zinc-deficient diet containing 35 mg/kg zinc for five weeks. Three littermates were fed the same basal diet plus zinc carbonate at the rate of 200 mg/kg diet. Puppies receiving the zinc-added diet received the same amount of food daily as was consumed by their zinc-deficient littermates. In other studies, increased levels of dietary calcium and constant low levels of zinc of 35 mg/kg, skin lesions increased in severity with increasing levels of calcium, indicating that zinc requirements based on maintaining skin integrity increase with increasing dietary calcium levels. Rankings of microscopic lesions of the skin are quantified in *Table 14.9*. The buccal mucosa was markedly thick in pups fed the zinc-deficient diet and the thymus of the zinc-deficient puppies were about half the size of puppies consuming the control diet. Lymph nodes of the zinc-deficient puppies were markedly atrophic (*Table 14.10*).

Calcium metabolism and skeletal development in young Great Dane dogs were studied by Hazewinkel *et al.* (1987) at various calcium-phosphorus levels on a dry-food basis. The following levels were studied: LC (5.5 g/kg Ca, 9.0 g/kg P); NC (11.0 g/kg Ca, 9.0 g/kg P); HC (33 g/kg Ca, 9.0 g/kg P); HCP (33 g/kg Ca, 30 g/kg P).

Table 14.8 AVERAGE WEEKLY SCORE OF GROSS EXTERNAL LESIONS IN PUPPIES FED A ZINC DEFICIENT DIET

Puppy no.	Feeding week										
	1	2	3	4	5	6	7	8	9	10	11
	Zinc-deficient diet					*Control diet*					
1	0	0.5	2.5	4.0	5	5	3	2	1.5	1	0
2	0	0.5	2.0	2.5	4	4.5	3	2	1.0	0	0
3	0	0.5	2.0	2.5	4	4.0	3	2	1.0	0	0
	Control diet					*Zinc-deficient diet*					
4	0	0	0	0	0	0	0	1	2.5	2.5	3
5	0	0	0	0	0	0	0	1	2.5	2.5	4

From Sanecki, Corbin and Forbes (1982)
Gross lesions in the epithelium were ranked according to the following criteria:
0, no clinical change
1, mild change consisting of erythema involving the pad, paws, tail, base of tail, perineal area, chin and mouth
2, eruptions of papules limited to the areas of the pads, paws, tail, perineal area, chin, and mouth, with patchy erythema over other areas of the body such as the ventral part of the abdomen, groin, axilla, flank, and thigh
3, eruptions over other areas, such as ventral part of the abdomen, groin, axilla, flank and thigh, and crusting of areas listed in No. 1, along with swelling of the paws
4, formation of exudate, crusts, small erosions, and mild alopecia over an area with progressively increased size of those areas listed in No. 1
5, formation of open sores in any area, with purulent exudate, crusts, papules, alopecia, and exudate over most of the body

Table 14.9 RANKING OF MICROSCOPIC SKIN LESIONS IN PUPPIES FED A ZINC DEFICIENT DIET WITH INCREASED LEVELS OF DIETARY CALCIUM

Puppy	Feeding week										
	1	2	3	4	5	6	7	8	9	10	11
	Zinc-deficient diet					Control diet					
1	0	0	1	2	4	4	2	1	1	0	0
2	0	0	1	1	4	4	3	2.5	1	0	0
3	0	0	1	3	3	3	2	0	0	0	0
	Control diet					Zinc-deficient diet					
4	0	0	0	0	0	0	0	1	1	1	2.5
5	0	0	0	0	0	0	1	1	2.5	3	4

From Sanecki, Corbin and Forbes (1985)
Histological changes were ranked in severity according to the following criteria:
0, no lesion seen
1, suggestion of thickening of any or all of the epithelial cell layers
2, definite thickening of the layers
3, severe thickening of the layers with retention of nuclear material in the stratum corneum and keratin layers
4, vacuolation of the cytoplasm of the cells in any of the layers, distortion of these layers, severe thickening, focal erosion, infiltration of neutrophils, lymphocytes or macrophages
Half steps in ranking index were used when the lesion did not fit into one or the other category

Calcium absorption coefficients were comparable in the NC, HC, and HCP groups and almost double those for the LC dogs. Histologically, severe irregularities in cartilage maturation with clinical and radiological appearances of radius curvus syndrome were most pronounced in the HC and HCP dogs, and were absent in the LC dogs. Pathological fractures appeared in the LC dogs. Hazewinkel et al. (1987) recommended avoidance of a high Ca intake with or without a constant Ca:P ratio to minimize abnormal skeletal development.

Iron requirement the growing English pointer puppies was determined by feeding a casein-based, fibre-free diet containing 5, 30, 55, 80, 105 or 130 mg Fe (supplied as $FeSO_4 \cdot 7H_2O$)/kg diet. These diets were fed to seven-week-old puppies with an initial weight of 2.62 kg for a period of 30 days. Based on the dose-response curve obtained, it was established that dietary Fe at a level of 84 mg/kg of diet maximized haemoglobin concentration and was more than the 81 mg/kg diet required for

Table 14.10 AVERAGE WEIGHT OF THE MESENTERIC LYMPH NODES PER PUP AND AVERAGE WEIGHT OF THE MESENTERIC LYMPH NODES/kg BODYWEIGHT FOR PUPS FED THE CONTROL OR Zn-DEFICIENT DIET

	Lymph node weight	
Diet	Per pup (g)	g/kg of body weight
Zn-deficient	6.4 ± 0.57[a]	2.6 ± 0.57[a]
Control	17.9 ± 0.29[b]	4.6 ± 0.29[b]

From Sanecki, Corbin and Forbes (1985)
[a,b]($0.001 < P < 0.01$) for numbers with different superscripts.
Data are expressed as average ± SD

maximizing haematocrit (Chausow and Czarnecki-Maulden, 1987). Inclusion of dietary phytate or fibre may decrease the efficiency of iron absorption, thus the iron requirement in conventional diets may be higher than levels determined with the casein-based, fibre free diet.

Methods of feeding influence behaviour. Alexander and Wood (1987) in comparing free-choice of self-feeding, time-restricted meal feeding and food-restricted meal feeding observed differences between each method. Free-choice feeding influences noise abatement by eliminating barking at mealtime. Boredom is reduced by frequent trips to the feeder and the dog consumes smaller amounts of food throughout the day, which also reduces coprophagy. Dogs housed together and fed free-choice may tend to overeat as they will be spurred by competition.

Conclusions

Dietary nutritional requirements of dogs are dependent on the dog's physiological status and the levels plus availability of other dietary nutrients. More is known about the nutrient requirements of dogs than is known about the nutrient requirements of humans. America's dogs receiving only good commercial diets and water consume a better balanced diet than is eaten by America's children.

References

ALEXANDER, J.E. and WODD, L.L. (1987). *Canine Practice*, **14**, March–April
ALLEN, S.E., FAHEY, G.C., CORBIN, J.E., PUGH, J.L. and FRANKLIN, R.A. (1981). *Journal of Animal Science*, **53**, 1538–1544
AMTSBERG, G. (1987). International Symposium, *Nutrition, Malnutrition and Diet of Dogs and Cats—1987*, p. 7. Institute for Animal Nutrition, Tierarztliche Hochschule Hannover
BLAZA, S.E., BURGER, I.H., HOLME, D.W. and KENDALL, P.T. (1982). *Journal of Nutrition*, **112**, 2033–2042
BOOLES, D. and RAINBIRD, A. (1987). International Symposium, *Nutrition, Malnutrition and Diet of Dogs and Cats—1987*, p. 23. Institute for Animal Nutrition, Tierarztliche Hochschule Hannover
BURNS, R.A. and MILNER, J.A. (1981). *Journal of Nutrition*, **111**, 2117–2124
BURNS, R.A. and MILNER, J.A. (1982). *Journal of Nutrition*, **112**, 447–452
BURROWS, C.F., KRONFELD, D.S., BANTA, C.A. and MERRITT, A.M. (1982). *Journal of Nutrition*, **112**, 1726–1732
CHAUSOW, D.G. and CZARNECKI-MAULDEN, G.L. (1987). *Journal of Nutrition*, **117**, 928–932
CZARNECKI, G.L. and BAKER, D.H. (1982). *Journal of Animal Science*, **55**, 1405–1410
CZARNECKI, G.L. and BAKER, D.H. (1984). *Journal of Nutrition*, **114**, 581–590
CZARNECKI, G.L., HIRAKAWA, D.A. and BAKER, D.H. (1985). *Journal of Nutrition*, **115**, 743–752
DONOGHUE, S., KRONFELD, D.S. and BATA, C.A. (1987). International Symposium, *Nutrition, Malnutrition and Diet of Dogs and Cats—1984*, p. 7. Institute for Animal Nutrition, Tierarztliche Hochschule Hannover
GRANDJEAN, D., PARAGON, B.M. and GRANDJEAN, R. (1987). International Symposium,

Nutrition, Malnutrition and Diet of Dogs and Cats—1987, p. 23. Institute for Animal Nutrition, Tierarztliche Hochschule Hannover
HAZEWINKEL, H.A.W., HOW, K.L., BOSCH, R., GOEDEGEBUURE, S.A. and VOORHORT, G. (1987a). International Symposium, *Nutrition, Malnutrition and Diet of Dogs and Cats—1987*, p. 35. Institute for Animal Nutrition, Tierarztliche Hochschule Hannover
HAZEWINKEL, H.A.W., VAN'T KLOOSTER, A.T., VOORHOUT, G. and GOEDEGEBUURE, S.A. (1987b). International Symposium, *Nutrition, Malnutrition and Diet of Dogs and Cats—1987*, p. 53. Institute for Animal Nutrition, Tierarztliche Hochschule Hannover
HIRAKAWA, D.A. and BAKER, D.H. (1985). *Nutrition Research*, **5**, 631–642
HIRAKAWA, D.A.S. (1986). The Role of Lysine and Sulfur Amino Acids in Canine Nutrition, PhD Thesis, University of Illinois
KLOCHE, B., INGWERSEN, M., MEYER, H. and SCHÜNEMANN, C. (1987). International Symposium, *Nutrition, Malnutrition and Diet of Dogs and Cats—1987*, p. 15. Institute for Animal Nutrition, Tierarztliche Hochschule Hannover
LAVELLE, R.B. (1987). International Symposium, *Nutrition, Malnutrition and Diet of Dogs and Cats—1987*, p. 53. Institute for Animal Nutrition, Tierarztliche Hochschule Hannover
LEIBETSEDER, J. (1987). International Symposium, *Nutrition, Malnutrition and Diet of Dogs and Cats—1987*, pp. 35–37. Institute for Animal Nutrition, Tierarztliche Hochschule Hannover
LOEW, E. (1987). *DVM*, **18** (11), 18
MALLUCHE, H.H., MATTHEWS, C., FAUGERE, M., FANTI, P., ENDRES, D.B. and FRIEDLER, R.M. (1986). *Endocrinology*, **119**, 1298–1304
MÄNNER, K., BRONSCH, K. and WAGNER, W. (1987). International Symposium, *Nutrition, Malnutrition and Diet of Dogs and Cats—1987*, p. 21. Institute for Animal Nutrition, Tierarztliche Hochschule Hannover
MEYER, H.C., SCHÜNEMANN, C., ELBERS, H. and JUNKER, S. (1987). International Symposium, *Nutrition, Malnutrition and Diet of Dogs and Cats—1987*, p. 13. Institute for Animal Nutrition, Tierarztliche Hochschule Hannover
MILNER, J.A. (1979). *Journal of Nutrition*, **109**, 1161–1167
MILNER, J.A. (1981). *Journal of Nutrition*, **111**, 40–45
MILNER, J.A. (1984). In *5th Annual Pet Food Institute Technical Symposium*, pp.25–37
MOORE, M.L., FOTTLER, H.J., FAHEY, G.C. and CORBIN, J.E. (1980). *Journal of Animal Science*, **50**, 892–896
NATIONAL RESEARCH COUNCIL (1974). *Nutrient Requirements of Dogs*. Revised 1974
NATIONAL RESEARCH COUNCIL (1985). *Nutrient Requirements of Dogs*. Revised 1985
ONTKO, J.A., WUTHIER, R.E. and PHILLIPS, P.H. (1957). *Journal Nutrition*, **62**, 163–169
RAINBIRD, A. (1987). International Symposium, *Nutrition, Malnutrition and Diet of Dogs and Cats—1987*, p. 25. Institute for Animal Nutrition, Tierarztliche Hochschule Hannover
ROBERTSON, B.T. and BURNS, M.J. (1963). *American Journal of Veterinary Research*, **24**, 997–1002
SANECKI, R.K., CORBIN, J.E. and FORBES, R.M. (1985). *American Journal of Veterinary Research*, **46**, 2120–2123
SANECKI, R.K., CORBIN, J.E. and FORBES, R.M. (1982). *American Journal of Veterinary Research*, **43**, 1642–1646
SCHÜNEMANN, C., MÜHIUM, A. and MEYER, H. (1987). International Symposium,

Nutrition, Malnutrition and Diet of Dogs and Cats—1987, p. 11. Institute for Animal Nutrition, Tierärztliche Hochschule Hannover

SCHÜNEMANN, C. and INGWERSEN, M. (1987). International Symposium, *Nutrition, Malnutrition and Diet of Dogs and Cats—1987*, pp. 17–19. Institute for Animal Nutrition, Tierärztliche Hochschule Hannover

WHEATLEY, V.R. and SHER, D.W. (1961). *Journal of Investigative Dermatology*, **36**, 169–170

LIST OF PARTICIPANTS

The twenty-second Feed Manufacturers Conference was organized by the following programme committee:

Mr J.W.C. Allen (David Patton Ltd)
Dr C. Brenninkmeijer (Hendrix Voeders, BV)
Dr L.G. Chubb (Private Consultant)
Mr P.D. Foxcroft (Prosper de Mulder, Ltd)
Mr J.J. Holmes (E.B. Bradshaw and Sons, Ltd)
Mr F.G. Perry (BP Nutrition (UK), Ltd)
Mr G. Phillips (Private Consultant)
Mr J.R. Pickford (Tecracon, Ltd)
Mr J.S.K. Round (Nitrovit, Ltd)
Mr M.H. Stranks (MAFF Bristol)
Dr A.J. Taylor (BOCM Silcock, Ltd)
Mr R.J. Thompson (Preston Farmers, Ltd)

Dr K.N. Boorman
Professor P.J. Buttery
Dr D.J.A. Cole (Chairman)
Dr P.C. Garnsworthy
Dr W. Haresign (Secretary) } University of Nottingham
Professor G.E. Lamming
Professor D. Lewis
Dr J. Wiseman

The twenty-second Conference was held at the School of Agriculture, Sutton Bonington, 6th–8th January 1988, and the Committee would like to thank the various authors for their valuable contributions. The following persons registered for the meeting:

Adams, Dr C.A.	Kemin Europa NV, Industriezone Wolfstee, 2410 Herentals, Belgium
Adamson, Mr A.H.	ADAS, Government Buildings, Westbury on Trym, Bristol BS10 6NJ
Aerts, Professor J.V.	Tech. State Univ. of Chemistry, Textile and Agriculture, Voskenslaan 270, 9,000 Gent, Belgium
Ainsworth, Ms H.	MAFF, Room 183, Great Westminster House, Horseferry Road, London
Alderman, Mr G.	Hunters Moon, Pearmans Glade, Shinfield Road, Shinfield, Reading RG2 9BE
Allder, Mr M.	Smith Kline Animal Health Ltd, Cavendish Road, Stevenage, Herts SG1 2EJ
Allen, Mr J.	Fisons plc, Animal Health, 12 Derby Road, Loughborough, Leics

List of participants

Allen, Dr J.D.	Eastman Chemical International AG, Kodak Ltd, Kirkby, Liverpool
Allen, Mr W.	David Patton Ltd, Milltown Mills, Dawson Street, Monaghan, Ireland
Arnott, Mr J.	Charles E. Ford Ltd, Old Dock, Avonmouth, Bristol BS11 9BU
Ashington, Mr B.	Peter Hand (GB) Ltd, Eastern Division, Tomo Building, Creeting Road, Stowmarket, Suffolk IP11 5AY
Ashley, Dr J.H.	AEC, 41 Avenue Bosquet, Paris 75007, France
Aspland, Mr P.	Aspland and James Ltd, 118 Bridge Street, Chatteris, Cambridge
Atherton, Dr D.	Thomson & Joseph Ltd, 119 Plumstead Road, Norwich, Norfolk
Barnes, Mr W.J.	BP Nutrition (UK) Ltd, Wincham, Northwich, Cheshire
Barrie, Mr M.	Elanco Products, Kingsclere Road, Basingstoke, Hants
Barrigan, Mr W.	Unichema Chemicals Ltd, Bebington, Wirral, Merseyside
Barron, Mr K.J.	USC (Industrial) Ltd, 8 Heddon Street, London W1R 8BP
Bartram, Dr C.	Feed Flavours (Europe) Ltd, Waterlip, Cranmore, Shepton Mallet, Somerset
Bates, Dr A.	Vitrition Ltd, Ryhall Road, Stamford, Lincs
Baxter, Mr A.	Smith Kline Animal Health Ltd, Cavendish Road, Stevenage, Herts SG1 2EJ
Beardsworth, Dr P.M.	Unilever Research, Colworth House, Sharnbrook, Bedfordshire MK44 1LQ
Beaumont, Mr D.	BP Nutrition (UK) Ltd, Wincham, Northwich, Cheshire
Beer, Mr J.H.	W & J Pye Ltd, Fleet Square, Lancaster LA1 1HA
Beer, Dr J.V.	The Game Conservancy, Fordingbridge, Hampshire SP6 1EF
Bell, Mr J.G.	Preston Farmers Ltd, Kinross, New Hall Lane, Preston, Lancs
Berkouwer, Ir C.	Kloek BV, PO Box 1147, 3300 BC Dordrecht, Holland
Birch, Ms V.	Cargill UK Ltd, Milling Division, Tilbury Docks, Tilbury, Essex
Bishop, Miss R.	University of Nottingham, School of Agriculture, Sutton Bonington, Loughborough, Leics LE12 5RD
Boorman, Dr K.N.	University of Nottingham, School of Agriculture, Sutton Bonington, Loughborough, Leics LE12 5RD
Booth, Miss A.	Page Feeds Ltd, Mill Lane, Tadcaster, N. Yorks
Bourne, Mr S.J.	Cranswick Mill Ltd, The Airfield, Cranswick, Driffield, N. Humberside
Bowtell, Mr M.G.	NUTEC Ltd, Greenhills Centre, Tallaght, Dublin 24, Ireland
Boyd, Dr J.	BOCM Silcock, Olympia Mills, Barlby Road, Selby, Yorkshire
Bradfield, Miss G.	University of Nottingham, School of Agriculture, Sutton Bonington, Loughborough, Leics LE12 5RD
Brenninkmeijer, Dr C.	Hendrix Voeders, Postbus 1, 5830 Boxmeer, Holland
Breukink, Ir L.M.	Orfam BV, Peppelkade U6, 3992 AK Houten, Netherlands
Brooking, Miss P.	W. J. Oldacre Ltd, Cleeve Hall, Bishops Cleeve, Cheltenham, Glos
Broom, Mr P.	French's (Feeds & Seeds) Ltd, Hennock Road, Marsh Barton, Exeter
Brosnan, Mr J.P.	Volac Ltd, Orwell, Royston, Herts
Brown, Mr G.	Colborn Dawes Nutrition, Heanor Gate, Heanor, Derbyshire

List of participants

Brown, Mr M.	R.J. Seaman & Sons Ltd, Egmere, Walsingham, Norfolk NR22 6BD
Brumby, Dr P.E.	WCF, North East Regional Office, Burn Lane, Hexham NE46 3HU
Buchanan, Ms A.G.	Elanco Products Ltd, Kingsclere Road, Basingstoke, Hants
Burt, Dr A.W.A.	Burt Research Ltd, 23 Stow Road, Kimbolton, Huntingdon
Bush, Mr T.	Colborn Dawes Nutrition, Heanor Gate, Heanor, Derbyshire
Butcher, Mr D.	Biocon (UK) Ltd, Eardiston, Tenbury, Wells, Worcester WR15 8JJ
Buttery, Professor P.J.	University of Nottingham, School of Agriculture, Sutton Bonington, Loughborough, Leics LE12 5RD
Buysing Damste, Ir B.A.	Trouw International BV, Research & Development, PO Box 50, 3880 AB Putten, Holland
Carlisle, Mr B.	ADAS, St Marys Manor, Beverley HU17 8DN
Carruthers, Miss S.	Eastern Counties Farmers, 136 Fore Street, Ipswich, Suffolk IP7 5QD
Carter, Mr T.	Kemin (UK) Ltd, Waddington, Lincoln LN5 9NT
Chubb, Dr L.G	'Koonunga', 39 Station Road, Harston, Cambridge CB2 5PP
Clarke, Mr A.N.	Four-F Nutrition, Darlington Road, Northallerton DL6 2NW
Clay, Mr J.	Alltech (UK), Units 16–17, Abenbury Way, Wrexham Industrial Estate, Wrexham, Clwyd LL13 9UY
Close, Dr W.H.	AFRC, Inst. for Grassland & Animal Production, Church Lane, Sheffield, Reading RG2 9AQ
Cogan, Ms D.	Butterworth Scientific Ltd, PO Box 63, Bury Street, Guildford, Surrey GU2 5BH
Cole, Dr D.J.A.	University of Nottingham, School of Agriculture, Sutton Bonington, Loughborough, Leics LE12 5RD
Colenso, Mr J.	BP Nutrition (UK) Ltd, Wincham, Northwich, Cheshire
Connolly, Mr J.C.	William Connolly & Sons Ltd, Red Mills, Goresbridge, Co. Kilkenny, Ireland
Cooke, Mr A.P.J.	Friendship Estates Ltd, Old House Farm, Stubbs Walden
Cooke, Dr B.C.	Dalgety Agriculture Ltd, Dalgety House, The Promenade, Clifton, Bristol BS8 3NJ
Corbet, Mr M.A.	J. Waring (Feeds) Ltd, Mission Mill, Musson, Doncaster
Corbin, Professor J.	University of Illinois, 1207 West Gregory Drive, Urbana, Illinois 61801, USA
Cornberg, Mr M.	Vitafoods Ltd, Riverside House, East St, Birkenhead L41 1BY
Courtin, Mr B.	EMC Belgium, Square de Meeusi, 1040 Brussels, Belgium
Cox, Mr N.	S C Associates Ltd, The Limes, Sowerby, Thirsk YO7 1HX
Crawford, Mr J.R.	Carrs Farm Foods, Old Croft, Stanwix, Carlisle, Cumbria CA3 9BA
Crehan, Mr M.P.	Nutec Ltd, Eastern Avenue, Lichfield WS13 7SE
Cullin, Mr A.W.	Forum Feeds, Forum House, Brighton Road, Redhill, Surrey
Dann, Mr R.	Rod Dann Marketing, Holme Farm, Cropton, Pickering, N. Yorks
Davies, Mr J.	Insta-Pro International (WE) Ltd, 11 Lypiatt Terrace, Cheltenham GL50 2SX
Davies, Mr T.	ADAS, Woodthorne, Wolverhampton WV6 8TQ

List of participants

Dawson, Dr R.	AHDA (UK) Ltd, 111, High Street, Tonbridge, Kent TN9 1DL
De Belder, Mr R.	EMC Belgium, Square de Meeusi, 1040 Brussels, Belgium
De Bruyne, Mr K.	EMC Belgium, Square de Meeusi, 1040 Brussels, Belgium
De Heus, Ir J.	Pricor BV, Postbus 51, 3420 DB Oudewater, Holland
De La Hunt, Mr T.E.	National Foods Ltd, PO Box 269, Harare, Zimbabwe
De Man, Dr T.H.J.	Kerkstraat 40, 3741 AK Baarn, Netherlands
Dean, Mr R.	Deans Agric. Assn, 39 Steeles, London WW3 4RG
Deverell, Mr P.	BASF (UK) Ltd, PO Box 4, Earl Road, Cheadle Hulme, Cheadle, Cheshire SK8 6QG
Dewhurst, Mr R.J.	University of Bristol, Dept Animal Husbandry, Langford House, Langford, Bristol BS18 7DU
Dixon, Mr D.H.	Brown & Gillmer Ltd, Seville Place, Dublin 1, Ireland
Dobson, Mr M	Elanco Products, Kingsclere Road, Basingstoke, Hants
Douglass, Mrs V.L.	Bernard Matthews plc, Gt Witchingham Hall, Norwich, Norfolk
Duncan, Dr M.S.	Holly Farms Poultry Ind. Inc., PO Box 88, Wilkesboro, North Carolina 28697, USA
Easter, Dr R.A.	University of Illinois, at Urbana-Champaign, Mumford Hall, 1301 West Gregory Drive, Urbana, Illinois 61801, USA
Ebbon, Dr G.P.	BP Research Centre, Chertsey Road, Sunbury on Thames, Middlesex TW16 7LN
Edmunds, Dr B.K.	Pauls Agriculture, Research & Advisory Dept, New Cut West, Ipswich, Suffolk
Edwards, Mr A.	Elanco Products Ltd, Kingsclere Road, Basingstoke, Hants
Edwards, Mr A.	Frank Wright Limited, Blenheim House, Blenheim Road, Ashbourne, Derbys. DE6 1HA
Ellis, Dr N.	71 Moat View, Roslin, Midlothian EH25 9NZ
Ellis-Jones, Miss T.	University of Nottingham School of Agriculture, Sutton Bonington, Loughborough, Leics LE12 5RD
Everington, Mrs J.M.	ADAS, Feed Evaluation Unit, Drayton Manor Drive, Alcester Road, Stratford upon Avon, Warks
Fallon, Dr R.J.	The Agricultural Institute, Grange, Dunsany, Co Meath, Ireland
Fawcett, Mr T.J.F.	Cumberland & Westmorland Farmers, Gilwill Estate, Penrith
Fawthrop, Mr G.	Smith Kline Animal Health Ltd, Cavendish Road, Stevenage, Herts SG1 2EJ
Fisher, Dr C.	IGAP (Poultry Dept), Roslin, Midlothian EH25 99PS
Fisher, Mr D.	Western Foods Ltd, Kines Court, High Street, Nailsea, Bristol BS19 1AW
Fitt, Dr T.	Colborn Dawes Nutrition, Heanor Gate, Heanor, Derbyshire
Flack, Mrs H.	BP Nutrition (UK) Ltd, Wincham, Northwich, Cheshire
Fletcher, Mr C.J.	Aynsome Laboratories, Grange over Sands, Cumbria
Fletcher, Dr J.M.	Unilever Research, Colworth House, Sharnbrook, Bedfordshire MK44 1LQ
Foxcroft, Dr P.D.	Prosper de Mulder Ltd, Ings Road, Doncaster
Franco, Dr M.	Sildamin Spa, 27010 Sostegno di Spessa, Pavia, Italy

List of participants

Frape, Dr D.L.	British Horse Feeds BHF Ltd, Victoria House, 50 Albert St, Rugby, Warks CV21 2RH
Fullarton, Mr P.J.	Finfeeds Ltd, Forum House, Brighton Road, Redhill
Gardner, Mr A.	Park Tonks Ltd, 104 High Street, Great Abington, Cambs
Garnsworthy, Dr P.C.	University of Nottingham, School of Agriculture, Sutton Bonington, Loughborough, Leics LE12 5RD
Geddes, Mr N.	Nutec Ltd, Eastern Avenue, Lichfield WS13 7SE
Gibson, Mr J.E	Park Nutrition Ltd, High Street Industrial Estate, Heckington, Sleaford, Lincs
Gilbert, Mr R.	Feed International, 18 Chapel Street, Petersfield, Hants GU32 3DZ
Gill, Dr R.D.	BOCM Silcock Ltd, Basing View, Basingstoke, Hants
Gillespie, Miss F.	Rumenco, UM Farm Liquids Division, Stretton House, Derby Road, Burton on Trent, Staffs
Gous, Dr R.M.	University of Natal, PO Box 375, Pietermaritzburg 3200, South Africa
Grant, Mr T.	PVA, Cambridge House, Woolpit, Bury St Edmunds, Suffolk
Gray, Mr B.	Boliden (UK) Ltd, Yorkshire House, East Parade, Leeds LS1 5SH
Green, Dr S.	AEC, 03600 Commentry, France
Grierson, Mr R.	Park Tonks Ltd, 48 North Road, Great Abington, Cambridge CB1 6AS
Haggar, Mr C.W.	Britphos Ltd, Rawdon House, Green Lane, Yeadon, Leeds LS19 7BY
Hall, Mr G.	Kemin (UK) Ltd, Waddington, Lincoln LN5 9NT
Hamilton, Mr G.A.	Agil Holdings Ltd, Fishponds Road, Wokingham, Berks RG11 2QL
Hanley, Mr B.	Biocon Ltd, Kilnagleary, Carrigaline, Co. Cork, Ireland
Hannagan, Mr M.J.	Dalgety Agriculture Ltd, Dalgety House, The Promenade, Clifton, Bristol BS8 3NJ
Harding, Mr G.	ABM Brewing & Food Group, Woodley, Stockport, Cheshire SK6 1PQ
Hardy, Dr B.	Dalgety Agriculture Ltd, Dalgety House, The Promenade, Clifton, Bristol BS8 3NJ
Haresign, Dr W.	University of Nottingham, School of Agriculture, Sutton Bonington, Loughborough, Leics LE12 5RD
Harker, Dr A.J.	Carrs Farm Foods, Old Croft, Stanwix, Carlisle, Cumbria CA3 9BA
Harland, Dr J.A.	British Sugar plc, PO Box 26, Oundle Road, Peterborough PE2 9QU
Harrington, Mr T.	ADAS, Block C. Government Buildings, Brooklands Avenue, Cambridge CB2 2DR
Harris, Mr W.J.	Vitafoods Ltd, East Street, Birkenhead L41 1BY
Harrison, J.M.	Peter Hand (GB) Ltd, 15–19 Church Road, Stanmore, Middlesex
Haythornwaite, Mr A.	Willow Lodge, Church Road, Warton, Preston, Lancs PR4 1BD
Hazzledine, Mr M.J.	Dalgety Agriculture Ltd, Dalgety House, The Promenade, Clifton, Bristol BS8 3NJ
Hegeman, Mr F.	Borculo Whey Products, Postbus 46, 7270 AA Borculo, Netherlands
Hemingway, Professor G.	University of Glasgow, Veterinary School, Bearsden, Glasgow
Henderson, Mr I.R.	Chapman & Frearson Ltd, Victoria Street, Grimsby DN31 1PX

List of participants

Henics, Dr Z.	University of Agriculture Keszthely, Hungary, Faculty of Animal Science, Kaposvar Pb. 16. h-7401, Hungary
Higgins, Mr L.	Pauls Agriculture Ltd, PO Box 39, 47 Key Street, Ipswich, Suffolk IP4 1BX
Higgins, Miss V.A.G.	Frank Wright Ltd, Blenheim House, Blenheim Road, Ashbourne, Derbys DE6 1HA
Hill, Dr R.	Royal Veterinary College, University of London, Dept of Animal Health & Production, Boltons Park, Potters Bar, Herts
Hine, Mr J.	BP Nutrition (UK) Ltd, Wincham, Northwich, Cheshire CW9 6DF
Hirst, Mr J.M.	John Hirst (Animal Feedstuffs) Ltd, Sworton Heath Farm, Swineyard Lane, High Leigh, Knutsford, Cheshire WH16 0RYY
Hirst, Mr M.	John Hirst (Animal Feedstuffs) Ltd, Sworton Heath Farm, Swineyard Lane, High Leigh, Knutsford, Cheshire WH16 0RY
Hitchins, Mr C.T.	Favor Parker Ltd, The Hall, Stoke Ferry, Kings Lynn, Norfolk FE33 9SE
Hockey, Mr R.	Smith Kline Animal Health Ltd, Cavendish Road, Stevenage, Herts SG1 2EJ
Hof, Ir J.B.	Windmill Holland, PO Box 58, 3130 AB Vlaardingen-NL, Holland
Hollows, Mr I.W.	Ian Hollows Feed Supplements, Wood Farm, Coppice Lane, Coton, Nr Whitchurch, Shropshire SY13 3LT
Holma, M.B.M.	Raision Tehtaat, PL 101, 21201 Raisio, Finland
Holmes, Mr J.J.	Laburnum Cottage, Thwing, Driffield, North Humberside
Hopkins, Mr J.R.	MAFF, Lawnswood, Leeds LS16 5PY
Houseman, Dr R.	EMC-Belgium, C/o Rawdon House, Green Lane, Yeadon, Leeds
Hudson, Mr K.A.	Beecham Animal Health, Gt West Road, Brentford, Middlesex
Hughes, Dr J.	Carrs Farm Foods, Old Croft, Stanwix, Carlisle CA3 9BA
Hyam, Mr J.M.	Nutral SA, Apartado 58, 28770 Colmenas, Viejo, Madrid, Spain
Inborr, Mr J.	Finfeeds Ltd, Forum House, 41–57 Brighton Road, Redhill, Surrey RH1 6YS
Ingham, Mr P.A.	A One Feed Supplements Ltd, North Hill, By RAF Dishforth, Thirsk, N. Yorks
Jagger, Dr S.	S C Associates Ltd, The Limes, Sowerby, Thirsk YO7 1HX
Jansegers, Dr L.	Kerselarendreef 82, B 2782 St Gillis Waas, Belgium
Jardine, Mr G.	Unitrition International Ltd, Basing View, Basingstoke, Hants
Jones, Mr E.J.	Format International Ltd, Owen House, Heathside Crescent, Woking, Surrey
Jones, Dr H.F.	Tate and Lyle, PO Box 68, Reading RB6
Jones, Mr R.E.	W. F. Tuck & Sons Ltd, Burston, Diss, Norfolk
Keith, Dr M.C.	7 Cliff Park, Cults, Aberdeen AB1 9JT
Kennedy, Mr D.A.	The Flavour Centre, Old Gorsey Lane, Wallasey, Merseyside L44 4AH
Kenyon, Mr P.J.	BOCM Silcock Ltd, Basing View, Basingstoke, Hants
Key, Mr D.S.	United Molasses Co, 167 Regent Road, Liverpool L20 8DD
Keys, Mr J.	J. E. Hemmings & Son Ltd, Barford Mills, Barford, Warwick CV35 8EJ
Kidd, Mr A.G.	ICI Biological Products, PO Box 1, Billingham, Cleveland TS23 1LB

Knight, Dr R.	BP Nutrition (UK) Ltd, Wincham, Northwich, Cheshire
Knox, Mr G.J.	Dalgety Agriculture Ltd, 102 Corporation St, Belfast, N. Ireland
Koch, Dr F.	Degussa AG ZN Wolfgang, Abt. IC-ATAV, Rodenbacher Chaussee 4, D-6450 Hanau 11
Lamming, Professor G.E.	University of Nottingham, School of Agriculture, Sutton Bonington, Loughborough, Leics LE12 5RD
Lane, Mr P.F.	Park Nutrition Ltd, High Street Industrial Estate, Heckington, Sleaford, Lincs
Law, Mr J.R.	Sheldon Jones plc, Priory Hill, West St, Wells, Somerset
Lee, Dr P.A.	IGAP, Shinfield Research Station, Reading RG2 9AQ
Lehtimaki, Mr T.	Suomen Rehu Oy, Kolmas Iinja 22, 00530 Helsinki, Finland
Lewis, Professor D.	University of Nottingham, School of Agriculture, Sutton Bonington, Loughborough, Leics LE12 5RD
Lewis, Dr M.	Edinburgh School of Agriculture, West Mains Road, Edinburgh EH9 3JG
Livingston, Mr D.H	Edward Baker Ltd, Cornard Mills, Sudbury, Suffolk CO10 0JA
Livingstone, Dr R.M.	Rowett Research Institute, Bucksburn, Aberdeen AB2 9SB
Longland, Dr A.C.	IGAP, Pig Department, Shinfield, Reading, Berks RG2 9AQ
Lonsdale, Dr C.R.	FAC (UK) Ltd, 11 Northfield Way, Appleton Roebuck, Yorks YO5 7EA
Low, Dr A.G.	IGAP, Shinfield, Reading, Berks RG2 9AQ
Lowe, Mr J.A.	Heygate & Sons Ltd, Bugbrooke Mills, Northampton NN7 3QH
Lyons, Dr T.P.	Alltech, Inc., 3031 Catnip Hill Pike, Nicholasville, KY 40356, USA
Mackey, Mr W.S.	G. E. Mclarnon & Sons Ltd, 126 Moneynick Road, Randalstown, Antrim BT41 3HU, N. Ireland
Malandra, Dr F.	Sildamin SPA, 27010 Sostegno di Spessa, Pavia, Italy
Marangos, Dr T.	Peter Hand (GB) Ltd, Southern Division, North Way, Walworth Industrial Estate, Andover, Hants SP10 5AZ
Mark, Mrs R.F.	Pauls Agriculture Ltd, Road One Industrial Estate, Winsford, Cheshire
Marriage, Mr P.	W & H Marriage & Sons Ltd, Chelmer Mills, Chelmsford CM1 1PN
Marsden, Dr M.	Peter Hand (GB) Ltd, Tomo Building, Creeting Road, Stowmarket, Suffolk IP11 5AY
Marsh, Mr R.	University of Nottingham, School of Agriculture, Sutton Bonington, Loughborough, Leics LE12 5RD
Martin, Mr W.S.D.	Curry Morrison & Co Ltd, Northern Road, Belfast Harbour Estate, Belfast, N. Ireland
McCollum, Mr I.	BP Nutrition (NI) Ltd, 8 Governor's Place, Carrick Fergus, Co. Antrim, N. Ireland
McDonald, Dr M.W.	IGAP, Poultry Division, Roslin, Midlothian EH25 9PS, Scotland
McFarquhar, Dr A.	Bernard Matthews plc, Gt Witchingham Hall, Norwich, Norfolk
McIlmoyle, Dr A.	Colborn Dawes Nutrition, Musgrave Park Industrial Estate, Stockman's Way, Belfast, N. Ireland
McLaughlin, Mr G.	International Additives Ltd, The Flavour Centre, Old Gorsey Lane, Wallasey L44 4AH, Merseyside
McLean, Dr A.F.	Volac Ltd, Orwell, Herts SG8 5QX

List of participants

McLean, Mr D.R.	R. J. Seaman & Sons Ltd, Egmere, Walsingham, Norfolk NR22 6BD
McNab, Dr J.M.	IGAP, Poultry Dept, Roslin Research Station, Roslin, Midlothian EH25 9PS
Meggison, Dr P.A.	Colborn-Dawes Aust (PTY) Ltd, PO Box 279, Wagga Wagga, NSW 2650, Australia
Merrin, Mr A.G.	A One Feed Supplements, North Hill, By RAF Dishforth, Thirsk
Miller, Mr C.	Waterford Co-op, Dungarvan, Co. Waterford, Ireland
Mills, Mr. C	University of Nottingham, School of Agriculture, Sutton Bonington, Loughborough, Leics LE12 5RD
Milner, Mr C.K.	Beecham Animal Health Res. Centre, Walton Oaks, Dorking Road, Tadworth, Surrey KT20 7NT
Moore, Mr D.R.	David Moore (Flavours) Ltd, 39 High Street, Harpenden, Herts AL5 2RU
Morgan, Dr J.T.	Four Gables, The Fosseway, Cheltenham, Glos GL54 1JU
Morrey, Mr A.	Farmore Farmers, Craven Arms, Shropshire
Morris, Professor T.R.	University of Reading, Dept of Agriculture, Earley Gate, Reading RG6 2AT
Mounsey, Mr H.	The Feed Compounder, Abney House, Baslow, Derbyshire DE4 1RZ
Mullan, Dr B.P.	IGAP, Shinfield, Reading RG2 9AQ
Murphy, Mr J.	Grinsted Products Ltd, Northern Way, Bury St. Edmunds, Suffolk
Murray, Mr A.G.	W C F, North East Regional Office, Barn Lane, Hexham NE46 3HJ
Murray, Mr F.	Kerry Foods Ltd, Hillingdon Hill, Uxbridge, Middlesex
Naylor, Mr P.	International Additives Ltd, The Flavour Centre, Old Gorsey Lane, Wallasey, Merseyside L44 4AH
Nelson, Mr R.	RMB Animal Health Ltd, Rainham Road South, Dagenham, Essex RM10 7XS
Newcombe, Mrs J.	University of Nottingham, School of Agriculture, Sutton Bonington, Loughborough, Leics LE12 5RD
Nixey, Mr C.	British United Turkeys Ltd, Hockenhall Hall, Tarvin, Chester CH3 8LE
Noordenbos, Ir H.U.	Orffa E.V., Houten, The Netherlands
Norcott, Mr G.	Inter-Lacto Ltd, 5 Garlinge Road, Tunbridge Wells, Kent TN4 0NR
O'Beirne, Mr P.	Cyanamid of Great Britain Ltd, Fareham Road, Gosport, Hants PO13 0A
O'Toole, Mr C.	Park Tonks Ltd, 48 North Road, Great Abington, Cambridge CB1 6AS
Oldham, Dr J.D.	Edinburgh School of Agriculture, Kings Buildings, West Mains Road, Edinburgh EH9 3JG
Owers, Dr M.	Pauls Agriculture Ltd, PO Box 39, 47 Key Street, Ipswich, Suffolk IP4 1BX
Paling, Mr I.	Brookside Metal Company Ltd, Bilston Lane, Willenhall, West Midlands
Pallister, Dr S.M.	Nutec Ltd, Eastern Avenue, Lichfield WS13 7SE
Palmer, Mr F.G.	ABR Foods Ltd, Hunters Road, Corby NN17 1JR
Papasolomontos, Dr S.	Dalgety Agriculture Ltd, Dalgety House, The Promenade, Clifton, Bristol BS8 3NJ
Partridge, Miss M.	Pauls Agriculture Ltd, Lords Meadow Mill, Crediton, Devon
Pass, Mr R.T.	Pentlands Scotch Whisky Research Ltd, 84 Slateford Road, Edinburgh EH11 1QU

Pattinson, Mr F.	Dow Agriculture, Catchmore Court, Brand Street, Hitchin, Hants SG5 1HZ
Pearce, Mr D.P.	Degussa Ltd, Paul Ungerer House, Earl Road, Stanley Green, Handforth, Wilmslow, Cheshire SK9 3RL
Perry, Mr F.G.	BP Nutrition (UK) Ltd, Wincham, Northwich, Cheshire
Pettigrew, Dr J.E.	University of Minnesota, Dept of Animal Science, St Paul, MN 55108, USA
Phillips, Mr G.	Greenway Farm, Greenway Lane, Charlton Kings, Cheltenham GL52 6PL
Pickford, Mr J.R.	Tecracon Ltd, Bocking Hall, Bocking Church Street, Braintree, Essex CM7 5JX
Pickles, Dr R.W.	Elanco Products Ltd, Kingsclere Road, Basingstoke, Hants
Pike, Dr I.H.	IAFM, Hoval House, Mutton Lane, Potters Bar EN6 3AR
Piva, Professor G.	Faculta Di Agraria, Piacenza, Ist Scienze Della Nutrizione, Via E. Parmense 84, 29100 Piacenza, Italy
Plowman, Mr G.B.	G. W. Plowman & Son Ltd, Selby House, High Street, Spalding PE11 1TW
Poornam, Mr P.K.	Format International Ltd, Owen House, Heathside Crescent, Woking, Surrey
Portsmouth, Mr J.I.	Peter Hand (GB) Ltd, 15–19 Church Road, Stanmore, Middlesex
Priest, Mr L.	Borculo Whey Products, Postbus 46, 7270 AA, Borculo, Netherlands
Putnam, Dr M.	Roche Products Ltd, PO Box 8, Welwyn Garden City, Herts AL7 3AY
Pye, Mr R.E.	W. & J. Pye Ltd, Fleet Square, Lancaster LA1 1HA
Queenborough, Miss R.	University of Nottingham, School of Agriculture, Sutton Bonington, Loughborough, Leics LE12 5RD
Rainbird, Dr A.	Waltham Centre for Pet Nutrition, Freeby Lane, Waltham on the Wolds, Melton Mowbray LE14 4RT
Raine, Dr H.	J. Bibby Agriculture Ltd, Oxford Road, Adderbury, Banbury, Oxon
Raper, Mr G.J.	Laboratories Pancosma (UK) Ltd, Anglia Industrial Estate, Saddlebow Road, Kings Lynn, Norfolk
Read, Mr M.	Smith Kline Animal Health Ltd, Cavendish Road, Stevenage, Herts SG1 2EJ
Read, Mr S.	Pauls Agriculture Ltd, PO Box 39, 47 Key Street, Ipswich IP4 1BX
Record, Mr S.J.	Fishers Nutrition Ltd, Cranswick, Driffield, N. Humberside
Reeve, Mr. J.	R.S. Feed Blocks Ltd, Orleigh Mill, Bideford, Devon
Rice, Dr D.A.	Dept of Agriculture for N. Ireland, Veterinary Research Laboratories, Stormont, Belfast, N. Ireland
Rigg, Mr G.	Elanco Products, Kingsclere Road, Basingstoke, Hants
Roberton, Mr D.	Roche Products Ltd, PO Box 8, Welwyn Garden City, Herts AL7 3AY
Robinson, Mrs G.	University of Nottingham, School of Agriculture, Sutton Bonington, Loughborough, Leics LE12 5RD
Roet, Mr R.	Monsanto Europe SA/NV, Avenue de Tervuren 270–272, Tervurenlaan 270–272, B-1150, Brussels, Belgium
Rosen, Dr G.	66 Bathgate Road, London SW19 5PH
Rosillo, Mr J.	University of Nottingham, School of Agriculture, Sutton Bonington, Loughborough, Leics LE12 5RD

List of participants

Round, Mr J.S.K.	Nitrovit (North), J. Bibby Agric. Ltd, Jubilee Mills, Copgrove, Harrogate HG3 3TB
Rowsell, Miss J.	Cyanamid of Great Britain Ltd, Fareham Road, Gosport, Hants PO13 0AS
Ryan, Mr T.	RMB Animal Health Ltd, Rainham Road South, Dagenham, Essex RM10 7XS
Salter, Dr D.	IGAP, Shinfield, Reading, Berks RG2 9AQ
Sandboel, Mr P.	Galenica A/S, Jacob Gades Alle 4, 1, DK-6600 Vejen, Denmark
Santoma, Dr G.	Cyanamid Iverica SA, 8 Apartado 471, 28080 Madrid, Spain
Schoener, Dr F.	BASF, Aktiengesellschaft, Finanzbuchhaltung ZLF/ZA, D-6700 Ludwigshafen/Rhein, W. Germany
Schoner, Dr F.J.	BASF, Tierernahrungsstation, Neumuhle 13, 6745 Offenbach, Federal Republic of Germany
Scott, Mr M.R.	Fergusson Wild & Co Ltd, 3 St Helen's Place, London EC3A 6BD
Segers, Ir L.	Orffa NV, Industriepark, Oudemanstraat 13, 2900 Londerzeel, Belgium
Shearn, Ms A.E.	Peter Hand (GB) Ltd, 15–19 Church Road, Stanmore, Middlesex
Shepperson, Dr N.P.G.	UFAC (UK) Ltd, Waterwitch House, Exeter Road, Newmarket, Suffolk CB7 5TR
Shipston, Mr A.H.	Dalgety Agriculture Ltd, Dalgety House, The Promenade, Clifton, Bristol BS8 3NJ
Shipton, Mr P.	T M Agricom Ltd, Bannister Hall Mill, Higher Walton, Preston, Lancs PR5 4DB
Silcock, Mr R.	Whitworth Bros Ltd, Agricultural Division, Fletton Mills, Peterborough PE2 8AD
Silvester, Miss L.M.	W J Oldacre Ltd, Cleeve Hall, Bishops Cleeve, Nr Cheltenham, Glos
Simpson, Mr A.	Eurofeed Hellas ABEE, Schimatari Viotias, Greece
Singer, Dr M.I.C.	Roche Products Ltd, PO Box 8, Broadwater Road, Welwyn Garden City AL7 3AY
Skov Larsen, Mr C.	Leo Pharmaceutical Products, 55 Industriparken, DK-2750 Ballerup, Denmark
Smith, Mr F.H.	University of Dublin, Faculty of Veterinary Medicine, Ballsbridge, Dublin 4
Snoek, Ir G.	Duphar BV, pob 900 1380 DA, Weesp, Holland
de Siqueira Castro e Solla, L.	Rua de Sao Marcal, 176, 1, 1200 Lisboa, Portugal
Speight, Mr D.	Nitrovit, Nitrovit House, Dalton, Thirsk YO7 3JE
Stainsby, Mr A.K.	BATA Ltd, Railway Street, Malton, N. Yorks YO17 0NU
Statham, Mr R.	Joseph Pyke & Son (Preston) Ltd, Harvey Lane, Golbourne, Warrington, Lancs
Strachan, Mrs P.J.	Unilever Research, Colworth Laboratory, Colworth House, Sharnbrook, Bedfordshire HK44 1LQ
Stranks, Mr M.H.	ADAS, Ministry of Agric. Fisheries & Food, Westbury on Trym, Bristol
Sunderland, Mr R.J.	BP Nutrition (UK) Ltd, Wincham, Northwich, Cheshire CW9 6DF
Taylor, Dr A.J.	BOCM Silcock Ltd, Basing View, Basingstoke, Hants
Taylor, Dr S.J.	VOLAC Ltd, Orwell, Royston, Herts

List of participants

Thacker, Dr P.A.	University of Saskatchewan, Dept of Animal & Poultry Science, Saskatoon, Canada
Thiele, Ir W.J.G.	Postbus 163, 2280 AD Ryswyk, Netherlands
Thomas, Professor P.C.	West of Scotland Agric. College, Auchincruive, Ayr, Scotland
Thompson, Mr D.	BP Nutrition (UK) Ltd, Castlegarde, Coppermore, Co. Limerick, Ireland
Thompson, Dr R.F.	Rumenco, Stretton House, Derby Road, Burton on Trent, Staffs DE13 0DW
Thompson, Mr J.	Feed Flavours Europe Ltd, Waterlip, Cranmore, Shepton Mallet, Somerset
Thompson, Mr R.J.	Preston Farmers Ltd, Kinross, New Hall Lane, Preston, Lancs
Todd, Mr R.	Inter-Lacto Ltd, 5 Garlinge Road, Tunbridge Wells, Kent TN4 0NR
Tolonen, Mr J.	Suomen Rehu Oy, Kolmas Linja 22, 00530 Helsinki, Finland
Tonks, Mr W.P.	Park Tonks Ltd, 48 North Road, Great Abington, Cambridge CB1 6AS
Toplis, Mr P.	Four F Nutrition, Darlington Road, Northallerton, N. Yorkshire DL6 2NW
Twigge, Mr J.R.	BP Nutrition (UK) Ltd, Wincham, Northwich, Cheshire CW9 6DF
Twyford, Mr I.	Brookside Metal Company Ltd, Bilston Lane, Willenhall, West Midlands WV13 2QE
Tyler, Mr A.	Colborn-Dawes Nutrition, Heanor Gate, Heanor, Derbyshire
Underwood, Mr N.	Cyanamid of Great Britain Ltd, Fareham Road, Gosport, Hants PO13 0AS
Van Aelton, Ir G.	Proefstation Aveve Molens, Mierdsedijk, 114, B-2391 Poppel, Belgium
Van Den Broecke, Ir J.	Eurolysine, Rue Ballu 16, 75009 Paris, France
Varley, Dr M.A.	University of Leeds, Dept of Animal Physiology & Nutrition, Leeds LS2 9JT
Vernon, Dr B.G.	Dalgety Agriculture Ltd, Dalgety House, The Promenade, Clifton, Bristol
Wakelam, Mr J.A.	George A. Palmer Ltd, Oxney Road Peterborough PE 15Y2
Wallace, Mr J.R.	Volac Ltd, Orwell, Royston, Herts
Wallbank, Mr D.	Vitafoods Ltd, Riverside House, East Street, Birkenhead L41 1BY
Ward, Mr J.H.	Nitrovit (South), Brook Mill, The Ham, Westbury, Wilts
Ward, Mr T.	J. N. Miller Ltd, Old Steam Mill, Corn Hill, Wolverhampton WV10 0BD
Waterworth, Mr D.G.	ICI Biological Products, PO Box 1, Billingham, Cleveland TS23 1LB
Webster, Professor A.J.F.	University of Bristol, Department of Animal Husbandry, School of Veterinary Science, Langford, Bristol BS18 7DUU
Webster, Mrs M.	BP Nutrition (UK) Ltd, Wincham, Northwich, Cheshire
Welsh, Mr R.	Hoechst UK Ltd, Walton Manor, Walton, Milton Keynes MK7 7AJ
Whiteoak, Mr R.A.	West Midlands Famers Assn Ltd, Llanthony Mills, Merchants Road, Gloucester
Wigger, Ing J.J.	NAFAG, Nahr-u. Futtermittel AG, 9202 Gosau SG, Switzerland
Wilding, Mr C.N.	Birch Hall Feeds Ltd, Waymills, Whitchurch, Shropshire
Williams, Mr C.	ABM Brewing & Food Group, Woodley, Stockport, Cheshire SK6 1PQ
Williams, Mr D.J.	International Molasses Ltd, Shell Road, Royal Edward Dock, Avonmouth, Bristol BS11 9BW

List of participants

Williams, Dr D.R.	BOCM Silcock Ltd, Basing View, Basingstoke, Hants RG21 2EQ
Williams, Mr W.T.	BP Nutrition (UK) Ltd, Wincham, Northwich, Cheshire
Wilson, Mr B.J.	Cherry Valley Farms, N. Kelsey Moor, Caister, Lincs LN7 6HH
Wilson, Mr S.	Pauls Agriculture Ltd, PO Box 39, 47 Key Street, Ipswich, Suffolk IP4 1BX
Wilson, Dr S.	Prosper de Mulder Ltd, Ings Road, Doncaster
Wiseman, Dr J.	University of Nottingham, School of Agriculture, Sutton Bonnington, Loughborough, Leics LE12 5RD
Woodgate, Mr S.	Prosper de Muller Ltd, Ings Road, Doncaster
Woolford, Dr M.K.	Agil Holdings Ltd, Fishponds Road, Wokingham, Berks RG11 2QL
Wright, Mr I.	Microbial Developments Ltd, Spring Lane North, Malvern Link, Worcester WR14 1AH
Youdan, Dr J.	Nutrimix, Boundary Industrial Estate, Boundary Road, Lytham, Lancs FY8 5HU
Zeller, Mr B.M.	Rhone-Poulenc (UK) Ltd, 271 High Street, Uxbridge, Middlesex

INDEX

Acidification of pig diets,
 by lactic acid producing microbes, 68
 response to inorganic acids, 66
 response to organic acids, 63
 interaction between diet type and acid response, 65
Acidosis, in poultry, 103
Alkalosis, in poultry, 103
Amino acids,
 content in legumes, 14–18
 efficiency of use of, in ruminants, 149
 effect of dietary supply of, on egg production, 112
 optimizing intake of, in laying hens, 119
 requirement for,
  dogs, 224
  gamebirds, 197
  turkeys, 88
 utilization by ruminants, 150
Antibiotics, pig industry use of, 73
Antinutritive factors,
 presence in legumes, 21–29
Autoxidation, of lipids, *see* Lipid peroxidation
Availability of nutrients,
 in legumes, 20

Basal endogenous nitrogen, 149
Best cost diet, 11
Beta-glucanase,
 addition to barley based diets, 80
Body tissue gain,
 energy costs associated with, 153

Cellulase digestible organic matter,
 as a predictor of feed ME, 145
Cell wall carbohydrates,
 proportions of in certain feeds, 186
Chemical analysis,
 prediction of feed ME from, 127–130, 137–145, 169–171
Compounded feeds, for ruminants,
 prediction of ME content of, 127–146

Dehulling of legumes, 16

Dietetic microangiopathy, 48
Dog,
 nutrition of,
  effects of advertisers on, 221
  effects on digestion and metabolism, 224
  research and publications on, 221–231
  scale of US market for dog food, 221
 recommended requirements for
  amino acid and protein, 224–228
  minerals and vitamins, 223, 228–231

Early weaned piglet,
 physiological difficulties for, 61–63
Endogenous amino acid utilization,
 by ruminants, 150
Egg output, effect of,
 amino acid supply on, 112
 dietary energy concentration on, 113–116
 environmental temperature on, 116
 nutrient density on, 114
Egg shell quality,
 influence of dietary calcium and phosphorus on, 101
Eicosanoids,
 formation of, 40
 nutritional modification of, 41
 role of, 40
Electrolyte balance,
 in poultry, 103–105
Energy retention,
 measurement by carcass analysis, 153
Enzyme supplementation,
 in piglet diets, 79
Essential amino acids,
 content in legumes, 17
Essential fatty acids,
 requirement for, 40

Favism factors,
 presence in legumes, 28
Feed evaluation systems (ruminants),
 historical development of, 168–172
 ideal requirements of, 167

Index

Feed evaluation systems (ruminants)—*cont.*
  types of system
    energy systems, 168–171
    models of ruminant digestion and
      metabolism, 173
    protein systems, 171, 172
    proximate analyses, 168
Feed intake,
  of leisure horse, 206
  of poultry,
    effect of amino acids on, 117
    effect of energy concentration on, 116
    effect of environmental temperature on, 116
Feeding standards,
  historical development of (ruminants), 148
Fibre,
  digestion of, by horse, 207, 208, 213
Field beans, *see* Legumes
Flatus,
  from fermentation of legumes, 33
Free radical production, 39

Gamebirds,
  as a viable industry, 195
  common species of, 195
  feed consumption and feed efficiency of, 200
  growth rates of, 200
  rearing programmes for, 195
  nutrition of,
    amino acid requirements, 197
    energy requirements, 200
    fat and fibre requirements, 200
    mineral and vitamin requirements, 200
    protein requirements, 197
Gastric pH,
  inability of young pig to maintain correctly, 62
  problems associated with increases in, 62
Growth hormone,
  as a growth promoter, 73
  control of secretion of, 75
  involvement in growth regulation, 73
Growth promoters,
  potential types of for pigs, 73–82

Horse, *see* Leisure horse

Immunocompetence of livestock,
  effect of vitamin E on, 51
Inorganic acids,
  as dietary acidifiers, 63

Lactic acid producing microbes,
  as dietary acidifiers, 68
Laying hens,
  effect of environmental temperature on, 116
  food intake in, 117
  optimizing amino acid intake of, 119
  optimizing feeding strategy of, 121
  response to energy and amino acids, 111–123
Lectins, *see* Phytohaemagglutinins

Legumes,
  antinutrative factors in, 21–29
  availability of nutrients in, 20
  carbohydrate content of, 18–20
  lipid content of, 20
  nutritive value of, for non-ruminants, 13–34
  processing of, 29–34
  protein and amino acid content of, 14–18
Leisure horse,
  diet-related ailments of, 213
  difficulties in rationing, 205
  feed intake of, 206
  fibre digestibility by, 207, 208, 213
  formulation of diets for, 215
  important response characteristics of, 205
  raw material feeds for, 216
  nutrient requirements,
    energy, 207–211
    protein, 211–213
    vitamins and minerals, 214
  effect of exercise on, 207–213
Lipid peroxidation,
  beneficial types of, 40
  pathological consequences of, 46–51
  prevention of, 44–46
  problems caused by, 44
Lupins, *see* Legumes

ME, efficiency of use of, for growth, 155
ME of feeds,
  comparison of values for sheep and cows, 137
  effect of fat and fibre sources on, 136
  prediction equations for (ruminants), 137–143, 170
  prediction from chemical composition, 127–130, 137–145, 169–171
MENTOR (model for ruminant feed evaluation), 173–188
  calculation of microbial synthesis by, 180–182
  description of, 173
  merits and shortcomings of, 185
  prediction of ME of feeds by, 174–179
  prediction of metabolizable protein of feeds by, 179–185
Metabolizability,
  of a diet, 152
Microbial protein synthesis (ruminants), 180–182
Milk constituent yield,
  nutritional manipulation of, 156–159
Mineral elements,
  requirements of,
    in dogs, 228–231
    in gamebirds, 200
    in poultry, 99–108

Nutrient allowances for ruminants,
  calculation of, 147–163
  response prediction for, 160–163
  safety margins in, 151–153
  manipulation of growth with, 153–156

Nutrient allowances for ruminants—*cont.*
  manipulation of milk constituent yield with, 156–159
Nutritional degenerative myopathy, 47

Oil content,
  as a predictor of feed ME, 145
Organic acids,
  as dietary acidifiers, 63
Oxygen free radicals,
  examples of, 42
  involvement in lipid peroxidation, 43
  production of, 42

Pancreas disease myopathy, 49
Partition of nutrients in ruminants, 153–162
Peas, *see* Legumes
Phytate,
  influence on poultry mineral metabolism, 100
Phytohaemagglutinins,
  presence in legumes, 27
Porcine stress syndrome, 49
Post weaning diarrhoea,
  effect of dietary acidification on, 63
Poultry,
  requirements for minerals and trace elements, 100–107
Prediction equations for feed energy (ruminants),
  comparison of predicted and database values, 141
  statistical assessment of accuracy of, 138
Probiotics,
  as porcine growth promoters, 77
  mode of action of, 78
Protease inhibitors,
  presence in legumes, 24–27
Protein requirements (ruminants), 149, 171
Proximate analysis,
  of feedstuffs, 168

Raw ingredient variability,
  effect on animal producer of, 8–11
  effect on feed manufacturer of, 4–8
  methods of minimizing effects of, 5–11
  problems of dealing with, 3–11
Reading model,
  prediction of amino acid requirements of,
    laying hens, 119
    turkeys, 89
Repartitioning agents,
  as growth promoters for pigs, 76

Rumen degradable nitrogen, 149
Rumen VFAs,
  influence on fat partition, 158

Somatostatin immunization,
  as a growth promoting technique in pigs, 75
Sulphur amino acids,
  content in legumes, 17
  growing puppy requirement for, 225
  turkey requirement for, 88

Tannins,
  content in legumes, 22–24
Trace elements,
  requirements for,
    dogs, 228
    gamebirds, 200
    leisure horse, 214
    poultry, 105–108
Turkeys,
  amino acid requirements of, 88–94
  bodyweight gain in, 90
  commercial dietary recommendations for, 93
  factors affecting food intake in, 94
  genetic progress in, 88
  leg problems in, 96
  market opportunities for, 87
  nutritional manipulation of meat yields in, 96
  subcutaneous fat cover in, 97

Undegraded dietary nitrogen, 149, 173

Vitamin E,
  as a free radical detoxifier, 39–45
  effect on immunocompetence of livestock, 51–55
Vitamins
  requirement for,
    dogs, 223
    gamebirds, 200
    leisure horse, 214
Volatile fatty acids,
  influence on fat partition, 158
  production from legumes, 33

Weende system,
  feedstuffs evaluation, 167
White flowered beans, *see* Legumes